Fundamental Crystallography, Powder X-ray Diffraction, and Transmission Electron Microscopy for Materials Scientists

Advances in Materials Science and Engineering
Series Editor
Sam Zhang

For more information about this series, please visit:
https://www.routledge.com/Advances-in-Materials-Science-and-Engineering/
book-series/CRCADVMATSCIENG

Fundamentals of Crystallography, Powder X-ray Diffraction, and Transmission Electron Microscopy for Materials Scientists

Dong ZhiLi

CRC Press
Taylor & Francis Group
Boca Raton London New York

CRC Press is an imprint of the
Taylor & Francis Group, an **informa** business

First edition published 2022
by CRC Press
6000 Broken Sound Parkway NW, Suite 300, Boca Raton, FL 33487-2742

and by CRC Press
4 Park Square, Milton Park, Abingdon, Oxon, OX14 4RN

© 2022 Taylor & Francis Group, LLC

CRC Press is an imprint of Taylor & Francis Group, LLC

Library of Congress Cataloging-in-Publication Data
Names: Dong, ZhiLi, author.
Title: Fundamentals of crystallography, powder X-ray diffraction, and transmission electron microscopy for materials scientists / Dong ZhiLi.
Description: First edition. | Boca Raton, FL : CRC Press, 2022. | Series: Advances in materials science and engineering | Includes bibliographical references and index. | Summary: "The goal of this textbook is to effectively equip readers with an in-depth understanding of crystallography, x-ray diffraction, and transmission electron microscopy theories as well as applications. Written as an introduction to the topic with minimal reliance on advanced mathematics, the book will appeal to a broad spectrum of readers, including students, engineers, and researchers in materials science and engineering, applied physics, and chemical engineering. It can be used in XRD and TEM lab training"– Provided by publisher.
Identifiers: LCCN 2021056285 (print) | LCCN 2021056286 (ebook) | ISBN 9780367357948 (hbk) | ISBN 9781032246802 (pbk) | ISBN 9780429351662 (ebk)
Subjects: LCSH: Materials--Analysis. | Crystallography. | X-ray diffraction imaging. | Transmission electron microscopy.
Classification: LCC QD131 .D66 2022 (print) | LCC QD131 (ebook) | DDC 543–dc23/eng20220207
LC record available at https://lccn.loc.gov/2021056285
LC ebook record available at https://lccn.loc.gov/2021056286

ISBN: 978-0-367-35794-8 (hbk)
ISBN: 978-1-032-24680-2 (pbk)
ISBN: 978-0-429-35166-2 (ebk)

DOI: 10.1201/9780429351662

Contents

PART I   Introduction to Crystallography

PART II   X-ray Diffraction of Materials

PART III   Transmission Electron Microscopy of Materials

Acknowledgments

I would like to take this opportunity to express my thanks to those who have shared with me their experiences in crystallography, X-ray diffraction, the Rietveld method, transmission electron microscopy and the HRTEM image simulations. I am deeply grateful to my teachers from Tsinghua University. I learned mathematical techniques for reciprocal lattice calculations as well as X-ray diffraction theories for crystal structure examination from Professor Lin Zi Wei. I further learned X-ray crystallography, experimental X-ray diffraction techniques and transmission electron microscopy from Professor Tao Kun and Professor Zhu Bao Liang. I received much help from my Ph.D. supervisor, Professor Hori Shigenori, and from Dr. Zhang Di and Mr. Fujitani Wataru while I was doing TEM experiments for my Ph.D. project at Osaka University during 1987 and 1988.

I am indebted to Professor Nutting who shared his invaluable experience on the interpretation of electron diffraction and the diffraction contrast formation in metallic and ceramic materials when I was a visiting professor at the University of Barcelona from June 1993 to November 1996. I would like to thank Dr. Tim White, who shared his Rietveld refinement, HRTEM multi-slice simulation experiences and helped our collaborative research projects on synthetic apatites when I was a senior industrial research scientist at the Environmental Technology Institute of Singapore.

I also want to express my special thanks to the following TEM experts. While I was preparing my lecture notes on transmission electron microscopy, many discussions with Dr. Pan Ming, of Gatan Inc., were very helpful. I learnt from Professor Andrew Johnson and Professor Ray Withers how to conduct CBED in a modern TEM during their visits to our TEM labs and during my visit to Andrew for further discussions at the University of Western Australia. I wish to thank Professor Lumin Wang and Dr. Kai Sun from the University of Michigan, Dr. Liu Binghai from Wintech Nano-Technology Services Pte Ltd, and Professor Jie Lian from Rensselaer Polytechnic Institute for their help in interpreting electron-beam-induced structure modification during TEM observations. I also wish to thank the following experts. For the Cs-corrected TEM analysis in some research projects, I received much help from Dr. Zhang Zaoli from the Erich Schmid Institute of Materials Science, from Dr. Christopher Brian Boothroyd and Dr. András Kovács from Forschungszentrum Julich GmbH in the Helmholtz Association. During the IAMNano Workshop 2015 at Hamburg, Professor Harald Rose shared a very encouraging message in his TEM research, and his experience of training scientists. I truly benefited from the excellent EELS analysis by Dr. Michel Bosmann from the Singapore Institute of Materials Research and Engineering when we were supervising a Ph.D. research project together.

In addition, tremendous thanks must go to my former research advisors Professor Chen Nanping and Professor Tang Xiangyun for their guidance on metallic materials research, to my previous research group directors, Professor Jose Maria Guilemany and Professor Khor Khiam Aik for giving me the opportunity to investigate the coating structures for their thermal spray projects, and to my former School Chair

Professor Freddy Boey for his help with our carbon reinforced polymer collaborative project. I would like to thank my past and present colleagues at Nanyang Technological University including Dr. Gan Chee Lip, Dr. Chen Zhong, Dr. Raju Ramanujan, Dr. Kong Ling Bing, Dr. Sun Xiaowei, Dr. Sun Changqing, Dr. Zhu Weiguang, Dr. Tang Dingyuan, Dr. Xu Chunxiang, Dr. Lim Teik Thye, Dr. Wang Rong, Dr. Li Changming, Dr. Chua Chee Kai, Dr. Liu Erjia, Dr. Li Lin, Dr. Yang En-Hua, Dr. Claude Guet, Dr. Sean Li, Dr. Tom Baikie, Dr. Sun Handong, Dr. Xiong Qihua and Dr. Tom Wu for their support in various ways in my X-ray diffraction and transmission electron microscopy analyses. I would like to express my sincere appreciation to Professor Nuria Llorca from the University of Barcelona, Professor Shi Ji from the Tokyo Institute of Technology, Professor Dr. Jozef Keckes from the Erich Schmid Institute of Materials Science, Professor Ding Jun, Professor Chen Jingshen and Dr. Yang Ping from the National University of Singapore, Dr. Pei Qing Xiang and Dr. Han Ming Yong from A*STAR for the beneficial discussions had on materials structure analysis.

Under the joint Ph.D. supervision programs between Nanyang Technological University and European universities, there were many meetings among supervisor teams, and I greatly acknowledge the help of the co-supervisors, especially Professor Yves Wouters from the University of Grenoble, Dr. Frederic Schuster from CEA, Professor Fanny Balbaud-Celerier from Sorbonne University, Professor Valeska Ting from the University of Bristol, Professor Sanka Gopinathan from UCL, Dr. Tim Hyde from Johnson Matthey and Professor Vadim V. Silberschmidt from the University of Loughborough.

I would also like to thank my past and present graduate students and postdoctoral research fellows, especially Shen Yiqiang, Li Zhipeng, Wang Jingxian, Li Ruitao and Yao Bingqing who had concentrated on the HRTEM analyses of metallic and ceramic materials.

My special thanks go to my wife Kriss Ding Ke and my church members Ang Se An and Jimmy Chew, who have been encouraging me with God's words from the holy Bible. Lastly, I am very thankful to my son, Brian Dong Bo Yuan, who patiently developed some of the graphics for my lecture notes.

Preface

One of the key topics in materials science and engineering is the structure–property relationship. To understand why a material displays certain behaviors, the first step is to resolve its crystal structure and reveal its characteristics. This is one of the reasons why I wish to share with young materials scientists my lecture notes on crystal structure description and crystal structure determination. This textbook covers (i) the fundamentals of crystallography, (ii) powder X-ray diffraction analysis (XRD) and (iii) the transmission electron microscopy (TEM) of materials, which are presented in three separate parts. I hope my lecture notes will help not only graduate students in the materials science and engineering field in their crystal structure examination studies, but also some undergraduates who are preparing themselves to do materials research.

The lecture notes presented are primarily based on the fundamental theories and technical skills I learnt when I was an undergraduate student at Tsinghua University and a research student during my PhD study at Osaka University. During my further academic advancement at Tsinghua University and the University of Barcelona, I used transmission electron microscopy intensively in my research projects for materials microstructure studies. It was exciting that these techniques entered a new era from the 1990s due to the development in spherical aberration corrections. After I transferred from the Thermal Spray Centre at the University of Barcelona to the Thermal Spray Research Group in the School of Mechanical and Production Engineering of Nanyang Technological University in 1996, especially after I joined the transmission electron microscopy laboratory at the Environmental Technology Institute of Singapore, we used more and more techniques associated with high resolution transmission electron microscopy (HRTEM), including imaging, diffraction and spectroscopy techniques in our apatite research projects, which enabled a better understanding of apatite structures and behaviors. Since I left the research institute and joined the School of Materials Engineering of Nanyang Technological University in 2004, I have been endeavoring to teach young students in the materials science and engineering field how to characterize materials using X-ray diffraction and transmission electron microscopy, and have summarized my lecture notes in a book chapter "X-ray Diffraction, Rietveld Crystal Structure Refinement and High-resolution transmission Electron Microscopy of Nanostructured Materials" in the *Handbook of Nanoceramics and Their Based Nanodevices* (Tseng and Nalwa, 2009). For this textbook, I have included more content that is suitable to my graduate course. I was also involved in teaching short graduate courses on TEM specimen preparation at the University of Barcelona when I was a visiting professor there. However, in this textbook, the TEM specimen preparation techniques have not been explained in detail. For those who are interested in understanding more specimen preparation techniques, you can find many useful information from other textbooks.

In my lectures and in some sections of this book, I have recommended some text-books rich in physics, such as *X-ray Diffraction* (Warren, 1990), *Elements of X-ray Diffraction* (Cullity and Stock, 2001), *X-ray Diffraction in Crystals, Imperfect Crystals and Amorphous Bodies* (Guinier, 1994), *Electron Microscopy of Thin Crystals* (Hirsch, 1977), *Transmission Electron Microscopy: A Textbook for Materials Science* (Williams, 1996) and *Diffraction Physics* (Cowley, 1995) to those students who are ready to continue their in-depth studies.

It is believed that, to study the topics of (i) imaging, (ii) diffraction and (iii) spectroscopy, a crystal structure course is a prerequisite. Therefore, this textbook introduces the fundamentals of crystallography before the explanations of X-ray diffraction and transmission electron microscopy theories. As this textbook covers very limited spectroscopy analysis, for those who need to conduct spectroscopy analysis, I would recommend them to read other spectroscopy-related textbooks. In some textbooks, the crystallography content is limited but the spectroscopy analysis is presented in detail. Therefore, this textbook can be used as a supplement to other textbooks for university students.

In X-ray diffraction and transmission electron microscopy areas, the relevant spectroscopy techniques include X-ray absorption fine structure (XAFS) analysis and electron energy loss spectroscopy (EELS) analysis. Energy dispersive X-ray spectroscopy (EDX) is a very common attachment in a modern transmission electron microscope and has become a routine chemical analysis technique during TEM operations. EDX is a quick and efficient method for composition analysis. XAFS can utilize a high brightness synchrotron light source and obtain the chemical composition information, the oxidation state of the elements and the coordination numbers. EELS can simply be installed in-column or post-column in a modern transmission electron microscope. EELS is a powerful tool in a microscope for the analysis of chemical compositions, bonding information and energy band information. Therefore, the understanding of atomic structures in quantum physics and the fundamental theories of band structures in solid state physics can help in the interpretation of the spectra collected.

I also feel that some fundamental techniques in mathematics and physics would be very helpful for reading this textbook. For example, Fourier series and Fourier transform analysis have been employed in the TEM imaging process. Diffraction theories have been used throughout X-ray diffraction and transmission electron microscopy analysis. The electron wavelength calculations need a relativistic correction as the velocities of electrons accelerated by high voltages are high. The atomic scattering factor calculation has employed some fundamental theories in quantum mechanics. In some cases, we just employ 1D treatments to deliver the concepts, although in real situations, 3D analyses are required. For example, to obtain the Mott relationship, we have used a 1D Fourier transform to correlate the atomic scattering factors for X-ray beams and electron beams. Some mathematical calculations employed in this textbook are only true under certain conditions. As we are materials scientists and materials engineers, but not mathematicians or physicists, we can always consult the latter in our neighboring schools when we encounter tough problems in mathematics and physics.

Again, I would like to clarify that I have presented some topics in more detail than others, as I have noticed that many students struggle with some of the topics. It is believed that the more detailed explanations of those topics in this textbook will guide students effectively during their studies.

Dong ZhiLi
Nanyang Technological University, Singapore

REFERENCES

Cowley, J. M. (John M.) (1995) *Diffraction physics [electronic resource]/John M. Cowley*. 3rd rev. ed. Amsterdam: Elsevier Science B.V. (North-Holland personal library).

Cullity, B. D. (Bernard D.) and Stock, S. R. (2001) *Elements of x-ray diffraction/B.D. Cullity, S.R. Stock*. 3rd ed. Upper Saddle River, NJ: Prentice Hall.

Guinier, A. (1994) *X-ray diffraction in crystals, imperfect crystals, and amorphous bodies/A. Guinier; translated by Paul Lorrain and Dorothée Sainte-Marie Lorrain*. New York: Dover.

Hirsch, P. B. (1977) *Electron microscopy of thin crystals/P. B. Hirsch, A. Howrie, R. B. Nicholson, D. W. Pashley and M. J. Whelan*. Malabar, FL: Krieger Pub Co.

Tseng, T.-Y. and Nalwa, H. S. (eds) (2009) *Handbook of nanoceramics and their based nanodevices*. Stevenson Ranch, CA: American Scientific Pub. (Nanotechnology Book Series; 24).

Warren, B. E. (Bertram E.) (1990) *X-ray diffraction*. Dover ed. New York: Dover Publications.

Williams, D. B. (1996) *Transmission Electron Microscopy [electronic resource]: A Textbook for Materials Science/by David B. Williams, C. Barry Carter*. Edited by C. B. Carter. Boston, MA: Springer US. doi:10.1007/978-1-4757-2519-3.

Author

Dr. **Dong ZhiLi** received his B.Eng. degree in metallic materials engineering from Tsinghua University in 1984. He obtained his Ph.D. from Tsinghua University in 1989 under the Joint Ph.D. Program of the Ministry of Education of China. He received a Japanese Government Scholarship and carried out experiments from his Ph.D. research at Osaka University in 1987 and 1988.

Dong developed his research in the areas of materials engineering, the synthesis of geo-mimetic materials, crystal structure/electronic structure-property relationships, and interface structure analysis. He has more than 30 years' experience in X-ray diffraction and the transmission electron microscopy of materials.

Dong is an associate professor in the School of Materials Science & Engineering at Nanyang Technological University (NTU). Prior to joining NTU, he worked at the Environmental Technology Institute of Singapore as a senior research scientist, the School of Mechanical and Production Engineering at NTU as a research fellow, the University of Barcelona as a visiting professor and Tsinghua University as a lecturer.

Symbols Used in This Book

$\vec{a}, \vec{b}, \vec{c}$	unit cell vectors of the direct lattice
$a, b, c, \alpha, \beta, \gamma$	lattice parameters
$\vec{a}^*, \vec{b}^*, \vec{c}$	unit cell vectors of the reciprocal lattice
$a*, b*, c*, \alpha*, \beta*, \gamma*$	reciprocal lattice parameters
$\vec{a}_1, \vec{a}_2, \vec{a}_3$	unit cell vectors of the direct lattice
$\vec{a}_1, \vec{a}_2, \vec{a}_3, \vec{c}$	unit cell vectors of hexagonal/trigonal system
$\vec{a}_1^*, \vec{a}_2^*, \vec{a}_3^*$	unit cell vectors of the reciprocal lattice
$\vec{b}_1, \vec{b}_2, \vec{b}_3$	unit cell vectors of the reciprocal lattice
$A(\theta)$	absorption correction factor
A	amplitude of wave
$A(u)$	aperture function
$\exp\left[-B\left(\dfrac{\sin\theta}{\lambda}\right)^2\right]$	isotropic function Debye–Waller factor, $B = 8\pi^2\bar{u}^2$
$B(u)$	aberration function
C_S, C_3	spherical aberration coefficient
d	resolution
d, d_{hkl}	interplanar spacing
$\vec{E}$	electric vector
E	energy
$E(u)$	envelope function
F, F_{hkl}, F_g	structure factor
FT	Fourier transform operator
FT^{-1}	inverse Fourier transform operator
f	focal length
f_e	atomic scattering factor for electron
f_X	atomic scattering factor for X-ray
$f(\theta)$	atomic scattering amplitude
$\vec{g} = h\vec{a}^* + k\vec{b}^* + l\vec{c}$	reciprocal lattice vector
(hkl)	Miller indices of a plane
$\{hkl\}$	indices of a set of all symmetrically equivalent planes
$(hkil)$	Miller-Bravais indices of a plane
$\{hkil\}$	indices of a set of all symmetrically equivalent planes
I	intensity
I_0	intensity of direct beam
I_g	int ensity of diffracted beam
i	inversion
k	wave number
$\vec{k}_0$	wave vector of incident wave
$\vec{k}$	wave vector of diffracted wave

L	camera length
m	mirror ref lection
m_e, m_0	mass of electron at rest
n	rotation operation
$\bar{n}$	roto-rotation operation
$\vec{q} = u\vec{a}^* + v\vec{b}^* + w\vec{c}^*$	reciprocal space vector/scattering vector
$\vec{r} = x\vec{a} + y\vec{b} + z\vec{c}$	position vector in real space
$\vec{r}^*$	position vector in reciprocal space
$\vec{r}'$	displacement vector
$\vec{R}_n = n_1\vec{a}_1 + n_2\vec{a}_2 + n_3\vec{a}_3$	position vector in direct lattice
R_B	R-Bragg factor
$R_{\exp}$	expected R-factor
R_{wp}	weighted profile R-factor
S	goodness of fit indicated
$\bar{s}$	excitation error or deviation parameter
s_{eff}	effective excitation error
s'	excitation error due to defect
S_n	roto-reflection
t	foil thickness
u, v	coordinates in back focal plane
$[uvw]$	indices of direction/zone axis
$<uvw>$	indices of set of symmetrically equivalent directions
$[uvtw]$	indices of direction/zone axis
$<uvtw>$	indices of a set of symmetrically equivalent directions
U	potential energy
V	volume
V_c, V_{cell}	volume of unit cell
V_c^*, V_{cell}^*	volume of unit cell in reciprocal lattice
V	accelerating voltage
$V(\vec{r})$	electrostatic potential
V_t	projected potential
δ	path difference
Δf	change of focus
Δ	spread of focus
ε	defocus
$\theta = \theta_s/2$	scattering semiangle
θ_B	Bragg angle
$\theta_S = 2\theta$	scattering angle
λ	wavelength
μ	absorption coefficient
ξ_g	extinction distance
ξ_g^{eff}	effective extinction distance
ϕ	phase difference/phase
ϕ	interplanar angle
ϕ_0	amplitude of incident beam

ϕ_g amplitude of diffracted beam
χ phase shift
$\bar{\chi}_0$ wave vector of incident wave in vacuum
$\bar{\chi}$ wave vector of diffracted wave in vacuum
ψ wave function
ψ_0 incident wave
ψ_g diffracted wave
$\otimes$ convolution operation

Introduction

The structure–property relationship is one of the key topics in materials research. To understand crystal structure description, and know how to use the theories and methods to determine crystal structures, is important to materials scientists. In the traditional definition, crystals are considered as a periodic arrangement of identical unit cells or building blocks. However, as more and more materials with ordered but aperiodic structures are discovered and synthesized, such as quasicrystals, the crystallography communities classify crystals into periodic crystals and aperiodic crystals, both of which have sharp diffraction peaks (Janssen, 2007). In these lecture notes, I only explain the periodic crystal structure, and I still use the expression "crystal" to represent "periodic crystal".

Although it is common to elucidate the properties of a material from its structural characteristics, in some cases we can also try to understand them from the material's chemical bonding characteristics or electronic structure. In this section, three examples are presented in order for students to learn how materials' properties are linked with chemical bonding characteristics, with crystal structures and with the electronic structures of crystals.

Our general understanding is that while we are looking into the behaviors of a crystalline material, we need to know its crystal structure characteristics. However, in some cases, we also need to understand the chemical bonding and the electronic structures of the material in order to fully clarify the material's behavior in relation to the structural characteristics.

In fact, crystal structure formation highly depends on the chemical bonding characteristics of the comprising elements. And both the resultant crystal structure and chemical bonding between the atoms or ions will determine the electronic structure of a crystal, which is well interpreted through two extreme cases: nearly free electron approximation and the tight-binding model (Omar, 1993).

It is also known that chemical bonding depends on the electronic structures of the atoms involved, in particular the valence electrons in the outermost shells. The bond lengths and bond angles associated with the atoms influence the packing of those atoms, and therefore the crystal structures, which is very well exhibited in diamond-type structures.

The main bonds governing inorganic material structures are metallic, ionic and covalent. A metallic bond is formed by the attraction between positively charged atomic nuclei and the delocalized electrons, and this bonding type is nondirectional, based on which the hard sphere packing model is often employed to describe the formation of metallic material structures. Face-centered cubic, body-centered cubic and hexagonal close-packed are the most common metal structure forms with higher symmetries. Ionic bonds are also nondirectional in nature, and the attractive forces manifest from all directions. The strong attraction between cations and anions creates a rigid framework, or lattice structure with higher symmetries. Ceramic materials are

DOI: 10.1201/9780429351662-1

often considered as ionic compounds, although the bonding characteristics are a mixture of ionic and covalent types. One of the Pauli principles tells us that a coordination polyhedron of anions is formed about each cation, the cation–anion distance equals the sum of their characteristic packing radii. The radius ratio of cation to anion determines both the nature of the coordination polyhedron and therefore the coordination number of the cation. Sodium chloride is a good example for explaining the above principle and the formation of the crystal structure type. Unlike ionic bonding, covalent bonding is highly directional. In diamonds, each carbon atom is sp^3-hybridized and forms four sigma bonds with neighboring carbon atoms, with a C-C-C bond angle of 109.5°. The basic tetrahedral units unite with one another to produce a cubic crystal structure.

Clearly, the above discussion is too simple to illustrate the formation of complicated crystal structures, as in some crystals there may be more than one bond type. Nonetheless, for materials scientists and engineers, the above simple discussion is useful for understanding crystal structure characteristics.

Having noticed the correlations between chemical bonding types, crystal structure characteristics and electronic structures for crystals, in some cases we can discuss the materials' properties not only based on the crystal structure characteristics, but also from the point of view of the bonding behaviors or electronic structures of crystals. Therefore, in teaching structure–property relationships in this Introduction, I have included the material property through its correlation with the chemical bonding and electronic structures of crystals. In the three main parts of the textbook, we focus more on crystal structure descriptions and examinations, than on the chemical bonding types and electronic structures of crystals.

The following three examples will show how (i) the chemical bonding, (ii) the crystal structure and (iii) the electronic structure of crystals influence material properties.

Example 1

It is reported that perovskite-type $BaTiO_3$ ferroelectric ceramic has a phase transition from a cubic to a tetragonal structure as it is cooled from a temperature above its Curie point (~120 °C) down to room temperature. The ferroelectric state at a certain temperature below the Curie point, with permanent dipoles, is shown in Figure 0.1. The reference planes are the top and bottom Ba^{2+} faces of the unit cell, and are halfway between the top and bottom Ba^{2+} faces. The unit cell constants are a = 3.98 Å and c = 4.03 Å (Barsoum, 2003). The displacement values of the ions are given as: (i) Ba^{2+} ions have no displacement; (ii) Ti^{4+} ion is displaced upwards by 0.06 Å; (iii) the midplane O^{2-} ions are displaced downwards by 0.06 Å; (iv) the top and bottom O^{2-} ions are displaced downwards by 0.09 Å. Calculate the polarization of ferroelectric $BaTiO_3$.

Notes: This is a typical example of crystal structure–property relationships. Due to the structure changes from the cubic phase to the tetragonal phase, the material varies from the paraelectric state to the ferroelectric state. Detailed discussion can be found in many materials science and engineering or ceramic materials books. For example, in the textbook *Introduction to Ceramics* (Kingery, 1976), the

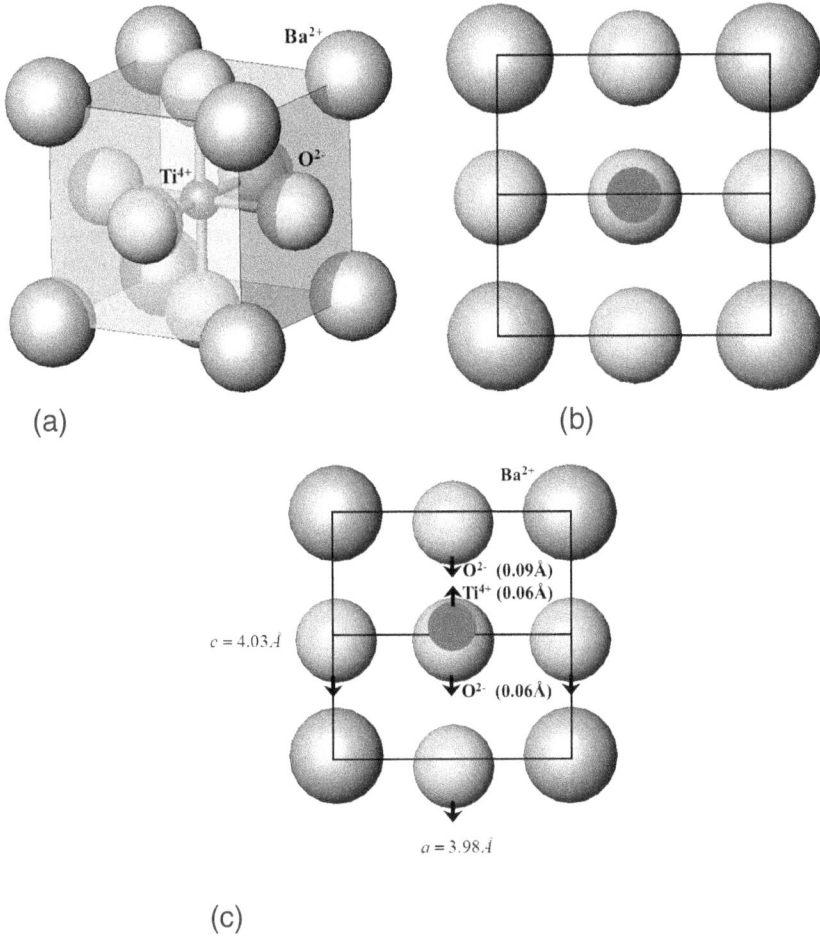

(a)   (b)

(c)

FIGURE 0.1 Crystal structures of $BaTiO_3$ ferroelectric ceramic. (a) Unpolarized cubic crystal structure with $a \approx 4.0\,\text{Å}$, (b) projection of the unpolarized structure, and (c) projection of the polarized structure. The figures were drawn using ATOMS.

representation of the polarized $BaTiO_3$ structure is based on the very original work published in 1955 by Shirane *et al.* There is also much other X-ray diffraction work reporting $BaTiO_3$ polarized crystal structures. Different textbooks may use different structure data, and the polarization value obtained can be different. Here we use the polarized structure data presented in materials textbooks (Callister, 2003; Barsoum, 2003) and we often use this model in our lectures to explain the crystal structure–property relationships. The paraelectric cubic phase is stable above 120 °C with the center of the positive charges coinciding with the center of the negative charges. Therefore, polarization $P = 0$. At room temperature, the material has a tetragonal structure, where the center of the cations is displaced relative to the O^{2-} ions, leading to the formation of electric dipoles.

In many cases, crystals may have defects, including point defects, line defects and plane defects. Defects inside a crystal can also affect the properties, for example, dislocations in metals can reduce the yield strengths compared with their counterparts free of dislocations. Sometimes, dislocations are purposely produced through deformation to increase the strengths of metal alloys. The strengths of metal alloys can also be increased through solid solution or through the decrease of grain size, resulting in the increase of grain boundaries.

Solution: For a system without net charge, the electric moment calculation is independent of the choice of origin. If the atoms are treated as point charges Q_i at positions r_i, then the electric moment can be obtained as:

$$\vec{p} = \sum_i Q_i \vec{r_i} \tag{0.1}$$

Polarization refers to the total electric dipole moment per unit volume of the material. For the tetragonal $BaTiO_3$, this polarization can be calculated based on the data from a unit cell as presented in Equation (0.2):

$$P = \frac{p_{cell}}{V_{cell}} = \frac{\sum_i Q_i d_i}{V_{cell}} \tag{0.2}$$

where Q_i is the ionic charge of each ion in a unit cell, d_i is its displacement with respect to the unpolarized structure and V_{cell} is the unit cell volume.

The first step is to set a moment direction. The second step is to calculate the dipole moment of ions in the unit cell.

If the upward direction is chosen as a positive direction, then we have the Ti^{4+} ion displaced by $d = +0.06\,\text{Å} = +0.06\times10^{-10}\,m$, $q = 4\times(1.602\times10^{-19}\,C)$, so that

$$qd = 4\times(1.602\times10^{-19}\,C)\times(+0.6\times10^{-10}\,m) = +3.84\times10^{-30}\,C\cdot m.$$

The top and bottom O^{2-} ions are each shared by two unit cells and thus count only as one ion, which is displaced by $d = -0.09\,\text{Å} = -0.09\times10^{-10}\,m$, $q = (-2)\times(1.602\times10^{-19}\,C)$, so that

$$qd = (-2)\times(1.602\times10^{-19}\,C)\times(-0.09\times10^{-10}\,m) = +2.88\times10^{-30}\,C\cdot m.$$

The four mid-plane O^{2-} ions are also each shared by two unit cells and thus count as two ions. Their displacement is $d = -0.66\,\text{Å} = -0.06\times10^{-10}\,m$. $q = (-2)\times2\times(1.602\times10^{-19}\,C)$. Therefore,

$$qd = (-2)\times2\times(1.602\times10^{-19}\,C)\times(-0.06\times10^{-10}\,m) = +3.84\times10^{-30}\,C\cdot m.$$

The total polarization is $P = (\Sigma qd)/V_{cell}$, where

$$\Sigma qd = (3.84 + 2.88 + 3.84) \times 10^{-30}\,C \cdot m = 10.56 \times 10^{-30}\,C \cdot m,$$

$$V_{cell} = (4.03 \times 3.98 \times 3.98) \times 10^{-30}\,m^3 = 63.83 \times 10^{-30}\,m^3.$$

As a result, $P = (10.56 \times 10^{-30}\,C \cdot m)/(63.83 \times 10^{-30}\,m^3) = 0.165 C \cdot m^{-2}$.

When the material is cooled down to room temperature, the spontaneous polarization for tetragonal $BaTiO_3$ is 0.165 C·m⁻².

Example 2 From the anharmonic curve for potential energy shown in Figure 0.2, derive the thermal expansion coefficient.

Note: This second example is on thermal expansion and its correlation with chemical bonding behavior. From the potential energy curve of chemical bonding, we can better understand the thermal expansion phenomenon and the coefficient of thermal expansion (Callister, 2003).

As discussed in many solid state physics textbooks, if we assume the vibration of each atom in a crystal is a classical harmonic oscillator, the curve of potential energy versus interatomic spacing is symmetrical, and there will be no thermal expansion. Therefore, we need to consider the effect of anharmonic terms in the potential energy on the mean separation of a pair of atoms at a temperature T (Kittel, 1996).

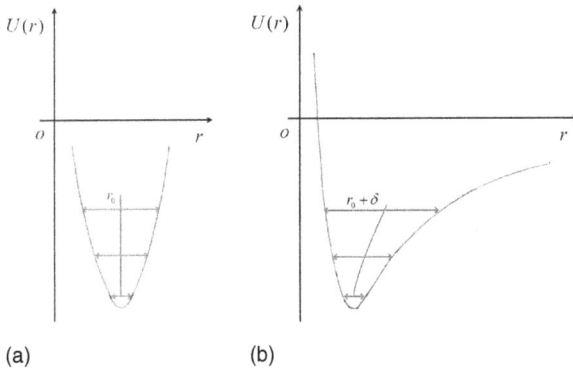

(a)   (b)

FIGURE 0.2 Potential energy as a function of interatomic spacing. (a) Symmetrical parabolic curve of potential energy corresponding to harmonic oscillator, (b) asymmetrical curve of potential energy, with the midpoint of oscillation moving away from the origin as energy increases.

Solution: In the asymmetrical curve of potential energy as a function of interatomic spacing, r_0 is the equilibrium interatomic spacing and δ is the displacement from r_0. When $U(r_0 + \delta)$ is used to represent the potential energy of the interaction, it can be expressed using a Taylor series:

$$U\left(r_0 + \delta\right) = U\left(r_0\right) + \left(\frac{\partial U}{\partial r}\right)_{r_0} \delta + \frac{1}{2!}\left(\frac{\partial^2 U}{\partial r^2}\right)_{r_0} \delta^2 + \frac{1}{3!}\left(\frac{\partial^3 U}{\partial r^3}\right)_{r_0} \delta^3 + \dots \tag{0.3}$$

It is known that $\left(\dfrac{\partial U}{\partial r}\right)_{r_0} = 0$ and the value of $U(r_0)$ has no influence on our discussions, then we just let $U(r_0) = 0$ to simplify the treatment. To continue the discussion, we only consider the Taylor series to third order as presented in some textbooks (Gu, 1989; Quéré, 1998; Wahab, 2015).

Let $\dfrac{1}{2!}\left(\dfrac{\partial^2 U}{\partial r^2}\right)_{r_0} = K$, $-\dfrac{1}{3!}\left(\dfrac{\partial^3 U}{\partial r^3}\right)_{r_0} = g$, then we can rewrite $U(r_0 + \delta)$, and

$$U\left(r_0 + \delta\right) = K\delta^2 - g\delta^3 \tag{0.4}$$

Imagine we only consider $U(r_0 + \delta) = K\delta^2$, which is the harmonic oscillation approximation, then we cannot observe any thermal expansion as the potential curve is symmetrical. In this case we can calculate the $\delta(T)$ value using Boltzmann distribution:

$$\bar{\delta} = \frac{\displaystyle\int_{-\infty}^{\infty} \delta \exp\left(-\frac{U}{k_B T}\right) d\delta}{\displaystyle\int_{-\infty}^{\infty} \exp\left(-\frac{U}{k_B T}\right) d\delta} = \frac{\displaystyle\int_{-\infty}^{\infty} \delta \exp\left(-\frac{K\delta^2}{k_B T}\right) d\delta}{\displaystyle\int_{-\infty}^{\infty} \exp\left(-\frac{K\delta^2}{k_B T}\right) d\delta} \tag{0.5}$$

As the integrand of the numerator is an odd function, the integral is 0. Therefore, we are not able to explain the thermal expansion phenomenon based on the harmonic oscillation approximation. We need to consider the asymmetrical effect using $U(r_0 + \delta) = K\delta^2 - g\delta^3$ for thermal expansion discussions. When T increases, the equilibrium position moves rightward, increasing the interatomic distance. In such a case thermal expansion occurs.

$$\bar{\delta} = \frac{\displaystyle\int_{-\infty}^{\infty} \delta \exp\left(-\frac{U}{k_B T}\right) d\delta}{\displaystyle\int_{-\infty}^{\infty} \exp\left(-\frac{U}{k_B T}\right) d\delta} = \frac{\displaystyle\int_{-\infty}^{\infty} \delta \exp\left(-\frac{K\delta^2 - g\delta^3}{k_B T}\right) d\delta}{\displaystyle\int_{-\infty}^{\infty} \exp\left(-\frac{K\delta^2 - g\delta^3}{k_B T}\right) d\delta} \tag{0.6}$$

When δ is small, the value of the δ^3 term is much smaller than that of δ^2, and the numerator is

$$\int_{-\infty}^{\infty} \delta \exp\left(-\frac{K\delta^2 - g\delta^3}{k_B T}\right) d\delta$$

$$= \int_{-\infty}^{\infty} \delta \exp\left(-\frac{K\delta^2}{k_B T}\right) \exp\left(\frac{g\delta^3}{k_B T}\right) d\delta$$

$$= \int_{-\infty}^{\infty} \delta \exp\left(-\frac{K\delta^2}{k_B T}\right)\left(1 + \frac{g\delta^3}{k_B T}\right) d\delta$$

For approximation, we use $e^x = 1 + x$ when x is small. The integral is further calculated as follows:

$$\int_{-\infty}^{\infty} \delta \exp\left(-\frac{K\delta^2}{k_B T}\right)\left(1 + \frac{g\delta^3}{k_B T}\right) d\delta$$

$$= \int_{-\infty}^{\infty} \delta \exp\left(-\frac{K\delta^2}{k_B T}\right) d\delta + \int_{-\infty}^{\infty} \frac{g\delta^4}{k_B T} \exp\left(-\frac{K\delta^2}{k_B T}\right) d\delta \qquad (0.7)$$

$$= \frac{g}{k_B T} \int_{-\infty}^{\infty} \delta^4 \exp\left(-\frac{K}{k_B T}\right) \delta^2 d\delta$$

$$= \frac{g}{k_B T}\left(\frac{3}{4}\pi^{\frac{1}{2}}\right)\left(\frac{k_B T}{K}\right)^{\frac{5}{2}}$$

The denominator is:

$$\int_{-\infty}^{\infty} \exp\left(-\frac{K\delta^2 - g\delta^3}{k_B T}\right) d\delta$$

$$= \int_{-\infty}^{\infty} \exp\left(-\frac{K\delta^2}{k_B T}\right)\left(1 + \frac{g\delta^3}{k_B T}\right) d\delta$$

$$= \int_{-\infty}^{\infty} \exp\left(-\frac{K\delta^2}{k_B T}\right) d\delta + \int_{-\infty}^{\infty} \frac{g\delta^3}{k_B T} \exp\left(-\frac{K\delta^2}{k_B T}\right) d\delta \qquad (0.8)$$

$$= \int_{-\infty}^{\infty} \exp\left(-\frac{K\delta^2}{k_B T}\right) d\delta$$

We can rewrite Equation (0.8) as:

$$\int_{-\infty}^{\infty} \exp\left(-\frac{K\delta^2}{k_BT}\right) d\delta$$

$$= \sqrt{\frac{k_BT}{K}} \int_{-\infty}^{\infty} \exp-\left(\sqrt{\frac{K}{k_BT}}\delta\right)^2 d\left(\sqrt{\frac{K}{k_BT}}\delta\right) \tag{0.9}$$

$$= \sqrt{\frac{k_BT}{K}}\sqrt{\pi} = \left(\frac{\pi k_BT}{K}\right)^{\frac{1}{2}}$$

Then we can obtain the average displacement from Equations (0.7) and (0.9):

$$\bar{\delta} = \frac{3}{4}\frac{g}{K^2}k_BT \tag{0.10}$$

From Equation. (0.10), the thermal expansion coefficient can be calculated. The expression is:

$$\alpha_l = \frac{1}{r_0}\frac{d\bar{\delta}}{dT} = \frac{3}{4}\frac{g}{K^2}\frac{k_B}{r_0} \tag{0.11}$$

For the above integral calculations, we use the following results in mathematics.

Let $\int_{-\infty}^{\infty} e^{-x^2}dx = I$, then

$$I^2 = \int_{-\infty}^{\infty} e^{-x^2}dx \int_{-\infty}^{\infty} e^{-y^2}dy$$

$$= \int_{-\infty}^{\infty}\int_{-\infty}^{\infty} e^{-(x^2+y^2)}dxdy$$

To continue the calculation through converting the Cartesian coordinates to polar coordinates, we have:

$$I^2 = \int_{0}^{\infty}\int_{0}^{2\pi} e^{-r^2}rd\theta dr$$

$$= 2\pi \int_{0}^{\infty} e^{-r^2}rdr$$

$$= \pi$$

Therefore:

$$\int_{-\infty}^{\infty} e^{-x^2} dx = I = \sqrt{\pi} \tag{0.12}$$

From Equation (0.12), the denominator is obtained.

The calculation of the numerator has employed the following integrations. From Equation (0.12), we know that $\int_{-\infty}^{\infty} e^{-x^2} dx = I = \sqrt{\pi}$.

Let $I(a) = \int_{-\infty}^{\infty} e^{-ax^2} dx$,

then $I(a) = \int_{-\infty}^{\infty} e^{-ax^2} dx = \frac{1}{\sqrt{a}} \int_{-\infty}^{\infty} e^{-(\sqrt{a}x)^2} d(\sqrt{a}x) = a^{-\frac{1}{2}}\sqrt{\pi}$, or we have

$$\int_{-\infty}^{\infty} e^{-ax^2} dx = a^{-\frac{1}{2}}\sqrt{\pi} \tag{0.13}$$

Now we differentiate both sides of Equation (0.13) with respect to the parameter a. The left side of the equation becomes:

$$\frac{d}{da}\left(\int_{-\infty}^{\infty} e^{-ax^2} dx\right) = \int_{-\infty}^{\infty} -x^2 e^{-ax^2} dx$$

And the right side of the equation becomes:

$$\frac{d}{da}\left(a^{-\frac{1}{2}}\sqrt{\pi}\right) = -\frac{1}{2}a^{-\frac{3}{2}}\sqrt{\pi}$$

We differentiate both sides with respect to a again. We now have the left side as:

$$\frac{d}{da}\left(\int_{-\infty}^{\infty} -x^2 e^{-ax^2} dx\right) = \int_{-\infty}^{\infty} x^4 e^{-ax^2} dx$$

And the right side as:

$$\frac{d}{da}\left(-\frac{1}{2}a^{-\frac{3}{2}}\sqrt{\pi}\right) = \frac{3}{4}a^{-\frac{5}{2}}\sqrt{\pi}$$

Therefore, we have:

$$\int_{-\infty}^{\infty} x^4 e^{-ax^2} dx = \frac{3}{4}a^{-\frac{5}{2}}\pi^{\frac{1}{2}} \tag{0.14}$$

Example 3 Based on the electronic structures of crystals, explain why some crystals are metals and others are insulators.

Note that this example is about the electronic structure–property relationship, and the topic is well discussed in textbooks in the solid state physics area (Kittel, 1996) and in the materials science area (Callister, 2003). The textbooks in solid state physics often propose the Nearly-Free Electron Model and the Tight Binding Model as two extreme cases for the calculations of the band structures of crystals (Omar, 1993). The wave functions for an electron in the periodic potential field are Bloch waves. In the Nearly-Free Electron Model, the band gap widths are related to the coefficients U_n in the Fourier series of the potential energy for an electron. For example, in a 1D periodic lattice with periodic potentials, the electron energies deviate from the values obtained by the free electron model near the Brillouin zone boundaries.

When $k = -\dfrac{\pi}{a}n$, $k' = \dfrac{\pi}{a}n$,

we have $E_+ = E_k^{(0)} + |U_n|$, $E_- = E_k^{(0)} - |U_n|$

Hence the band gap, $E_g = E_+ - E_- = 2|U_n|$.

For the Nearly-Free Electron Model, detailed discussion can be found in Appendix A2.

Solution: Based on the band structure presented in Kittel (1996) and Callister (2003), if the valence electrons exactly fill one or more bands as shown in Figure 0.3(a), leaving the high energy bands empty, the crystal will be an insulator. If the band gap is relatively narrow, it is a semiconductor; for example, the band gap of a silicon semiconductor is about 1.11 eV at 300 K. Doping P or B to Si can result in an n-type or a p-type semiconductor. The energy state of the donor or acceptor is within the band gap, which allows relatively large numbers of charge carriers to be created, leading to relatively high electrical conductivities compared with the corresponding intrinsic semiconductor. If the bands overlap in energy, the two partially filled bands will result in a metal character as presented in Figure 0.3(b). Figure 0.3(c) shows a partially filled band, and the material will be a conductor.

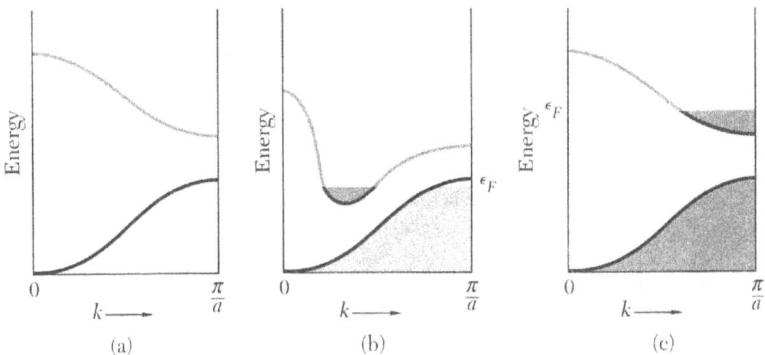

FIGURE 0.3 Occupied state and band structures giving (a) an insulator, (b) a metal or semimetal because of band overlap, and (c) a metal because of electron concentration.

(Source: (Kittel, 1996). Wiley, with the permission of John Wiley & Sons, Inc.)

REFERENCES

Barsoum, M. W. (2003) '*Fundamentals of ceramics/Michel W. Barsoum*,' Bristol, UK: Institute of Physics Pub. (McGraw-Hill series in materials science and engineering).

Callister, W. D. (2003) *Materials science and engineering: an introduction/William D. Callister, Jr.* 6th ed. New York: Wiley.

Gu, B. L. (1989). *Solid State Physics/Gu, B. L., Wang, X. K.*, Beijing: Tsinghua University Press.

Janssen, T. (2007) *Aperiodic crystals: from modulated phases to quasicrystals/Ted Janssen, Gervais Chapuis and Marc de Boissieu.* Edited by G. Chapuis and M. de. Boissieu. Oxford: Oxford University Press (International Union of Crystallography monographs on crystallography; 20).

Kingery, W. D. (1976) *Introduction to ceramics/W. D. Kingery, H. K. Bowen, D. R. Uhlmann.* 2nd ed. Edited by H. K. (Harvey K. Bowen and D. R. (Donald R. Uhlmann). New York: Wiley (Wiley series on the science and technology of materials).

Kittel, C. (1996) *Introduction to solid state physics/Charles Kittel.* 7th ed. New York: Wiley.

Omar, M. A. (1993) *Elementary solid state physics: principles and applications/M.A. Omar.* Rev. print. Reading, Mass: Addison-Wesley Pub. Co. (Addison-Wesley series in solid state sciences).

Quéré, Y. (1998) *Physics of materials/Yves Quéré; translated by Stephen S. Wilson.* Amsterdam: Gordon and Breach Science Publishers.

Wahab, M. A. (Mohammad A.) (2015) *Solid state physics : structure and properties of materials/M.A. Wahab.* 3rd ed. Oxford U.K: Alpha Science International.

Part I

Introduction to Crystallography

Natural or synthetic solid state materials can have amorphous, aperiodic crystalline or periodic crystalline structures; for periodic crystalline materials, the crystal size can be at the nanometer, micrometer, millimeter scale or even larger. A polycrystalline material contains many crystallites or grains oriented differently, whereas a single crystal only has one grain, therefore it has no grain boundaries. In our surroundings, we can find many useful materials that are in periodic crystalline forms; their properties are directly related to their chemical compositions, chemical bonding types, crystal structures and electronic structures, as presented in the examples in the Introduction.

From about the seventeenth century, it was noticed that the beauty of the symmetry of crystals suggested the existence of underlying orders of some kind. From their external symmetries and shapes, it was argued that crystals must consist of ordered arrangements of tiny particles, which are known as atoms and ions today. From the beginning of the nineteenth century, there was a real explosion of crystallography research (Lima-de-Faria and Buerger, 1990). For example, in 1801, the French scientist Rene Just Hauy produced a crystal drawing based on polyhedral molecules. Also in the early nineteenth century, the German scientist Christian Samuel Weiss derived crystal forms from crystallographic axes, and the English scientist William Hallowes Miller proposed the *hkl* notation.

The 32 crystallographic point groups were derived in 1830 by the German scientist Johann Friedrich Christian Hessel. Axel Gadolin, independently of Hessel, also derived the 32 crystal classes. We know that the 32 point groups can be classified into 7 crystal systems, the combination of which with 4 lattice types generate 14 Bravais lattices. In 1848, the French physicist Auguste Bravais derived 14 possible arrangements of points in space. Bravais lattices are named after him to remind us of his contribution.

The Russian crystallographer Evgraph Stepanovich Fedorov, and the German mathematician Arthur Schoenflies, derived independently in 1890–1891 the 230 possible space groups that symmetrically arrange the building units (atoms, ions, molecules) inside crystals. In 1894, the English geologist William Barlow, using a sphere packing approach, independently derived the 230 space groups. In the International Tables for Crystallography A (Hahn, 2005), both Hermann–Mauguin notation and Schoenflies's notation are employed in the descriptions of space groups and point groups.

As the course syllabus for students in materials science and engineering is different from that in solid state physics or solid state chemistry, we aim at teaching materials science and engineering students in a way that matches their levels in mathematics and physics. For some topics, especially those related to the powder X-ray diffraction and transmission electron microscopy of materials, we include more basic mathematical treatments and physics, following which our materials science and engineering students can better understand the overall picture.

In Part I of these lectures, the translational symmetry operations are introduced first. Then the ten unique point symmetry operations, followed by combinations of point symmetry operations, are discussed. After that the classification of 7 crystal systems from the 32 crystallographic point groups is described. Some students may think further and try to understand why there are 14 Bravais lattices but 28 derived from the combination of these 14 Bravais lattices with four lattice types.

My experience gained during teaching materials scientists suggests that many students are confused by the trigonal crystal system and hexagonal system representation. I believe that it is necessary to emphasize the differences between these systems, and further emphasize the differences between the rhombohedral cell and the non-rhombohedral cell within the trigonal system. The above concepts must be emphasized when we teach materials scientists and materials engineers.

For the derivation of the 32 crystallographic point groups, we follow the approaches taken in the textbook *The 23rd Edition of the Manual of Mineral Science* by Klein *et al.* (2008) and the textbook *Principles of Symmetry* by Tang Youqi (1977).

In the space group section, the two new types of symmetry elements, the glade planes and screw axis, are introduced first, after which the space groups made up from lattice symmetry (translational), point symmetry (not translational), and glide-reflection and screw symmetry (a translational component) are briefly explained.

The mathematical treatments based on direct latices and reciprocal lattices are powerful techniques for the study of X-ray diffraction and transmission electron microscopy theories. The calculations associated with reciprocal lattices are explained after discussion of the symmetries of crystals.

Now it is helpful to highlight again the methodology and approach of teaching crystallography to materials science and engineering students, which is presented in the flow chart below.

Concepts of translational symmetry and point symmetry
⇓
Translational symmetry: Concept of periodicity, and 4 lattice types; 7 crystal systems (the 32 crystallographic point groups can be categorized into 7 crystal systems); 14 Bravais lattices based on combinations of 7 crystal systems with 4 lattice types.
⇓
More detailed explanation on trigonal system: Rhombohedral cell of the trigonal system can be represented using rhombohedral axes as well as hexagonal axes; Non-rhombohedral cell of the trigonal system can only be represented using hexagonal axes.
⇓
Miller indices and Miller–Bravais indices: For rhombohedral cell in the trigonal system: transformation of indices from rhombohedral axes to hexagonal axes, and from hexagonal axes to rhombohedral axes; For hexagonal system: Description of Miller–Bravais indices.
⇓
Point symmetry: Ten unique crystallographic symmetry elements of rotation axes and improper rotation axes (constraints of translational symmetry); Descriptions of symmetry elements.
⇓
Combination of point symmetry operations: Theorems illustrating crystallographic symmetry operation combinations; Group theory.
⇓
Combination of symmetry operations (non-cubic systems)
⇓
Combination of symmetry operations (cubic system)
⇓
32 crystallographic point groups
⇓
7 crystal systems through classifying 32 point groups
⇓
14 Bravais lattices generated from 7 crystal systems and 4 lattice types (details are presented in the translational symmetry section).
⇓
Space group concepts
⇓
Differences between P3 (143) and R3 (P146) (refer to the translational symmetry section).
⇓
Direct lattice and reciprocal lattice: Definition of reciprocal lattice; Properties and associated calculations of reciprocal lattices.

REFERENCES

Hahn, T. (2005) *International tables for crystallography. Volume A, space-group symmetry/ Edited by Theo Hahn.* 5th ed. re, *International tables for crystallography [electronic resource] Volume A, Space-group symmetry.* 5th ed. re. Dordrecht: Published for the International Union of Crystallography by Springer.

Klein, C. et al. (2008) *The 23rd edition of the manual of mineral science*, 23rd ed. Hoboken, NJ: J. Wiley.

Lima-de-Faria, J. (José) and Buerger, M. J. (eds) (1990) *Historical atlas of crystallography/ edited by J. Lima-de-Faria; with the collaboration of M.J. Buerger ... [et al.].* Dordrecht; Published for International Union of Crystallography by Kluwer Academic Publishers.

Tang, Y. Q. (1977) *Principles of symmetry.* Beijing: Chinese Scientific Publishers.

1 Periodicity of Crystals and Bravais Lattices

In this chapter, we will describe how to represent the periodicity of crystals through a discussion on lattices, basis (or motif) and a unit cell. For the seven crystal systems and fourteen Bravais lattices, we will focus more on the trigonal system and hexagonal system. The purpose is to clear up the uncertainties of some students from materials science or an engineering background.

1.1 CRYSTALS, LATTICES AND BASIS

A crystal can be considered as a translationally periodic array of atoms in three dimensions. To effectively explain this characteristic, that is translational periodicity, it is essential to introduce the concepts of the lattice, unit cell and Bravais lattice.

The lattice is defined as a translationally periodic array of points in space, that is, to match with the translationally periodic array of atoms. A lattice is an imaginary pattern of points associated with the crystal, in which every point in the pattern has an environment that is identical to that of any other point. In the real space of a crystal, the origin of the lattice is an arbitrary point.

The basis or motif is the building block, which when placed on each lattice point, creates the physical structure of the crystal, that is, a translationally periodic array of atoms in three dimensions. The basis may correspond to a single atom, an ion or a molecule, but usually it consists of a group of atoms or ions. We can describe the relationship of the crystal, the lattice and the basis as follows:

$$\text{Crystal} = \text{Lattice} + \text{Basis}$$

This relationship is schematically shown in Figure 1.1.

In mathematics, a crystal structure can be expressed as a convolution of lattice and basis when the lattice is treated as a periodic array of delta functions:

$$\text{Crystal} = \text{Lattice} \otimes \text{Basis} \tag{1.1}$$

This is due to the fact that

$$f(x) \otimes \delta(x) = \int_{-\infty}^{+\infty} f(x')\delta(x-x')dx' = f(x)$$

and

$$f(x) \otimes \delta(x-na) = \int_{-\infty}^{+\infty} f(x')\delta\big[(x-na)-x'\big]dx' = f(x-na)$$

DOI: 10.1201/9780429351662-3

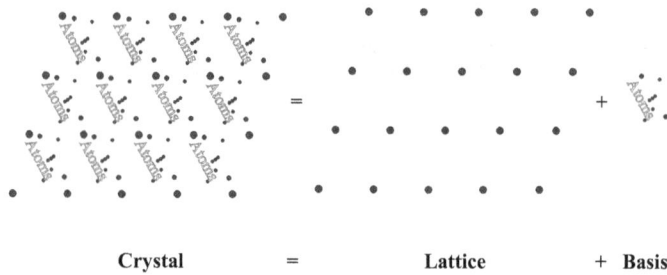

| Crystal | = | Lattice | + Basis |

FIGURE 1.1 Crystal, lattice and basis.

1.2 UNIT CELL TYPES, CRYSTAL SYSTEMS AND BRAVAIS LATTICES

In fact, every crystal structure has a certain symmetry (see Chapter 2), as does the lattice. In a three-dimensional lattice, a small repeat unit with representative symmetries is defined and called a unit cell, which contains the symmetrical characteristics of the entire lattice. By stacking identical unit cells in three-dimensional space, the entire lattice can be constructed. The unit cell is chosen in the following way:

1. The edges of the unit cell should coincide with the symmetry of the lattice.
2. The smallest possible cell that contains all elements should be chosen.

There are four types of unit cells in three-dimensional lattices as shown in Figure 1.2, which are primitive, body-centered, face-centered and one-face-centered (A-centered, B-centered or C-centered). These four types of unit cells, when constructing the whole lattice, satisfy the condition of "having identical surroundings". The primitive unit cell only contains one lattice point. Body-centered, face-centered and one-face-centered unit cells are not primitive as they contain more than one lattice point. In solid state physics textbooks, the Wigner–Seitz primitive unit cell is often used, which in the reciprocal lattice defined in physics textbooks is the first Brillouin zone. I often remind my students that the 2π factor is commonly used in solid state physics textbooks in order to simplify many expressions. There is only one Wigner–Seitz cell for any given lattice. In the International Table for Crystallography (Authier, 2010), there is a description of the Wigner–Seitz cell used in crystallography. The domain, or the polyhedron, of a particular lattice point consists of all points in space that are closer to this lattice point than to any other lattice point, or at most equidistant

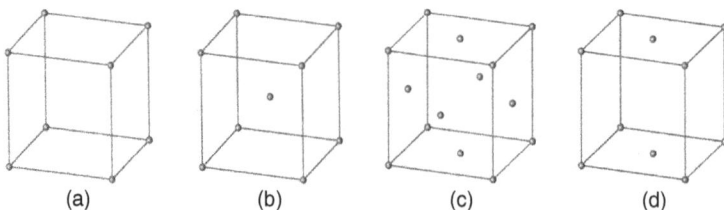

FIGURE 1.2 Four lattice types that have translational symmetries. (a) Primitive lattice, (b) body-centered lattice, (c) face-centered lattice and (d) one-face-centered lattice.

to it. The Wigner–Seitz cell is constructed by connecting the selected lattice origin to all other lattice points and drawing bisecting planes, which are perpendicular to these connecting lines and passing through their midpoints. The convex polyhedron enclosed by these planes is the Wigner–Seitz cell, In mathematics, it is called the Voronoi cell or Dirichlet domain.

The reason for choosing a conventional unit cell, instead of a primitive one, for some lattices is that we want to display the symmetry elements in a unit cell. Lattice parameters corresponding to a unit cell are shown in Figure 1.3.

Based on the symmetry characteristics of the 7 crystal systems and the 4 lattice types, not 28 lattices, but only 14, exist. For example, the single-side face-centered tetragonal does not exist and can be described using a smaller tetragonal primitive unit cell (see Figure 1.4(a)). For a cubic system, there is no single-side face-centered cell. The single-side centering can break the cubic symmetry, and the three-fold axes along the diagonal directions will disappear (see Figure 1.4(b)).

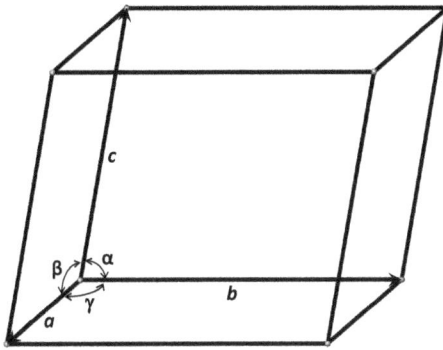

FIGURE 1.3   Unit cell and lattice parameters.

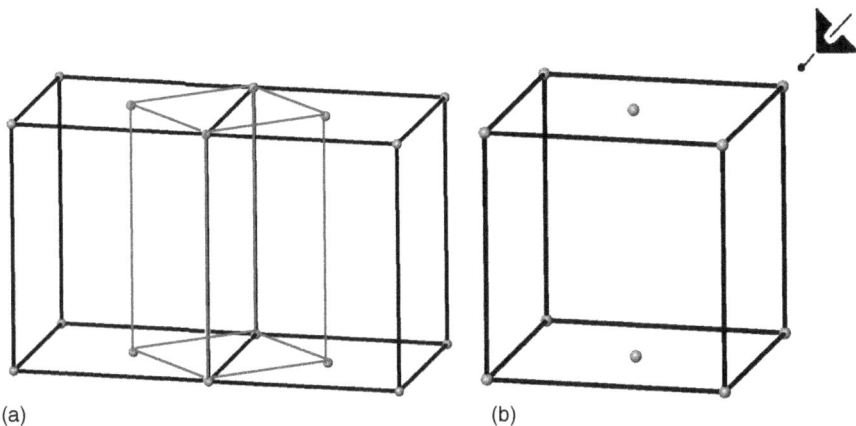

(a)   (b)

FIGURE 1.4   (a) A C-centering tetragonal unit cell does not exist as the lattice can be described using a smaller primitive unit cell, (b) C-centering destroys the three-fold symmetry about the four body diagonal axes in the cubic system.

The 14 Bravais lattices are shown in Figure 1.5.

In a Bravais lattice, we can find the coordinates of a point, the indices of a direction vector and the Miller indices of a crystallographic plane.

To determine the indices of a crystallographic direction, we just need to know the projections of the vector along the three axes, in terms of unit cell edge lengths. The three numbers representing the three projections are multiplied or divided by a common factor so that they are reduced to the smallest absolute integer values, which are enclosed in square brackets.

The position of any point located within a unit cell is specified in terms of its coordinates as a fractional multiple of the unit cell edge lengths.

To determine the Miller indices of a crystallographic plane, we need to (i) determine the intercepts of the face along the crystallographic axes, in terms of unit cell edge lengths; (ii) take the reciprocals; and (iii) multiply or divide the three numbers

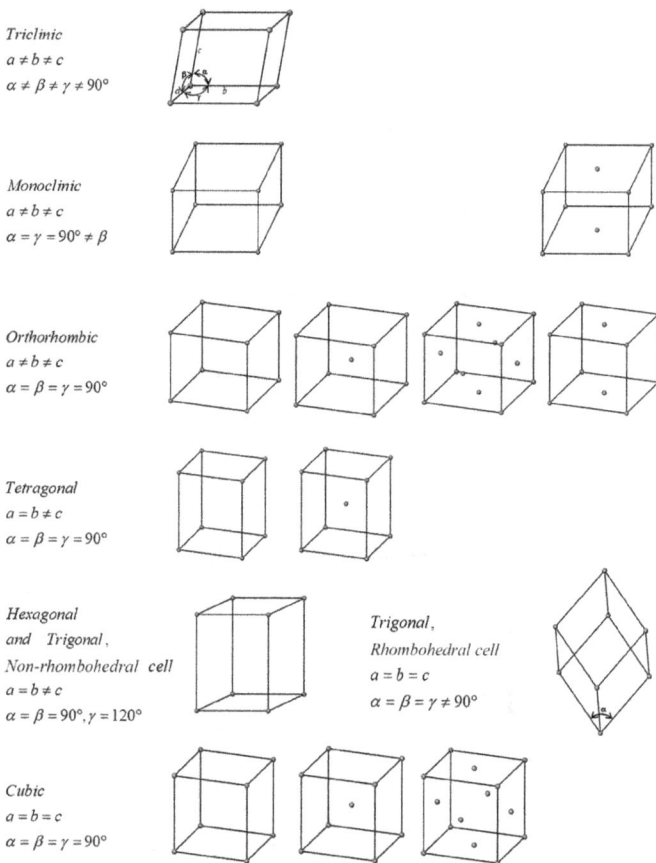

Triclinic
$a \neq b \neq c$
$\alpha \neq \beta \neq \gamma \neq 90°$

Monoclinic
$a \neq b \neq c$
$\alpha = \gamma = 90° \neq \beta$

Orthorhombic
$a \neq b \neq c$
$\alpha = \beta = \gamma = 90°$

Tetragonal
$a = b \neq c$
$\alpha = \beta = \gamma = 90°$

Hexagonal
and Trigonal,
Non-rhombohedral cell
$a = b \neq c$
$\alpha = \beta = 90°, \gamma = 120°$

Trigonal,
Rhombohedral cell
$a = b = c$
$\alpha = \beta = \gamma \neq 90°$

Cubic
$a = b = c$
$\alpha = \beta = \gamma = 90°$

FIGURE 1.5 Fourteen Bravais lattices derived from seven crystal systems and four lattice types. Note that the non-rhombohedral cell of the trigonal system can only be represented using hexagonal axes. The rhombohedral cell of the trigonal system can be represented by either hexagonal axes or rhombohedral axes.

by a common factor to reduce them to the smallest absolute integer values. These integral numbers are then parenthetically enclosed and are employed to designate that specific crystallographic plane within the lattice.

1.3 RHOMBOHEDRAL CELLS AND NON-RHOMBOHEDRAL CELLS IN TRIGONAL SYSTEMS

From my discussions with students and materials scientists on crystallography theories and applications, I have noticed that the trigonal system can cause some confusions to them.

We need to emphasize that the trigonal system includes rhombohedral cells and non-rhombohedral cells.

1. The non-rhombohedral cells of a trigonal system can only be represented using hexagonal axes.
2. The rhombohedral cells of a trigonal system can be represented by either hexagonal axes or rhombohedral axes. The unit cell represented by hexagonal axes is a non-primitive one as it contains three lattice points per cell, which are at $000, \dfrac{2}{3}\dfrac{1}{3}\dfrac{1}{3}, \dfrac{1}{3}\dfrac{2}{3}\dfrac{2}{3}$ for the obverse setting. The three lattice points are at $000, \dfrac{1}{3}\dfrac{2}{3}\dfrac{1}{3}, \dfrac{2}{3}\dfrac{1}{3}\dfrac{2}{3}$ for the reverse setting (Hahn, 2005). Both obverse setting and reverse setting are well presented in the International Table for Crystallography A. For the obverse setting, the unit cell vector relationship for the hexagonal non-primitive cell description and the rhombohedral primitive cell description is well described in Cullity and Stock (2001).

In fact, the symmetry elements for the rhombohedral cells and non-rhombohedral cells in the trigonal system are different, and their differences can be found in the graphic descriptions in space group No. 143 with symmetry P3 (Figure 1.6) and

$P3$ C_3^1 3   Trigonal

No. 143 $P3$ Patterson symmetry $P3$

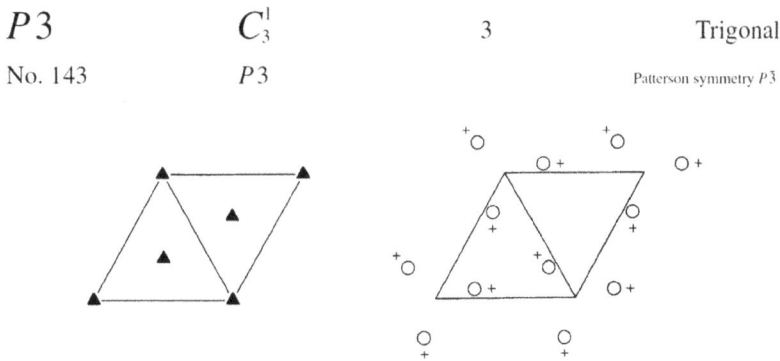

FIGURE 1.6   Symmetry elements for trigonal space group No. 143 P3. The arrangement of the three-fold rotation axes along the c-direction is shown.

(Source: Reproduced from Hahn (2005), with the permission of Springer Nature.)

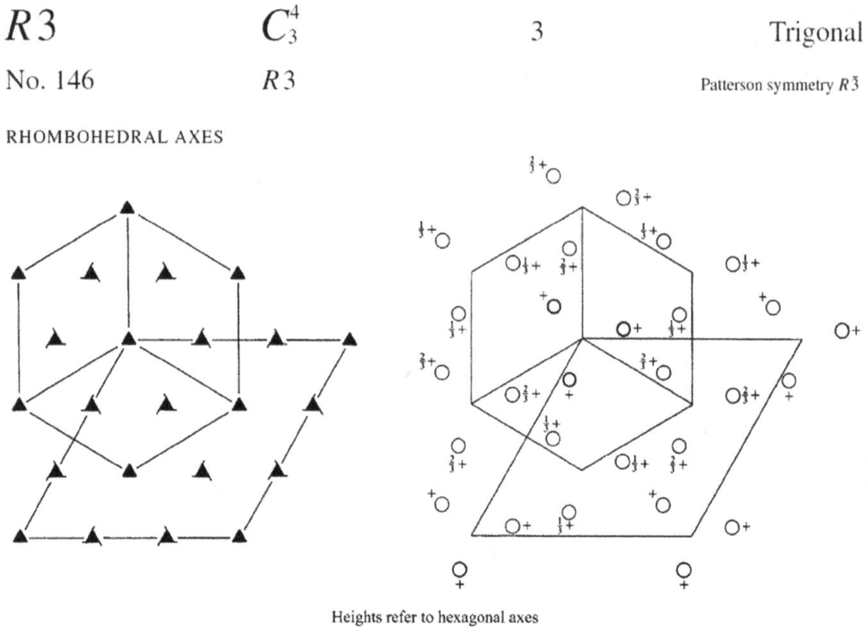

$R3$ C_3^4 3   Trigonal

No. 146 $R3$ Patterson symmetry $R\overline{3}$

RHOMBOHEDRAL AXES

Heights refer to hexagonal axes

FIGURE 1.7   Symmetry elements for trigonal space group No. 146 R3. In comparison with space group P3, R3 has both three-fold rotation and three-fold screw axes.

(Source: Reproduced from Hahn (2005, with the permission of Springer Nature.)

space group No. 146 with symmetry R3 (Figure 1.7) (Hahn, 2005). For R3 as presented in Figure 1.7, there are both three-fold rotation axes and three-fold screw axes. The coordinates of the general Wyckoff positions for P3 and R3 are different as well.

Although No. 143 P3 can only be presented using hexagonal axes, the representative symmetry element is a three-fold rotation axis, not a six-fold axis. To compare the symmetry characteristics of space group P3 and space group P6 when projected along the c-axis of the hexagonal lattice as displayed in Figures 1.6 and 1.8, it is not difficult to find their differences in symmetry element arrangements.

Based on the descriptions for the obverse setting and the reverse setting in Hahn (2005), the unit cell vector relationship between the hexagonal axes and rhombohedral axes for the obverse setting can be represented as in Figure 1.9. The relationship can also be expressed by Equation (1.2) (Cullity and Stock, 2001).

$$\begin{aligned}
\vec{a}_1(H) &= \vec{a}_1(R) - \vec{a}_2(R) \\
\vec{a}_2(H) &= \vec{a}_2(R) - \vec{a}_3(R) \\
\vec{a}_3(H) &= \vec{a}_3(R) - \vec{a}_1(R) \\
\vec{c}(H) &= \vec{a}_1(R) + \vec{a}_2(R) + \vec{a}_3(R)
\end{aligned} \tag{1.2}$$

$P6$ C_6^1 6   Hexagonal

No. 168 $P6$ Patterson symmetry $P6/m$

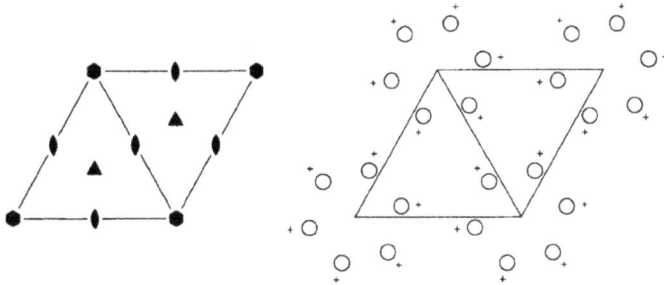

FIGURE 1.8 Symmetry elements for hexagonal space group No 168 P6. The arrangement of the six-fold, three-fold and two-fold rotation axes along the c-direction is shown.

(Source: Reproduced from Hahn (2005), with the permission of Springer Nature.)

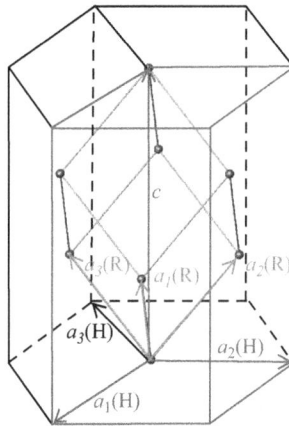

FIGURE 1.9 Representations of rhombohedral cells in the trigonal system using hexagonal axes and rhombohedral axes for the obverse setting. The unit cell represented by the hexagonal axes is a non-primitive one as it contains three lattice points per cell, which are at $000, \frac{2}{3}\frac{1}{3}\frac{1}{3}, \frac{1}{3}\frac{2}{3}\frac{2}{3}$ for the obverse setting. The three lattice points are at $000, \frac{1}{3}\frac{2}{3}\frac{1}{3}, \frac{2}{3}\frac{1}{3}\frac{2}{3}$ for the reverse setting.

The indices (*HKL*), referred to the hexagonal axes, of a plane and the indices (*hkl*), referred to the rhombohedral axes, are related in the following equations (Cullity and Stock, 2001):

$$H = h - k$$
$$K = k - l \qquad (1.3)$$
$$L = h + k + l$$

From Equation (1.3), it follows that:

$$-H + K + L = 3k \qquad (1.4)$$

Equation (1.4) tells us that $-H + K + L$ is always an integral multiple of 3. If this condition is not satisfied, the lattice is hexagonal (Cullity and Stock, 2001).

Transformation equations from hexagonal axes to rhombohedral axes:

$$h = \frac{1}{3}\left(2H + K + L\right)$$

$$k = \frac{1}{3}\left(-H + K + L\right) \qquad (1.5)$$

$$l = \frac{1}{3}\left(-H - 2K + L\right)$$

The unit cell parameters a_R and α of the rhombohedral unit cell are related to a_H and c by Equation (1.6):

$$a_R = \frac{1}{3}\sqrt{3a_H^2 + c_H^{\,2}}$$

$$\sin\frac{\alpha}{2} = \frac{3}{2\sqrt{3 + \left(c_H / a_H\right)^2}} \qquad (1.6)$$

The volumes of the unit cells described using rhombohedral axes and hexagonal axes can be calculated using Equation (1.7):

$$V_R = a_R^3\sqrt{1 - 3\cos^2\alpha + 2\cos^3\alpha}$$

$$V_H = \frac{\sqrt{3}a_H^2 c_H}{2} \qquad (1.7)$$

1.4   MILLER–BRAVAIS INDICES IN HEXAGONAL SYSTEMS

For the hexagonal axes, there is a three-index and a four-index scheme to represent the crystallographic directions and planes; it is more convenient to use the Miller–Bravais four-index system ($hkil$) and [$uvtw$] to demonstrate crystallographic planes and directions. This four-index scheme for labeling planes in a hexagonal system makes permutation symmetries apparent. Now the planes in the same family are identified by permutations of the (first three) indices, as with Miller indices for the other systems. For example, the similarity between (110) ≡ (11−20) and (−210) ≡ (−2110) is more obvious when employing a four-index scheme, or when the redundant index "i" is provided.

For the four-index scheme, the third index "i" ($hkil$) for a plane is obtained from $i = - (h + k)$ or from Equation (1.8):

$$h + k + i = 0 \qquad (1.8)$$

The four-index scheme can better represent the family of planes for the hexagonal system. For any l value, if the first three indices are the same, irrespective of the order and sign, they are equivalent and belong to the same family. For example, the $\{11\bar{2}0\}$ family consists of six equivalent planes, which are $(11\bar{2}0), (\bar{1}2\bar{1}0), (\bar{2}110), (\bar{1}\bar{1}20), (1\bar{2}10)$ and $(2\bar{1}\bar{1}0)$.

In the three-index system, the six planes in the same family are displayed as $(110), (\bar{1}20), (\bar{2}10), (\bar{1}\bar{1}0), (1\bar{2}0)$ and $(2\bar{1}0)$; some crystallographic equivalent planes do not have the same set of indices.

If the above planes in a trigonal system are labeled using the four-index scheme for hexagonal axes, $(\bar{1}\bar{1}20), (2\bar{1}\bar{1}0)$ and $(\bar{1}2\bar{1}0)$ is one set of equivalent planes, and $(11\bar{2}0), (\bar{2}110)$ and $(1\bar{2}10)$ is another set of equivalent planes. Therefore, the Miller–Bravais scheme is more suitable for labeling the planes and directions of the hexagonal system. Similarly, we know that in a hexagonal system, the $\{10\bar{1}0\}$ family consists of $(10\bar{1}0), (01\bar{1}0), (\bar{1}100), (\bar{1}010), (0\bar{1}10)$ and $(1\bar{1}00)$ planes.

From Figure 1.10 we can find the advantages of using the four-index scheme compared to the three-index scheme. The comparisons for some typical planes and directions are demonstrated.

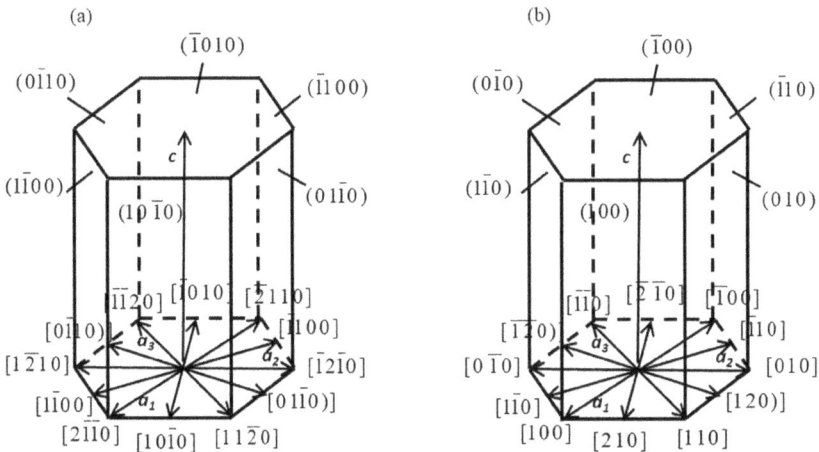

FIGURE 1.10 Planes and directions represented using four-index and three-index systems. (a) Indices of prism planes and some typical directions are displayed using a four-index system, (b) indices of prism planes and some typical directions are displayed using a three-index system.

For a crystallographic direction in the hexagonal system, the conversion from the three-index system to the four-index system, or from $[u'v'w']$ to $[uvtw]$, is not that straightforward in comparison with the index conversion for a crystallographic plane. We need to use the following conditions:

$$u'\vec{a}_1 + v'\vec{a}_2 + w'\vec{c} = u\vec{a}_1 + v\vec{a}_2 + t\vec{a}_3 + w\vec{c}$$

and

$\vec{a}_3 = -\left(\vec{a}_1 + \vec{a}_2\right)$, and further to let

$$t = -\left(u + v\right)$$

The obtained relationships are shown in Equation (1.9):

$$u = \frac{n}{3}\left(2u' - v'\right)$$
$$v = \frac{n}{3}\left(2v' - u'\right) \tag{1.9}$$
$$t = -\left(u + v\right)$$
$$w = nw'$$

n is a factor that is required to make u, v, t and w into the smallest absolute integer values.

The conversion from the four-index system to the three-index system for crystallographic directions $[uvtw] \Rightarrow [u'v'w']$ is accompanied by the following formulas:

$$u' = u - t$$
$$v' = v - t \tag{1.10}$$
$$w' = w$$

Then the common factor for u', v' and w' needs to be removed.

Zinc oxide has a wurtzite-type structure with the hexagonal $P6_3mc$ space group. Through a vapor-phase transport method, zinc oxide hexagram whiskers with uniform size and morphology, as shown in Figure 1.11, were fabricated by heating a mixture source of zinc oxide, indium oxide and graphite powders in air (Xu et al., 2006). In Figure 1.12, a four-index scheme was used to index the electron diffraction patterns and to show the equivalent planes and directions.

FIGURE 1.11 SEM images of ZnO hexagram whiskers (a) with two enlarged hexagrams (b), enlarged side rods (c) and enlarged center rod array (d).

(Source: Reproduced from Xu *et al.* (2006), with the permission of AIP Publishing.)

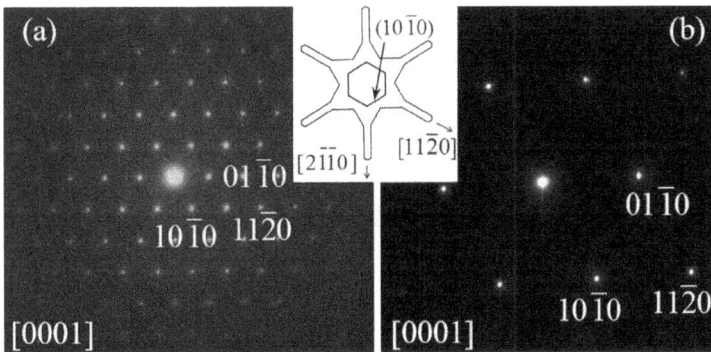

FIGURE 1.12 Selected area electron diffraction pattern taken from the edge of the core disk (a) and the side nanorod (b) of a ZnO hexagram from a [0001] zone axis. The insert shows schematically the crystallographic planes and directions of the core disk and side nanorods.

(Source: Reproduced from Xu *et al.* (2006), with the permission of AIP Publishing.)

SUMMARY

The understanding of the periodicity of crystals is the foundation for the further studies of the symmetry of periodic crystals, because the periodic characteristics of a periodic crystal impose constraints on symmetry operations.

A periodic crystal is a translationally periodic array of atoms in three dimensions. The lattice is a translationally periodic array of points in space. If we express the lattice, or the periodic array of points, using delta functions, then a periodic crystal can be represented on a lattice $\otimes$ basis.

In total, there are 14 Bravais lattices. However, the types of basis or motif are abundant, which results in a variety of crystals with different lattice constant values.

One of the most confusing parts for students is the trigonal system. Students need to know the differences between the rhombohedral cell and the non-rhombohedral cell in the trigonal system. The non-rhombohedral cell of the trigonal system can only be represented using hexagonal axes although there is no six-fold axis in the trigonal system. The rhombohedral cells of the trigonal system can be represented by either hexagonal axes or rhombohedral axes; the unit cell represented by the hexagonal axes is a non-primitive one as it contains three lattice points per cell. However, in many cases, we just use the hexagonal axes to describe the rhombohedral cells as it is easier to do mathematical treatments with hexagonal coordinates.

Students need to understand how to determine the Miller indices of a crystallographic plane, the indices of a crystallographic direction, and the coordinates of a position in the unit cell. For a hexagonal crystal, the four-index scheme can better represent the symmetries of crystallographic planes and directions.

Crystal defects including point defects, linear defects, planar defects and volume defects have not been discussed in this chapter.

We need to bear in mind that we often use the term "crystal" to represent the "periodic crystal". However, when we need to differentiate between periodic crystal and aperiodic crystal, we use the term "periodic crystal".

REFERENCES

Authier, A. (ed.) (2010) *International tables for crystallography. Volume D, physical properties of crystals/edited by A. Authier.* 1st ed., *Physical properties of crystals.* 1st ed. Chichester, West Sussex, UK: John Wiley & Sons.

Cullity, B. D. (Bernard D.) and Stock, S. R. (2001) *Elements of x-ray diffraction/B.D. Cullity, S.R. Stock.* 3rd ed. Upper Saddle River, NJ: Prentice Hall.

Hahn, T. (2005) *International tables for crystallography. Volume A, Space-group symmetry/ Edited by Theo Hahn.* 5th ed. re, *International tables for crystallography [electronic resource] Volume A, Space-group symmetry.* 5th ed. re. Dordrecht: Published for the International Union of Crystallography by Springer.

Lima-de-Faria, J. (José) and Buerger, M. J. (eds) (1990) *Historical atlas of crystallography/ edited by J. Lima-de-Faria; with the collaboration of M.J. Buerger ... [et al.].* Dordrecht; Published for International Union of Crystallography by Kluwer Academic Publishers.

Xu, C. X. et al. (2006) 'Zinc oxide hexagram whiskers', *Applied Physics Letters*, 88(9), pp. 23–25. doi:10.1063/1.2179133.

2 Symmetry of Crystals, Point Groups and Space Groups

Based on the concept of the translational symmetries discussed previously, a crystal structure can be represented using the combination of a lattice and the motif that is composed of atoms, ions or molecules. Our everyday experience also suggests that crystals exhibit various external symmetries, similar to flowers and the leaves of plants. For example, diamond crystals show beautiful external symmetries, and so do the crystals of ruby and quartz. In crystallography, symmetries include macroscopic symmetry (32 crystallographic point group symmetries) and microscopic symmetry (symmetry elements involving translations: lattices, screw axes and glide planes). We can use the characteristic symmetry features to describe macroscopic symmetry. For example, a snow flake has a six-fold axis perpendicular to the plane of the flake. An octahedron shaped diamond crystal shows four three-fold axes perpendicular to the octahedral cleavage surfaces. In crystallography, symmetry can be described as "invariance to an operation".

2.1 SYMMETRY ELEMENTS AND REPRESENTATIONS

Translational symmetry describes the periodic repetition of a motif or basis across a length or through an area or volume. Crystallographic point symmetry, on the other hand, describes the periodic repetition of atoms (ions) around a point. As presented in many textbooks, the symmetry elements in point symmetry are discussed first. By definition, a symmetry element is a geometrical entity about which a symmetry operation is performed.

In three-dimensional space, a point group symmetry element can be a point, an axis or a plane. For a rotation axis (international notation: n; Schoenflies notation: C_n), mirror reflection plane (international notation: m; Schoenflies notation: σ) and an inversion center (international notation: $\bar{1}$; Schoenflies notation: i), the corresponding symmetry operations are rotation, reflection and inversion as shown in Figure 2.1. Other than rotation, reflection and inversion, there are also symmetry operations with the combined action of rotation by $360/n$ degrees together with inversion through a point on the rotation axis (international notation: $\bar{n}$), or rotation by $360/n$ degrees together with reflection through a plane perpendicular to the rotation axis (Schoenflies notation: S_n). In the following discussion, we will notice that roto-inversion operations and roto-reflection operations are equivalent.

Since translational symmetry imposes restrictions on rotation symmetry, the five-fold axis, seven-fold axis and higher-order axes do not exist in periodic crystals.

DOI: 10.1201/9780429351662-4

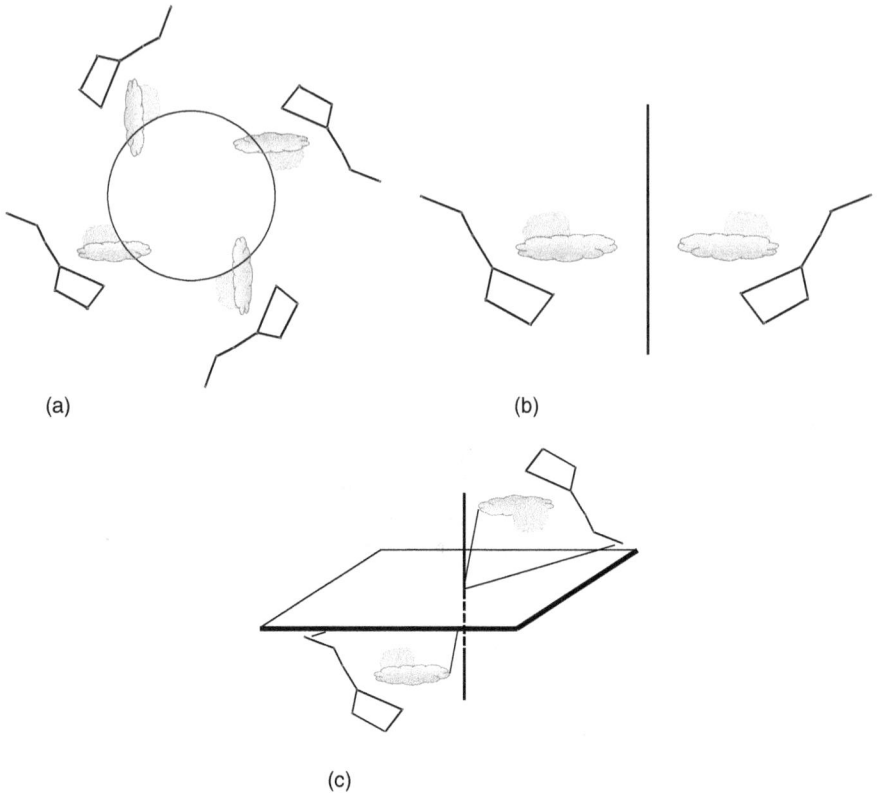

(a)   (b)

(c)

FIGURE 2.1 Schematic representations of (a) rotation, (b) reflection and (c) inversion operations.

However, in a paper published in November 1984, selected area electron diffraction patterns from rapidly solidified alloys of Al with 10–14% Mn exhibited icosahedral symmetry (Shechtman *et al.*, 1984). Since then, quasicrystals have been discovered in many alloy systems. In a research project, I analyzed Al-Cr-Fe quasicrystals and collected the electron diffraction patterns of five-fold and ten-fold symmetries as shown in Figure 2.2 (Li *et al.*, 2016).

   As just mentioned, in periodic crystals, only one-fold, two-fold, three-fold, four-fold and six-fold rotation axes exist. The rotation operation 1 is often called the identity. This symmetry element is included because it is a requirement in the definition of a group in group theory. A schematic representation of rotation symmetry in periodic crystals is given in Figure 2.3.

   The roto-inversion axes $\bar{n}$ (international notation) associated with the roto-inversion operations include $\bar{1}, \bar{2}, \bar{3}, \bar{4}$ and $\bar{6}$. For the corresponding operations, $\bar{1}$ is the inversion i, $\bar{2}$ is the mirror reflection m and $\bar{3}$ contains a three-fold rotation and inversion i. Operation $\bar{4}$ itself is an independent one, which is not equal to $4 + i$ as shown in Figure 2.4. $\bar{6}$ contains a three-fold rotation and reflection m $(3 \perp m)$. If a

FIGURE 2.2 (a) Five-fold rotation axis exhibited in a flower, (b) five-fold rotation axis present in an electron diffraction pattern from a quasicrystal in an Al-Cr-Fe alloy.

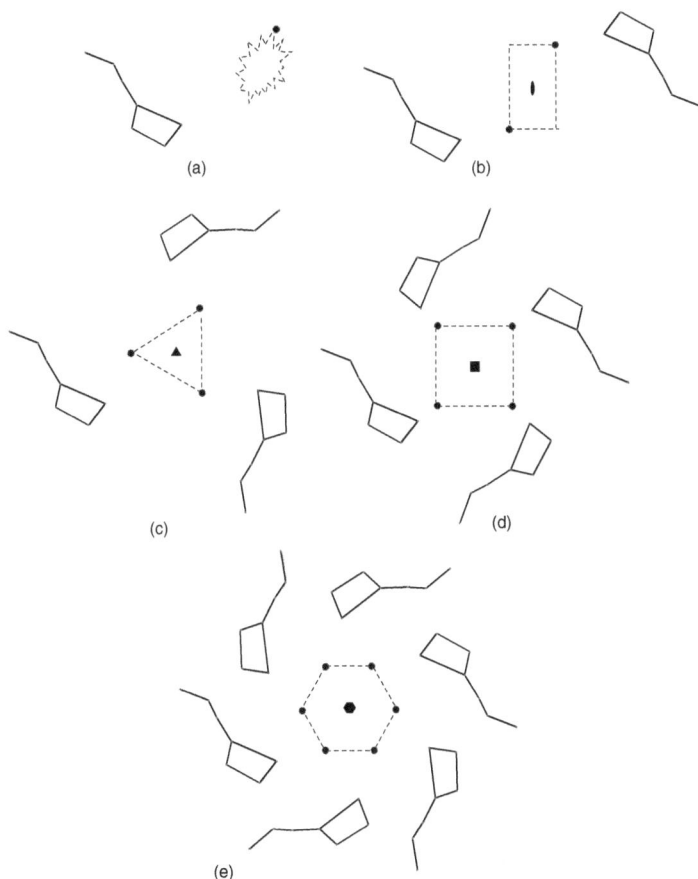

FIGURE 2.3 Rotation axes in periodic crystals. (a) One-fold rotation axis, (b) two-fold rotation axis, (c) three-fold rotation axis, (d) four-fold rotation axis and (e) six-fold rotation axis.

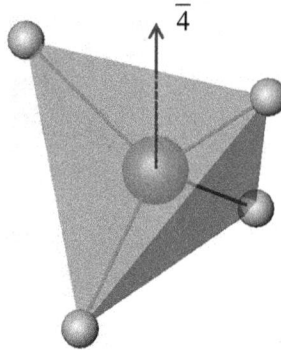

FIGURE 2.4   Schematic representation of $\bar{4}$ symmetry.

TABLE 2.1
Relationship Between Roto-Inversions $\bar{n}$ and Roto-reflections S_n

The roto-inversion operations $\bar{n}$	$\bar{1} = i$	$\bar{2} = m$	$\bar{3} = 3+i$	$\bar{4}$	$\bar{6} = \dfrac{3}{m}$
Roto-reflection operation S_n	$S_1 = m$	$S_2 = i$	$S_3 = \dfrac{3}{m}$	$S_4 = \bar{4}$	$S_6 = 3 + i$

---

crystal contains $3/m$, or $\bar{6}$, it belongs to a hexagonal system, not a trigonal system, because in crystallography, we use $\bar{6}$ and not $3/m$ to identify the crystal system.

Roto-reflection operation S_n (Schoenflies notation) means rotation about an axis followed by a reflection through a plane perpendicular to the axis, and this is equivalent to a roto-inversion operation $\bar{n}$ (international notation). Table 2.1 shows the relationships between $\bar{n}$ and S_n.

To summarize the previous discussions on symmetry operations, which include (i) identity, (ii) rotation, (iii) reflection, (iv) inversion and (v) roto-inversion (or roto-reflection), we find that there are ten unique three-dimensional crystallographic symmetry operations:

$$1, 2, 3, 4, 6, \bar{1}(=i), \bar{2}(=m), \bar{3}(=3+i), \bar{4}, \bar{6}(=3/m)$$

There are different ways to represent symmetry operations, for example, by matrix and by stereographic projection. In the textbook *Essentials of Crystallography* (Wahab, 2009), there is a detailed explanation of the matrix representation of symmetry operations. Stereographic projection can represent a three-dimensional crystal on two-dimensional paper. This is a projection of points from the surface of a sphere onto its equatorial plane as shown in Figure 2.5. The poles representing planes'

(a)   (b)

Reference sphere

N

P'

ϕ

Equator

O

P

Q

$OP = X = R\,\mathrm{tg}\dfrac{\phi}{2}$

O

P

Q

Equator - Projection plane

S - Point of projection

(c)

N

P'

ϕ

Equator

O

P

Q

S - Point of projection

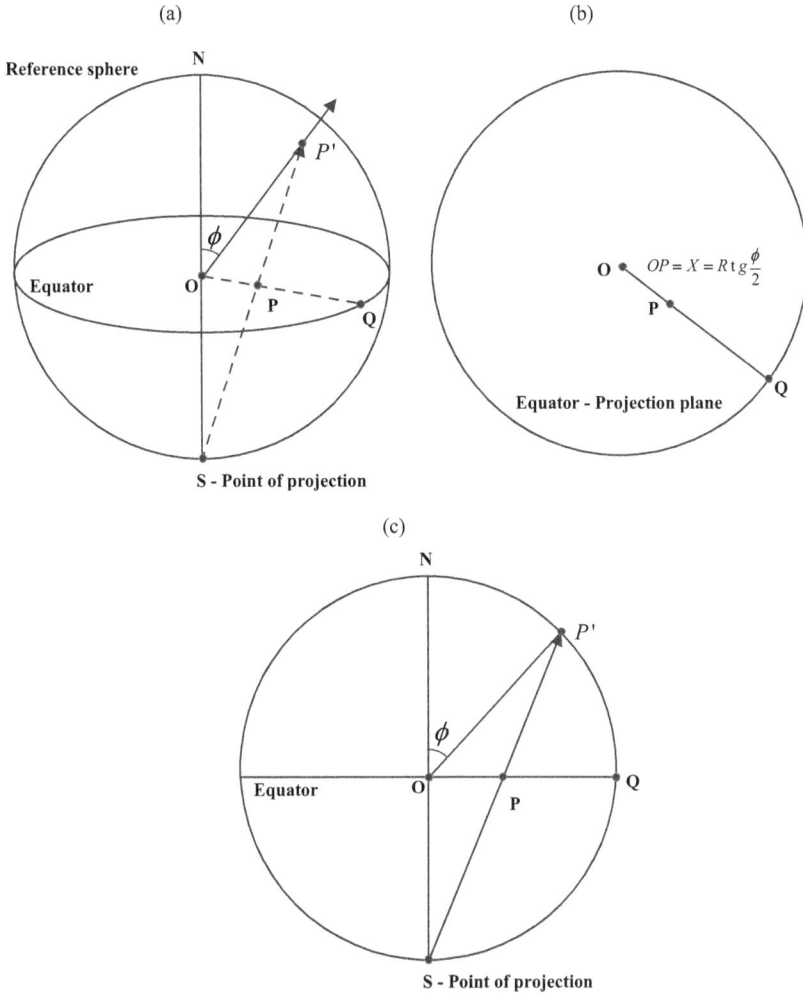

FIGURE 2.5 (a) Stereographic projection showing a point P' from the surface of a sphere onto its equatorial plane. A pole representing a plane normal in the northern (or southern) hemisphere appears as a point on the equatorial plane. (b) Top view showing the projection plane. (c) Vertical section through NS and OP'.

normals in the northern and southern hemispheres appear as points on the equatorial plane. There is an explanation as to why we do not use the parallel beams, but instead the beams connecting the points on the north or south hemisphere to the south or north pole (Hammond, 2015). Analysis shows that the stereographic projection of a circle is a circle (Ladd, 2014). It can also be seen that for distortions along the radial direction, equal angles are represented by smaller distances as we move out from the center, whereas in the case of stereographic projection, the distortion is different (Hammond, 2015). For stereographic projection, the distance from the center to a

TABLE 2.2

Example Demonstration of Symmetry Operations

Graphic Representation of a Four-fold Axis	Matrix Representation of a Four-fold Axis	Stereographic Projection of a Four-fold Axis
Axis along the z direction	For position X = x, y, z in Cartesian coordinates, there is a symmetry equivalent position at X' = −y, x, z after a 90° rotation about the z-axis. Therefore, we can use the expression	Axis along the z direction

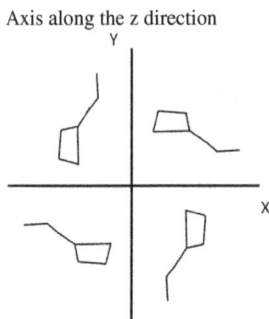

For position X = x, y, z in Cartesian coordinates, there is a symmetry equivalent position at X' = −y, x, z after a 90° rotation about the z-axis. Therefore, we can use the expression

$$\begin{pmatrix} -y \\ x \\ z \end{pmatrix} = \begin{pmatrix} 0 & -1 & 0 \\ 1 & 0 & 0 \\ 0 & 0 & 1 \end{pmatrix} \begin{pmatrix} x \\ y \\ z \end{pmatrix} \text{ or}$$

$$X' = R_4 X$$

where

$$R_4 = \begin{pmatrix} 0 & -1 & 0 \\ 1 & 0 & 0 \\ 0 & 0 & 1 \end{pmatrix} \text{ is}$$

the rotation matrix in this particular orientation.

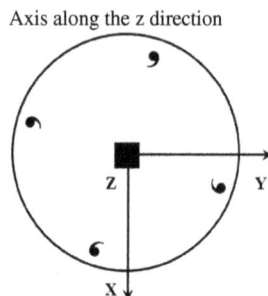

point is $OP = X = Rtg\dfrac{\phi}{2}$, and the equal angles $\Delta\phi$ are represented by larger distances ΔX as we move out from the center. The importance of the stereographic projection in crystallography derives from the fact that a set of points on the surface of the sphere provides a complete representation of a set of directions in three-dimensional space. This method is employed in crystallographic point group analysis to represent the existence of symmetry elements, including the rotation axes, the mirror reflection planes, the inversion center and the roto-inversion axes. Some textbooks draw a projection plane in contact with the north pole or south pole, and the projections of the plane and directions are demonstrated in a projection circle with a radius twice that of the reference sphere (Barrett and Massalski, 1980; Cullity, 1978).

In Table 2.2, we just use a four-fold rotation axis as an example to demonstrate the stereographic projection and matrix representations.

The ten unique symmetry elements can be combined and generate more symmetry features that can be observed in some crystals. In mathematics, the point symmetry operations are summarized using group theory, and the combinations of crystallographic symmetry elements follow the rules of this group theory. For crystallographic point symmetry operations, there are 32 possible combinations, which define the 32 crystallographic point groups. Every periodic crystal belongs to one of these 32 crystallographic point groups.

2.2   COMBINATIONS OF SYMMETRY ELEMENTS

The ten unique crystallographic point groups, each of which only has one rotation axis or improper rotation axis, are

$$1, 2, 3, 4, 6, \bar{1}\left(=i\right), \bar{2}\left(=m\right), \bar{3}\left(=3+i\right), \bar{4}, \bar{6}\left(=3/m\right).$$

The above ten unique symmetry operations can combine and form new point groups. There are rules for symmetry element combinations and they are well described in *Fundamentals of Crystals* (Vainsthein, 1994). In point group symmetry, theorems (I) and (III) are employed, whereas in space group symmetry, theorem (II) is also needed:

(I) The line of intersection of two planes m and m' at an angle $\alpha/2$ is a rotation axis n_α, which is demonstrated in Figure 2.6(a).

(II) Translation t can be obtained by two reflections in planes m spaced by $t/2$, parallel to one another and perpendicular to the translation axis (Figure 2.6(b)).

(III) Euler's theorem tells us that a rotation about two intersecting axes n_{α_1} and n_{α_2} is equivalent to a rotation about a third axis n_{α_3} as shown in Figure 2.7.

For Euler's theorem, the textbook *Essentials of Crystallography* (Wahab, 2009) provides a detailed explanation on how to obtain the α_3 value based on spherical triangle calculation.

In my lectures on the combinations of symmetry elements, the focus is more on the crystallographic point groups, in which case only theorem (I) and theorem (III) are used, as theorem (II) includes translation operations.

In mathematics, a set of elements (operations) is a group if the following properties hold, as explained in the textbook *Symmetry of Crystals and Molecules* (Ladd, 2014):

(I) Closure: The product of any two elements of the group gives a third element of the group.

(II) Association: For any three elements of the group (AB)C = A(BC).

(III) Identity: The group contains an identity element E, such that AE = EA = A for any element of the group.

(IV) Inverses: For each element A, there is an inverse A^{-1} in the group, such that $AA^{-1} = E = A^{-1}A$.

The crystallographic symmetry elements and corresponding operations satisfy group theory, which can be demonstrated from the ten unique symmetry elements and their combinations. There are only 22 possible unique combinations for the 10 rotation and improper rotation axes. Therefore, in crystallography, there are 32 crystallographic point groups in total based on the 10 rotation and improper rotation axes along with 22 combinations of them. A very comprehensive discussion is presented

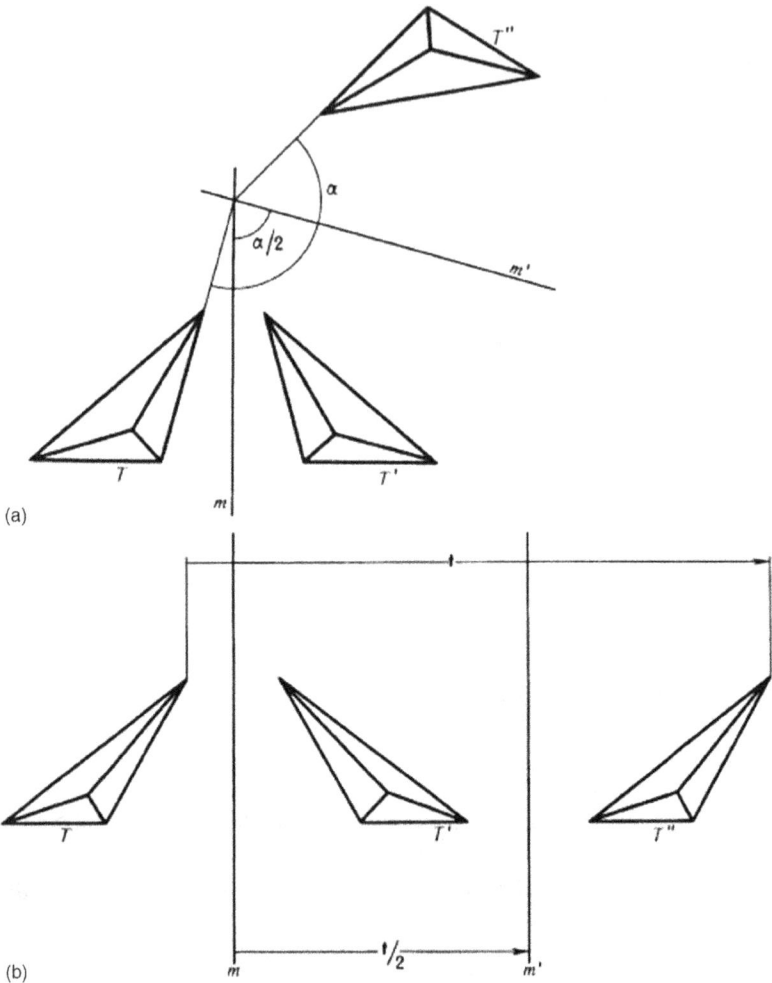

FIGURE 2.6 Motion operation as the result of successive reflections in two planes m and m'. (a) Rotation (planes m and m' intersect, the line of intersection is the rotation axis). The reflection in m transforms figure T to its mirror reflection T', and the reflection in m' gives a figure T'' congruent to T and rotated through angle α. (b) Translation (planes m and m' are parallel).

(Source: Reproduced from Vainsthein (1994), with the permission of Springer Nature.)

in the textbook *Structure of Materials: An Introduction to Crystallography, Diffraction and Symmetry* (De Graef and McHenry, 2012).

To teach materials scientists the concepts of 32 crystallographic point groups, the approaches in the textbook *The 23rd Edition of the Manual of Mineral Science* by Klein *et al.* (2008) and in the textbook *Principles of Symmetry* by Tang (1977) are helpful.

It is easy to understand the non-cubic crystallographic point groups based on the combinations described in Klein *et al.* The key characteristic of the 27 non-cubic

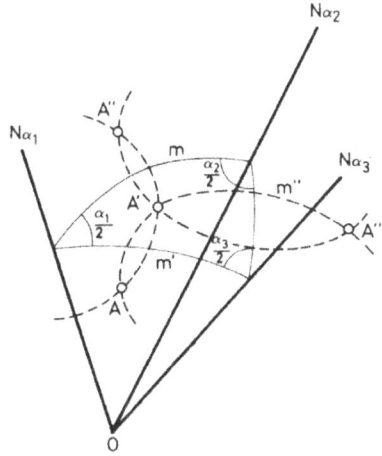

FIGURE 2.7 The derivation of Euler's theorem. The continuous arcs show the intersection of the symmetry planes with a sphere of arbitrary radius with center at the point of intersection of the axes. In the figure, the action of n_{α_1} is replaced by that of planes m' and m, and the action of n_{α_2} is replaced by that of m and m''. The successive reflections here are as follows: A in m' gives A', A' in m gives A'', and this is equivalent to a rotation of n_{α_1} . Similar explanations apply to n_{α_2} . A''' is congruent to A and rotated about axis n_{α_3} , which is the line of intersection of planes m' and m''. The value of angle α_3 depends on α_1, α_2 and the angle between axes n_{α_1} and n_{α_2}.

(Source: Reproduced from Vainsthein (1994), with the permission of Springer Nature.)

point groups is that either there is no higher-order axis (three-fold, four-fold or six-fold) or there is only one higher-order axis. For cubic point groups, several higher-order axes exist.

(I) Rotation axis only.

In each point group, there is only one rotation axis n.

n can be 1, 2, 3, 4 or 6. There are five point groups. Rotation axis 1 is just the identity (E).

(II) Roto-inversion axis only.

In each point group, there is only one $\bar{n}$.

$\bar{n}$ can be $\bar{1}, \bar{2}(=m), \bar{3}(=3+\bar{1}), \bar{4}, \bar{6}(=3/m)$. There are five point groups.

You can see that each of the above ten symmetry operations is only related to a single symmetry element. For the 22 point groups to be discussed below, each is associated with more than one symmetry element.

(III) A combination of rotation axes.

There are four point groups, which are 222, 32, 422, 622.

For rotation axis combination, 32 is used for the trigonal system, and 622 is used for the hexagonal system. Such notation reflects the fundamental facts that the point symmetry of the trigonal system contains one primary axis and three secondary axes, but there are no tertiary axes in the hexagonal system

(Donnay, 1969). In trigonal space groups, the lattice letter R or P is followed by two symmetry-element symbols, whereas in hexagonal space-group symbols the lattice letter P is followed by three symbols. Euler's Theorem can be used to interpret the combination of axes.

(IV)   One rotation axis with a perpendicular mirror.

In point groups $\dfrac{n}{m}$, axis n is normal to mirror m.

Although there are four point groups $\dfrac{2}{m}, \dfrac{3}{m}, \dfrac{4}{m}$ and $\dfrac{6}{m}$ for $\dfrac{n}{m}$. $\dfrac{3}{m}$ is the same as $\bar{6}$, and we use symbol $\bar{6}$ instead of $\dfrac{3}{m}$ in the 32 crystallographic point group list. $\bar{6}$ belongs to the hexagonal crystal system, not to the trigonal system. For $\dfrac{2}{m}, \dfrac{4}{m}$ and $\dfrac{6}{m}$, an inversion center is created due to the combination of an even-fold rotation axis with a mirror plane perpendicular to it.

(V)   One rotation axis with parallel mirrors.

The point groups include 2 mm, 3 m, 4 mm, 6 mm.

In the trigonal system, two symmetry element symbols 3 m are used, and in the hexagonal system three symmetry element symbols 6 mm are used. Theorem (I) can be used to interpret the combinations of one rotation axis with parallel mirror planes.

(VI)   Roto-inversion with rotation and mirror.

There are three point groups, which are $\bar{3}\dfrac{2}{m}$, $\bar{4}2m$ and $\bar{6}2m$.

$\bar{2}$ combined with a rotation perpendicular to it or combined with a mirror parallel to it is the same as 2 mm in (V) (Girolami, 2016).

$\bar{3}\dfrac{2}{m}$, $\bar{4}2m$ and $\bar{6}2m$ are new point groups.

(VII)   Three rotation axes and perpendicular mirrors.

There are three point groups, which are $\dfrac{2}{m}\dfrac{2}{m}\dfrac{2}{m}, \dfrac{4}{m}\dfrac{2}{m}\dfrac{2}{m}, \dfrac{6}{m}\dfrac{2}{m}\dfrac{2}{m}$.

An inversion center is generated due to the combination of an even-fold rotation axis with a mirror plane perpendicular to it.

(VIII)   Isometric patterns containing several higher-order axes.

The presentation on the point groups with several higher-order axes is very different, and we discuss them through a similar approach as presented in Tang (1977).

2.3   POINT GROUPS FOR THE CUBIC SYSTEM

Every point group in the cubic system contains several higher-order axes. The polyhedron exhibits such external symmetries and can be derived as follows.

Imagine that we draw a polyhedron having at least two higher-order rotation axes L^n_1 and L^m_1, and that the two axes intercept at point "O". We assume that L^n_1 is an n-fold rotation axis and L^m_1 is an m-fold rotation axis, with the angle between

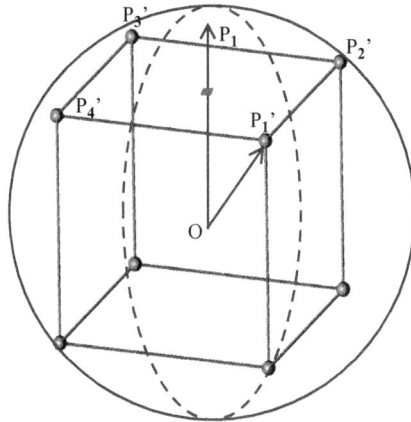

FIGURE 2.8 An example of cubic symmetry showing the combinations of three four-fold rotation axes and four three-fold rotation axes.

L^n_1 and L^m_1 being not greater than the angle between any two other higher-order rotation axes present in the group. We can imagine that there is a sphere with a center at "O", and the intercepts of axes L^n_1 and L^m_1 with the sphere are points P_1 and P_1' (Figure 2.8).

As L^n_1 is an n-fold rotation axis, around L^n_1 there should be other m-fold axes L^m_2, L^m_3, ..., L^m_n along with L^m_1. The intercepts of those (n−1) m-fold rotation axes with the sphere are P_2', P_3', ... and P_n'. We can draw lines between P_1' and P_2', between P_2' and P_3', ..., between P_{n-1}' and P_n', and between P_n' and P_1', which generates a regular polygon of n sides. P_1', P_2', ... and P_n' are the convex corners of the regular polygon. The intercept of the axis L^n_1 with the polygon mentioned above is located in the center of the polygon. Within the polygon, only axis L^n_1 has an intercept with it, and there is no other intercept from any other higher-order rotation axis.

Since L^m_1, L^m_2, ..., L^m_n are m-fold rotation axes, around each of the m-fold axis there should be m n-fold axes. Therefore, each of the convex corners P_1', P_2', ... and P_n' is surrounded by m of the same sized regular polygons (Figure 2.8).

To summarize the above. From the sphere we drew, a regular polyhedron composed of regular polygons of n-sides is obtained. The polyhedron is inscribed in the sphere, and each vertex of the polyhedron is surrounded by m n-sided regular polygons.

At any vertex of the polyhedron, there are m interior angles from m polygons. Each of the interior angles from an n-sided regular polygon is $\left(\dfrac{n\pi - 2\pi}{n} \right)$. So the sum of m interior angles is $m\left(\dfrac{n\pi - 2\pi}{n} \right) = \dfrac{m(n-2)}{n}\pi$, which should be less than 2π.

Figure 2.8 is just an example of when a cube is generated, in which case m = 3 and n = 4.

Therefore, the possible n and m value combinations and shapes of regular polyhedra are shown in Table 2.3.

TABLE 2.3

Rotation Axes and Shapes of Polyhedra in a Cubic System

$\dfrac{m(n-2)}{n}\pi < 2\pi;$

n: polygon
of n sides;
m: vertex
surrounded by m
n-sided polygons

n	m	Shape of Polyhedron	Rotation Axes	Point Group Using Shoeflies Notation
3	3	Regular tetrahedron 	Regular tetrahedron Having three two-fold axes, and four three-fold axes: $3L^2 4L^3$ (Vainsthein, 1994)	T
4	3	Cube 	Cube Having three four-fold axes, four three-fold axes and six two-fold axes: $3L^4 4L^3 6L^2$	O
3	4	Regular octahedron 	Regular octahedron Having three four-fold axes, four three-fold axes and six two-fold axes: $3L^4 4L^3 6L^2$	O

(Continued)

TABLE 2.3 (Continued)

$$\frac{m(n-2)}{n}\pi < 2\pi;$$

n: polygon of n sides;
m: vertex surrounded by m n-sided polygons

n	m	Shape of Polyhedron	Rotation Axes	Point Group Using Shoeflies Notation
5	3	Regular dodecahedron	Regular dodecahedron Not in periodic crystals	I
3	5	Regular icosahedron	Regular icosahedron Not in periodic crystals	I

The first three groups belong to the cubic system, and the last two groups containing a five-fold axis belong to aperiodic crystal systems. A quasicrystal description is not within the scope of our discussions.

The symmetry elements in the tetrahedron are 3L2 4L3, and in the cube and the octahedron are 3L4 4L3 6L2. In the symmetry formulas, symbols L, P and C are used to represent the rotation axis, the center of symmetry and the mirror plane following Table 2.3 in Vainsthein (1994).

We can add mirror planes or inversion centers to the above symmetry element combinations, which will not result in additional higher-order rotation axes. And in total there are three "T" point groups and two "O" point groups.

TABLE 2.4

Thirty-two Crystallographic Point Groups in Periodic Crystals (Full Symbol) (Klein *et al.*, 2008)

	Increasing Rotational Symmetry				
	1	2	3	4	6
Rotation Axis Only	1	2	3	4	6
Roto-inversion axis only	$\bar{1}(=i)$	$\bar{2}(=m)$	$\bar{3}$	$\bar{4}$	$\bar{6}\left(=\dfrac{3}{m}\right)$
Combination of rotation axes		222	32	422	622
One rotation with perpendicular mirror		$\dfrac{2}{m}$	$\dfrac{3}{m}(=\bar{6})$	$\dfrac{4}{m}$	$\dfrac{6}{m}$
One rotation with parallel mirrors		2mm	3m	4mm	6mm
Roto-inversion with rotation and mirror		(The same as 2mm)	$\bar{3}\dfrac{2}{m}$	$\bar{4}\,2m$	$\bar{6}\,2m$
Three rotation axes with perpendicular mirrors		$\dfrac{2}{m}\dfrac{2}{m}\dfrac{2}{m}$		$\dfrac{4}{m}\dfrac{2}{m}\dfrac{2}{m}$	$\dfrac{6}{m}\dfrac{2}{m}\dfrac{2}{m}$
Combination of several higher-order axes			$23,\ \dfrac{2}{m}\bar{3},\ 432,$ $\bar{4}\,3m,\ \dfrac{4}{m}\bar{3}\dfrac{2}{m}$ (Four three-fold axes are characteristics)		

(I) Symmetry elements associated with tetrahedrons are 3L2 4L3
 Add m → 3L2 4L3 6P, which is T$_d$
 Add inversion center $\bar{1}$ → 3L2 4L3 3P C, which is T$_h$
(II) Symmetry elements associated with the cube and the octahedron are 3L4 4L3 6L2
 Add mirror m or inversion center $\bar{1}$ → 3L4 4L3 6L2 9P C, which is O$_h$
 Therefore, there are five cubic point groups in total.

Table 2.4 lists the 32 crystallographic point groups according to the method presented in Klein *et al.* (2008).

2.4 THIRTY-TWO CRYSTALLOGRAPHIC POINT GROUPS AND 230 SPACE GROUPS

Based on the symmetry characteristics, the 32 crystallographic point groups can be classified into seven crystal systems as shown in Figure 1.5. Each crystal system has its lattice constraints due to the symmetry elements in the crystal. The order of the point group symbols for each crystal system highly reflects the symmetry characteristics of that system, and they are presented in Table 2.5.

TABLE 2.5

Crystal Systems and the 32 Crystallographic Point Groups.

Crystal System	Point Groups (Short Symbol) (Hahn, 2011)	Symbol Meaning for Each Position (Ladd & Palmer, 2003)			Symmetry Constraints on Cell Parameters
		1st Position	2nd Position	3rd Position	
Triclinic	$1, \bar{1}$	All directions in crystal			None
Monoclinic	$2, m, 2/m$	2 and/or $\bar{2}$ along y			$\alpha = \gamma = 90°$ (b-unique setting)
Orthorhombic	$222, mm\,2, mmm$	2 and/or $\bar{2}$ along x	2 and/or $\bar{2}$ along y	2 and/or $\bar{2}$ along z	$\alpha = \beta = \gamma = 90°$
Tetragonal	$4, \bar{4}, 4/m$	4 and/or $\bar{4}$, along z			$a = b$
	$422, 4mm, \bar{4}\,2m, 4/mmm$	4 and/or $\bar{4}$, along z	2 and/or $\bar{2}$ along x, y	2 and/or $\bar{2}$ along <110>	$\alpha = \beta = \gamma = 90°$
Trigonal	$3, \bar{3}$	3 and/or $\bar{3}$, along z			$a = b$
	$32, 3m, \bar{3}\,m$	3 and/or $\bar{3}$, along z	2 and/or $\bar{2}$ along x, y, u		$\alpha = \beta = 90°$ $\gamma = 120°$ (non-rhombohedral cell) $a = b = c$ $\alpha = \beta = \gamma \ne 90°$ (rhombohedral cell)
Hexagonal	$6, \bar{6}, 6/m$	6 and/or $\bar{6}$, along z			$a = b$
	$622, 6mm, \bar{6}\,m\,2, 6/mmm$	6 and/or $\bar{6}$, along z	2 and/or $\bar{2}$ along x, y, u	2 and/or $\bar{2}$ perpendicular to x, y, u and in xy plane	$\alpha = \beta = 90°$ $\gamma = 120°$
Cubic	$23, m\bar{3}$	2 and/or $\bar{2}$ along x, y, z	3 and/or $\bar{3}$, along <11>		$a = b = c$
	$432, \bar{4}\,3m, m\bar{3}\,m$	4 and/or $\bar{4}$, along x, y, z	3 and/or $\bar{3}$, along <11>	2 and/or $\bar{2}$ along <110>	$\alpha = \beta = \gamma = 90°$

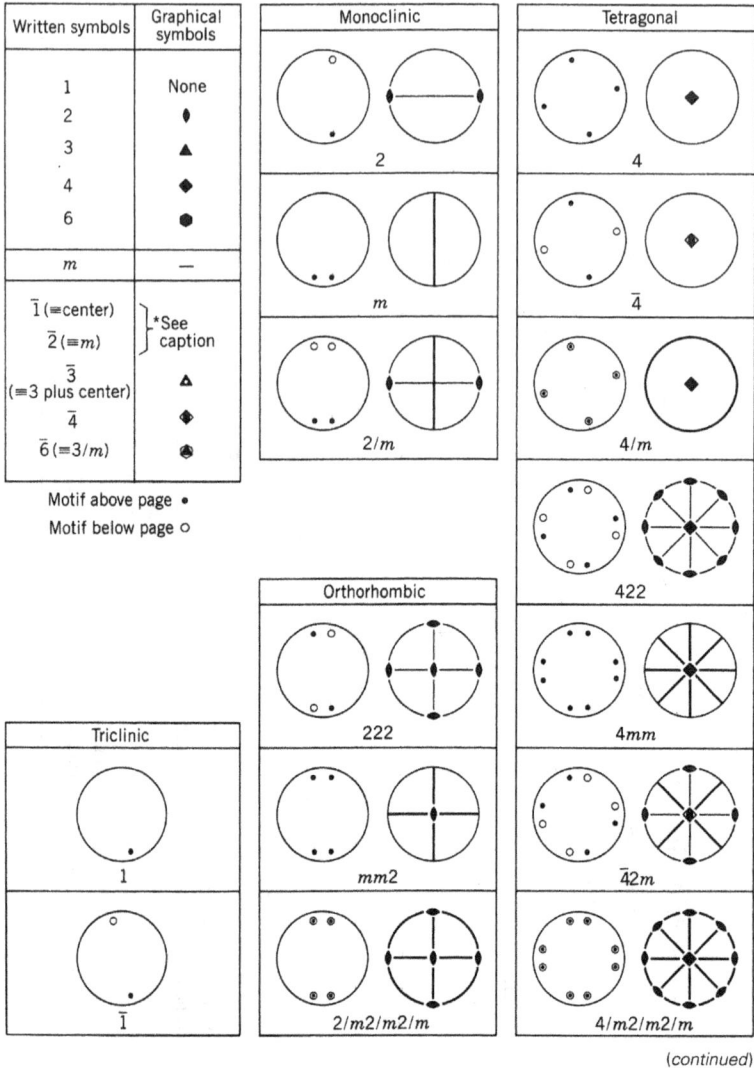

FIGURE 2.9 Stereographic projection representations of the 32 point groups.

(Source: Reproduced from Klein *et al.* (2008), with the permission of John Wiley & Sons, Inc.)

The stereographic projection representations of the 32 point groups are shown in Figure 2.9.

The seven crystal systems, combined with the four lattice types, can generate fourteen Bravais lattices, as discussed in Chapter 1. Some centering types are not allowed for some crystal systems because they would lower the symmetry of the unit cell. For example, the one-face-centered cubic lattice does not exist, as the three-fold

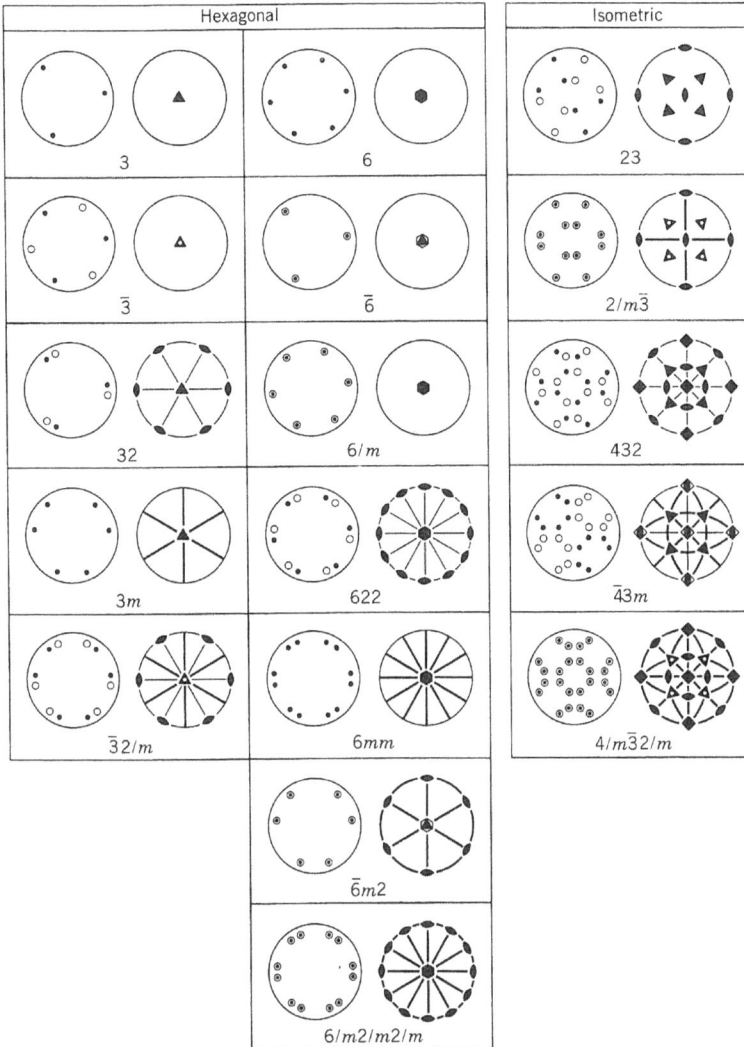

	Hexagonal		Isometric

(Hexagonal column diagrams: 3, 6; 3̄, 6̄; 32, 6/m; 3m, 622; 3̄2/m, 6mm; 6̄m2; 6/m2/m2/m)

(Isometric column diagrams: 23; 2/m3̄; 432; 4̄3m; 4/m3̄2/m)

FIGURE 2.9 (*Continued*)

axes along the diagonal directions do not allow it. Some centering types are unnecessary. For example, the one-face-centered tetragonal can be described using a smaller tetragonal primitive unit cell. It is worthwhile mentioning that, in the trigonal system with a rhombohedral cell, R is used for both the rhombohedral description (primitive cell) and the hexagonal description (triple cell).

In X-ray diffraction, only the 11 centro-symmetric crystallographic point groups, or Laue groups, can be distinguished due to the phase problem. Table 2.6 lists the Laue groups that are associated with the crystallographic point groups.

TABLE 2.6

Thirty-two Crystallographic Point Groups and Eleven Laue Groups

Crystal System	Point Group (Hahn, 2011)	Laue Group
Triclinic	1, $\bar{1}$	$\bar{1}$
Monoclinic	2, m, 2/m	2/m
Orthorhombic	222, mm2, mmm	mmm
Tetragonal	4, $\bar{4}$, 4/m	4/m
	422, 4mm, $\bar{4}$ 2m, 4/mmm	4/mmm
Trigonal	3, $\bar{3}$	$\bar{3}$
	32, 3m, $\bar{3}$ m	$\bar{3}$ m
Hexagonal	6, $\bar{6}$, 6/m	6/m
	622, 6mm, $\bar{6}$ m2, 6/mmm	6/mmm
Cubic	23, m3	m $\bar{3}$
	432, $\bar{4}$ 3m, m $\bar{3}$ m	m $\bar{3}$ m

In the 32 crystallographic point groups, only identity, rotation, reflection, inversion and roto-inversion (or roto-reflection) operations are included, and this kind of symmetry is called macroscopic symmetry. When translation components are incorporated, the group is called a space group. The following is a basic discussion of space group operations.

If a rotation is followed by a translation operation, it becomes a screw operation, as shown in the example in Figure 2.10. In total, there are 11 screw operations: $2_1, 3_1, 3_2, 4_1, 4_2, 4_3, 6_1, 6_2, 6_3, 6_4$ and 6_5. If a reflection is followed by a translation operation, then it is called a glide operation. Figure 2.11 is a schematic representation of glide operation b.

There are 230 possible combinations among all the available crystallographic symmetry operations, including point symmetry operations, screw operations, glide operations and Bravais lattice translations. The 230 combinations are called 230 space groups. In the projection descriptions, some graphical symbols are employed for symmetry elements. Students can find a complete list of the graphical symbols for the symmetry elements in Volume A of the *International Tables for Crystallography*.

We can use apatite materials as an example to explain the space group symmetry elements and symbols. Apatite materials have a general formula, $A^I_4 A^{II}_6 (BO_4)_6 X_2$, where A^I and A^{II} are cation sites with different symmetries; BO_4 is the tetrahedral site and X is the anion site. Many of the apatite materials belong to the space group No.176, $P6_3/m$, which is the Short Hermann–Mauguin space group symbol. P refers to the primitive unit cell; 6_3 refers to the screw axis along c of the hexagonal lattice; m represents the mirror plane perpendicular to the screw axis 6_3. C^2_{6h} is the Shoenflies symbol. The space-group diagram consists of the projections of symmetry elements. The space-group tables (Hahn, 2011) also contain general and special positions, as well as their reflection conditions.

Translational component

c/2

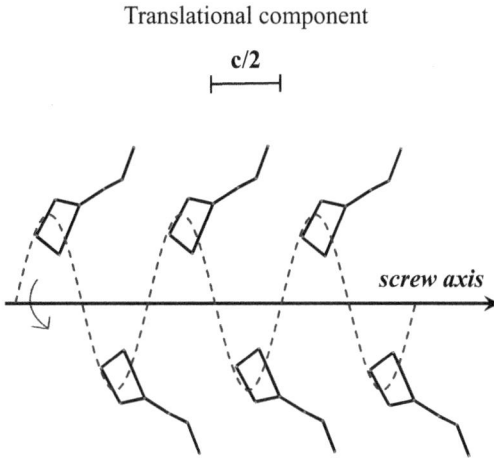

screw axis

FIGURE 2.10   Schematic representation of a 2_1 screw axis and associated screw operation.

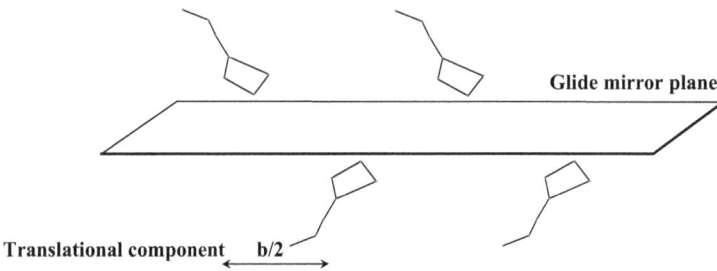

Glide mirror plane

Translational component   b/2

FIGURE 2.11   Schematic representation of glide mirror plane and b-glide operation.

SUMMARY

Translational symmetry in three-dimensional periodic crystals imposes limitations to rotation axes. Only two, three, four and six-fold rotation axes allow for space filling translational symmetry.

The permitted combinations of point symmetry elements, or external symmetry elements, form crystallographic point groups. There are 32 crystallographic point groups, which can be classified into 7 crystal systems based on the presence of the characteristic symmetry elements.

The combinations of all possible crystallographic symmetry operations, including 32 point symmetry operations, glide operations and screw operations, together with the Bravais lattice translations, have exactly 230 types, which are called the 230 space groups.

REFERENCES

Barrett, C. S. (Charles S.) and Massalski, T. B. (1980) *Structure of metals: crystallographic methods, principles and data/C.S. Barrett, T.B. Massalski*. 3rd rev. ed. Oxford: Pergamon (International series on materials science and technology; v. 35).

Cullity, B. D. (Bernard D.) (1978) *Elements of x-ray diffraction*. 2nd ed. Reading, MA: Addison-Wesley Pub. Co. (Addison-Wesley series in metallurgy and materials).

Donnay, J. D. H. (1969) 'Symbolism of rhombohedral space groups in Miller axes', *Acta Crystallographica Section A*. International Union of Crystallography (IUCr), 25(6), pp. 715–716. doi:10.1107/S056773946900163X.

Girolami, G. S. (2016) *X-ray crystallography/Gregory S. Girolami (Professor of Chemistry, University of Illinois at Urbana-Champaign)*. Mill Valley, CA: University Science Books.

De Graef, M. and McHenry, M. E. (2012) *Structure of materials: an introduction to crystallography, diffraction and symmetry/Marc De Graef, Carnegie Mellon University, Pittsburgh, Michael E. McHenry, Carnegie Mellon University, Pittsburgh*. 2nd ed. Cambridge: Cambridge University Press.

Hahn, T. (2011) *International tables for crystallography. Volume A, Space-group symmetry/ edited by Theo Hahn*. 5th ed. re, *Space-group symmetry*. 5th ed. re. Chichester, West Sussex: Published for the International Union of Crystallography by John Wiley & Sons.

Hammond, C. (2015) *The basics of crystallography and diffraction/Christopher Hammond*. 4th ed. Oxford: Oxford University Press (International Union of Crystallography monographs on crystallography; 21).

Klein, C. et al. (2008) *The 23rd edition of the manual of mineral science: (after James D. Dana)/Cornelis Klein, Barbara Dutrow*. 23rd ed., *Mineral science*. 23rd ed. Hoboken, N.J: J. Wiley.

Ladd, M. F. C. (2014) *Symmetry of Crystals and Molecules*, 1st ed. Oxford: Oxford University Press

Ladd, M. F. C. & Palmer, R. A. (2003). *Structure determination by X-ray crystallography*, 4thed. New York: Kluwer Academic/Plenum Publishers

Li, R. T., Dong, Z. L. and Khor, K. A. (2016) 'Al-Cr-Fe quasicrystals as novel reinforcements in Ti based composites consolidated using high pressure spark plasma sintering', *Materials and Design*, 102, pp. 255–263. doi:10.1016/j.matdes.2016.04.040.

Shechtman, D. et al. (1984) 'Metallic phase with long-range orientational order and no translational symmetry', *Physical Review Letters*. American Physical Society, 53(20), pp. 1951–1953. doi:10.1103/PhysRevLett.53.1951.

Tang, Y. Q. (1977) *Principles of symmetry/Tang, Y. Q*. Beijing: Chinese Scientific Publishers.

Vainsthein, B. K. (1994) *Fundamentals of crystals: symmetry, and methods of structural crystallography*. 2nd ed. Berlin, Heidelberg: Springer Berlin/Heidelberg. doi:10.1007/978-3-662-02975-6.

Wahab, M. A. (2009) *Essentials of crystallography/M.A. Wahab*. Oxford, UK: Alpha Science International.

3 Reciprocal Lattices

A reciprocal lattice is mathematically defined based on a direct lattice. It is an effective tool in resolving crystallography and diffraction problems. It is also employed in quantum mechanics and solid state physics.

3.1 DEFINITION AND LATTICE PARAMETER RELATIONSHIPS

If the unit cell vectors of a direct lattice $\vec{a}, \vec{b}, \vec{c}$, and the cell parameters a, b, c and α, β, γ, are given, the unit cell vectors of the corresponding reciprocal lattice, $\vec{a}^*, \vec{b}^*, \vec{c}^*$, and cell parameters, a^*, b^*, c^* and $\alpha^*, \beta^*, \gamma^*$, are defined by the following equations:

$$\vec{a}^* = \frac{\vec{b} \times \vec{c}}{\vec{a} \cdot \vec{b} \times \vec{c}} = \frac{\vec{b} \times \vec{c}}{V}$$

$$\vec{b}^* = \frac{\vec{c} \times \vec{a}}{\vec{b} \cdot \vec{c} \times \vec{a}} = \frac{\vec{c} \times \vec{a}}{V} \qquad (3.1)$$

$$\vec{c}^* = \frac{\vec{a} \times \vec{b}}{\vec{c} \cdot \vec{a} \times \vec{b}} = \frac{\vec{a} \times \vec{b}}{V}$$

where $\vec{a} \cdot \vec{b} \times \vec{c} = \vec{b} \cdot \vec{c} \times \vec{a} = \vec{c} \cdot \vec{a} \times \vec{b} = V$ and V is the unit cell volume in the direct lattice. The above definition is equivalent to:

$$\vec{a}^* \cdot \vec{b} = \vec{a}^* \cdot \vec{c} = \vec{b}^* \cdot \vec{a} = \vec{b}^* \cdot \vec{c} = \vec{c}^* \cdot \vec{a} = \vec{c}^* \cdot \vec{b} = 0$$

$$\vec{a}^* \cdot \vec{a} = \vec{b}^* \cdot \vec{b} = \vec{c}^* \cdot \vec{c} = 1 \qquad (3.2)$$

It is obvious that the reciprocal lattice of a reciprocal lattice is the direct lattice.

Based on the definition, the unit cell parameters between the direct lattice and the reciprocal lattice can be correlated by vector operations.

The relationships between direct and reciprocal lattice cell parameters are summarized in Table 3.1. Many crystallography textbooks have discussed the relationship between direct and reciprocal lattice parameters, for example *Essentials of Crystallography* (Wahab, 2009) and *Fundamentals of Crystallography* (Giacovazzo, 2011).

The formula for the volume of the parallelepiped of a unit cell can be found in some textbooks. As $V = \vec{a} \cdot \vec{b} \times \vec{c} = \vec{b} \cdot \vec{c} \times \vec{a} = \vec{c} \cdot \vec{a} \times \vec{b}$ for a parallelepiped, there is no

DOI: 10.1201/9780429351662-5

TABLE 3.1

Cell Parameters of the Reciprocal Lattice in Relation to the Direct Lattice

$$a^* = \frac{bc\sin\alpha}{V} \qquad a = \frac{b^*c^*\sin\alpha^*}{V^*}$$

$$b^* = \frac{ca\sin\beta}{V} \qquad b = \frac{c^*a^*\sin\beta^*}{V^*}$$

$$c^* = \frac{ab\sin\gamma}{V} \qquad c = \frac{a^*b^*\sin\gamma^*}{V^*}$$

$$V^* = \frac{1}{V} \qquad V = \frac{1}{V^*}$$

$$\sin\alpha^* = \frac{V}{abc\sin\beta\sin\gamma} \qquad \sin\alpha = \frac{V^*}{a^*b^*c^*\sin\beta^*\sin\gamma^*}$$

$$\sin\beta^* = \frac{V}{abc\sin\gamma\sin\alpha} \qquad \sin\beta = \frac{V^*}{a^*b^*c^*\sin\gamma^*\sin\alpha^*}$$

$$\sin\gamma^* = \frac{V}{abc\sin\alpha\sin\beta} \qquad \sin\gamma = \frac{V^*}{a^*b^*c^*\sin\alpha^*\sin\beta^*}$$

$$\cos\alpha^* = \frac{\cos\beta\cos\gamma - \cos\alpha}{\sin\beta\sin\gamma} \qquad \cos\alpha = \frac{\cos\beta^*\cos\gamma^* - \cos\alpha^*}{\sin\beta^*\sin\gamma^*}$$

$$\cos\beta^* = \frac{\cos\gamma\cos\alpha - \cos\beta}{\sin\gamma\sin\alpha} \qquad \cos\beta = \frac{\cos\gamma^*\cos\alpha^* - \cos\beta^*}{\sin\gamma^*\sin\alpha^*}$$

$$\cos\gamma^* = \frac{\cos\alpha\cos\beta - \cos\gamma}{\sin\alpha\sin\beta} \qquad \cos\gamma = \frac{\cos\alpha^*\cos\beta^* - \cos\gamma^*}{\sin\alpha^*\sin\beta^*}$$

need to use the definition of the reciprocal lattice in unit cell volume calculation, or to use the associated properties. The expressions are given below.

$$V = abc\sqrt{1-\cos^2\alpha-\cos^2\beta-\cos^2\gamma+2\cos\alpha\cos\beta\cos\gamma} \qquad (3.3)$$

$$V^* = a^*b^*c^*\sqrt{1-\cos^2\alpha^*-\cos^2\beta^*-\cos^2\gamma^*+2\cos\alpha^*\cos\beta^*\cos\gamma^*} \qquad (3.4)$$

As explained in Wahab (2009), the unit cell vectors $\vec{a}$, $\vec{b}$ and $\vec{c}$ are expressed using the Cartesian system:

$$\vec{a} = a_x\vec{i} + a_y\vec{j} + a_z\vec{k}$$
$$\vec{b} = b_x\vec{i} + b_y\vec{j} + b_z\vec{k}$$
$$\vec{c} = c_x\vec{i} + c_y\vec{j} + c_z\vec{k}$$

In this case, the unit cell volume can be written as

$$V_{cell} = \vec{a} \cdot \vec{b} \times \vec{c} = \begin{vmatrix} a_x & a_y & a_z \\ b_x & b_y & b_z \\ c_x & c_y & c_z \end{vmatrix}$$

As the value of the determinant remains unchanged when its rows and columns are interchanged, it follows that

$$V_{cell}^2 = \begin{vmatrix} a_x & a_y & a_z \\ b_x & b_y & b_z \\ c_x & c_y & c_z \end{vmatrix} \times \begin{vmatrix} a_x & a_y & a_z \\ b_x & b_y & b_z \\ c_x & c_y & c_z \end{vmatrix}$$

$$= \begin{vmatrix} a_x & a_y & a_z \\ b_x & b_y & b_z \\ c_x & c_y & c_z \end{vmatrix} \times \begin{vmatrix} a_x & b_x & c_x \\ a_y & b_y & c_y \\ a_z & b_z & c_z \end{vmatrix} = \begin{vmatrix} \vec{a}\cdot\vec{a} & \vec{a}\cdot\vec{b} & \vec{a}\cdot\vec{c} \\ \vec{b}\cdot\vec{a} & \vec{b}\cdot\vec{b} & \vec{b}\cdot\vec{c} \\ \vec{c}\cdot\vec{a} & \vec{c}\cdot\vec{b} & \vec{c}\cdot\vec{c} \end{vmatrix}$$

From the above V_{cell}^2 expression, the unit cell volume equation can be obtained. Now, students are encouraged to complete the above V_{cell}^2 calculation and obtain the following V_{cell} expression:

$$V_{cell} = abc\sqrt{1 - \cos^2\alpha - \cos^2\beta - \cos^2\gamma + 2\cos\alpha\cos\beta\cos\gamma}$$

The reciprocal lattice of the reciprocal lattice is the direct lattice, which can be observed from the crystallographic reciprocal lattice definition. From the definitions of the reciprocal lattice, the reciprocal lattice constants a*, b* and c* can be calculated from the direct lattice parameters. For example, from $\vec{a}* = \dfrac{\vec{b}\times\vec{c}}{\vec{a}\cdot\vec{b}\times\vec{c}}$, we have $a^* = \dfrac{bc\sin\alpha}{V}$. Similarly, we can obtain expressions for b* and c*. In many crystallography textbooks, there are detailed mathematical derivations of reciprocal lattice parameters, and in the following we provide some examples of derivations with similar or different methods.

Example 3.1 Prove that the following two definitions are equivalent.

Definition I.

$$\vec{a}^*\cdot\vec{b} = \vec{a}^*\cdot\vec{c} = \vec{b}^*\cdot\vec{a} = \vec{b}^*\cdot\vec{c} = \vec{c}^*\cdot\vec{a} = \vec{c}^*\cdot\vec{b} = 0$$
$$\vec{a}^*\cdot\vec{a} = \vec{b}^*\cdot\vec{b} = \vec{c}^*\cdot\vec{c} = 1$$

Definition II.

$$\vec{a}* = \frac{\vec{b}\times\vec{c}}{\vec{a}\cdot\vec{b}\times\vec{c}}$$
$$\vec{b}* = \frac{\vec{c}\times\vec{a}}{\vec{b}\cdot\vec{c}\times\vec{a}}$$
$$\vec{c}* = \frac{\vec{a}\times\vec{b}}{\vec{c}\cdot\vec{a}\times\vec{b}}$$

where $\vec{a} \cdot \vec{b} \times \vec{c} = \vec{b} \cdot \vec{c} \times \vec{a} = \vec{c} \cdot \vec{a} \times \vec{b} = V$.

Solution: From Definition I, we know that $\vec{a}^* \cdot \vec{b} = \vec{a}^* \cdot \vec{c} = 0$, or $\vec{a}^*$ is perpendicular to both $\vec{b}$ and $\vec{c}$. Therefore, we can write $\vec{a}^* = k \cdot \vec{b} \times \vec{c}$, where k is a constant.

From $\vec{a} \cdot \vec{a}^* = 1$, we have $k \vec{a} \cdot \vec{b} \times \vec{c} = 1$, or $k = \dfrac{1}{\vec{a} \cdot \vec{b} \times \vec{c}} = \dfrac{1}{V}$

Therefore:

$$\vec{a}* = \frac{\vec{b} \times \vec{c}}{\vec{a} \cdot \vec{b} \times \vec{c}} = \frac{\vec{b} \times \vec{c}}{V}.$$

Similarly, we have

$$\vec{b}* = \frac{\vec{c} \times \vec{a}}{\vec{b} \cdot \vec{c} \times \vec{a}}$$

and

$$\vec{c}* = \frac{\vec{a} \times \vec{b}}{\vec{c} \cdot \vec{a} \times \vec{b}}$$

From Definition I, Definition II can be obtained as shown above, indicating that the two definitions are equivalent.

Example 3.2

Prove that $V^* = 1/V$.

Solution: From $\vec{a}* = \dfrac{\vec{b} \times \vec{c}}{V}$ and $\vec{a} = \dfrac{\vec{b}^* \times \vec{c}^*}{V^*}$, we have

$$\vec{a}^* \cdot \vec{a} = \frac{1}{V^*V}\left[\left(\vec{b} \times \vec{c}\right) \cdot \left(\vec{b}^* \times \vec{c}^*\right)\right] = \frac{1}{V^*V}\left[\left(\vec{b} \cdot \vec{b}^*\right) \cdot \left(\vec{c} \cdot \vec{c}^*\right) - \left(\vec{b} \cdot \vec{c}^*\right) \cdot \left(\vec{c} \cdot \vec{b}^*\right)\right] = \frac{1}{V^*V}.$$

As $\vec{a}^* \cdot \vec{a} = 1$, we can obtain the volume of the reciprocal unit cell from the volume of the direct cell, or $V^* = 1/V$.

Example 3.3

Use direct lattice parameters to express $\sin\alpha^*$, $\sin\beta^*$, and $\sin\gamma^*$.

Solution: Since we know that $\vec{a} = \dfrac{\vec{b}^* \times \vec{c}^*}{V^*}$, or $a = \dfrac{b^* c^* \sin\alpha^*}{V^*}$, we have $\sin\alpha^* = \dfrac{aV^*}{b^* c^*}$.

We also know that $b^* = \dfrac{ca\sin\beta}{V}$, $c^* = \dfrac{ab\sin\alpha}{V}$ and $V^* = \dfrac{1}{V}$ can be substituted into $\sin\alpha^* = \dfrac{aV^*}{b^* c^*}$.

Therefore:

$$\sin\alpha^* = \frac{aV^*}{b^* c^*} = \frac{a\big/V}{(ca\sin\beta/V)(ab\sin\gamma/V)} = \frac{V}{abc\sin\beta\sin\gamma}.$$

Similarly, we can get expressions for $\sin\beta^*$ and $\sin\gamma^*$.
Therefore:

$$\sin\alpha^* = \frac{V}{abc\sin\beta\sin\gamma}$$

$$\sin\beta^* = \frac{V}{abc\sin\gamma\sin\alpha}$$

$$\sin\gamma^* = \frac{V}{abc\sin\alpha\sin\beta}$$

Example 3.4

Use direct lattice parameters to express $\cos\alpha^*$, $\cos\beta^*$ and $\cos\gamma^*$.

Solution: Since $\vec{b}^* \cdot \vec{c}^* = \left|\vec{b}^*\right|\left|\vec{c}^*\right|\cos\alpha^*$, we have

$$\cos\alpha^* = \frac{\vec{b}^* \cdot \vec{c}^*}{\left|\vec{b}^*\right|\left|\vec{c}^*\right|} = \frac{\left(\dfrac{\vec{c}\times\vec{a}}{V}\right)\cdot\left(\dfrac{\vec{a}\times\vec{b}}{V}\right)}{\left|\dfrac{\vec{c}\times\vec{a}}{V}\right|\left|\dfrac{\vec{a}\times\vec{b}}{V}\right|}$$

$$= \frac{(\vec{c}\cdot\vec{a})\cdot(\vec{a}\cdot\vec{b})-(\vec{c}\cdot\vec{b})\cdot(\vec{a}\cdot\vec{a})}{ca\sin\beta\cdot ab\sin\gamma} = \frac{ca\cos\beta\cdot ab\cos\gamma-cb\cos\alpha\cdot a^2}{ca\sin\beta\cdot ab\sin\gamma},$$

$$= \frac{\cos\beta\cdot\cos\gamma-\cos\alpha\cdot}{\sin\beta\cdot\sin\gamma}$$

Similarly, we can get expressions for $\cos \beta^*$ and $\cos \gamma^*$.
Therefore:

$$\cos \alpha^* = \frac{\cos \beta \cos \gamma - \cos \alpha}{\sin \beta \sin \gamma}$$

$$\cos \beta^* = \frac{\cos \gamma \cos \alpha - \cos \beta}{\sin \gamma \sin \alpha}$$

$$\cos \gamma^* = \frac{\cos \alpha \cos \beta - \cos \gamma}{\sin \alpha \sin \beta}$$

It should be mentioned that in solid state physics and quantum mechanics, the definition includes a 2π factor. However, its reciprocity is lost, which means the reciprocal lattice of the reciprocal lattice is no longer the direct lattice (Authier, 2010).

If the unit cell vectors of the direct lattice are $\vec{a}_1, \vec{a}_2, \vec{a}_3$, then the unit cell vectors of the reciprocal lattice, $\vec{b}_1, \vec{b}_2, \vec{b}_3$, are defined as:

$$\vec{b}_1 = 2\pi \frac{\vec{a}_2 \times \vec{a}_3}{\vec{a}_1 \cdot \vec{a}_2 \times \vec{a}_3}$$

$$\vec{b}_2 = 2\pi \frac{\vec{a}_3 \times \vec{a}_1}{\vec{a}_2 \cdot \vec{a}_3 \times \vec{a}_1} \qquad (3.5)$$

$$\vec{b}_1 = 2\pi \frac{\vec{a}_1 \times \vec{a}_2}{\vec{a}_3 \cdot \vec{a}_1 \times \vec{a}_2}$$

$$\vec{a}_1 \cdot \vec{a}_2 \times \vec{a}_3 = \vec{a}_2 \cdot \vec{a}_3 \times \vec{a}_1 = \vec{a}_3 \cdot \vec{a}_1 \times \vec{a}_2 = V,$$

Or:

$$\vec{b}_i \cdot \vec{a}_j = 2\pi \delta_{ij} \qquad i, j = 1, 2, 3 \qquad (3.6)$$

In physics, a Brillouin zone is a particular choice of the unit cell. The first Brillouin zone is defined as the Wigner–Seitz cell in the reciprocal lattice. The nth Brillouin zone is the set of points that can be reached with a straight line from the origin by passing through $n-1$ Bragg diffraction planes, and within n Bragg diffraction planes. As the nth Brillouin zone is a shell around lower Brillouin zones, its shape becomes rather complicated for higher values of n. For a lattice, all Brillouin zones have the same volume.

In Part II, we will study X-ray diffraction and Bragg's law. The vector form of the Bragg condition is

$$\vec{k} - \vec{k}_0 = \vec{g}_{hkl}, \text{ or } \vec{k} = \vec{k}_0 + \vec{g}_{hkl} = \vec{k}_0 - \vec{g}_{\overline{hkl}}$$

$\vec{g}_{hkl}$ and $\vec{g}_{\overline{hkl}}$ are antiparallel reciprocal lattice vectors, and we can just use $\vec{g}$ to represent any reciprocal lattice vector, for instance, $\vec{k} = \vec{k}_0 - \vec{g}$.

If we square the left side and right side of the equation, we have

$$\vec{k}^2 = \vec{k}_0^2 - 2\vec{k}_0 \cdot \vec{g} + \vec{g}^2.$$

As $k^2 = k_0^2$ or $\vec{k}^2 = \vec{k}_0^2$

we have $\vec{k}_0 \cdot \vec{g} = \dfrac{1}{2} g^2$, or

$$\vec{k}_0 \cdot \frac{\vec{g}}{|\vec{g}|} = \frac{1}{2} |\vec{g}| \tag{3.7}$$

Eq. (3.7) can be used to represent the boundary of Brillouin zones.

Eq. (3.7) is deduced from the vector form of the Bragg condition, and tells us that if the projection of the incident wave vector along any reciprocal lattice vector is just half of the reciprocal lattice vector length, the Bragg condition is satisfied.

3.2   SOME IMPORTANT PROPERTIES OF THE RECIPROCAL LATTICE AND ASSOCIATED CALCULATIONS

The reciprocal lattice is a useful tool for crystallographic calculations; some useful relationships are shown below.

(I) The Fourier transform of the real lattice is the reciprocal lattice.

The expression below is just for a one-dimensional infinite point row (the lattice function):

$$f(x) = \sum_{n=-\infty}^{\infty} \delta(x - na)$$

$$F(u) = \frac{1}{a} \sum_{h=-\infty}^{\infty} \delta(u - h/a)$$

(II) The reciprocal lattice vector, $\vec{g} = h\vec{a}^* + k\vec{b}^* + l\vec{c}^*$, is normal to the real lattice plane (hkl), and is of magnitude $1/d_{hkl}$.

(III) The interplanar spacing, d_{hkl}, can be calculated based on:

$$\left| \vec{g}_{hkl} \right|^2 = \frac{1}{d_{hkl}^2} = \left(h\vec{a}^* + k\vec{b}^* + l\vec{c}^* \right) \cdot \left(h\vec{a}^* + k\vec{b}^* + l\vec{c}^* \right) \tag{3.8}$$

(IV) The interplanar angle, φ, can be calculated based on:

$$\vec{g}_1 \cdot \vec{g}_2 = g_1 g_2 \cos\phi$$

$$\cos\phi = \frac{\vec{g}_1 \cdot \vec{g}_2}{g_1 g_2} = \frac{1}{g_1 g_2} \left(h_1\vec{a}^* + k_1\vec{b}^* + l_1\vec{c}^* \right) \cdot \left(h_2\vec{a}^* + k_2\vec{b}^* + l_2\vec{c}^* \right) \tag{3.9}$$

(V) The reciprocity of the different lattices is:

$$P(\text{primitive}) \Leftrightarrow P(\text{primitive})$$

$$F(\text{face – centered}) \Leftrightarrow I(\text{body – centered})$$

$$I(\text{body – centered}) \Leftrightarrow F(\text{face – centered})$$

$$C(\text{one – face – centered}) \Leftrightarrow C(\text{one – face – centered})$$

The reciprocity between the *F* and *I* lattices (for both cubic and non-cubic systems) is derived as follows.

We know from Figure 3.1 that basic vectors of the primitive cell of a face-centered lattice are:

$$\vec{a}' = \frac{\vec{b} + \vec{c}}{2}$$
$$\vec{b}' = \frac{\vec{c} + \vec{a}}{2} \qquad\qquad (3.10)$$
$$\vec{c}' = \frac{\vec{a} + \vec{b}}{2}$$

(a)   (b)

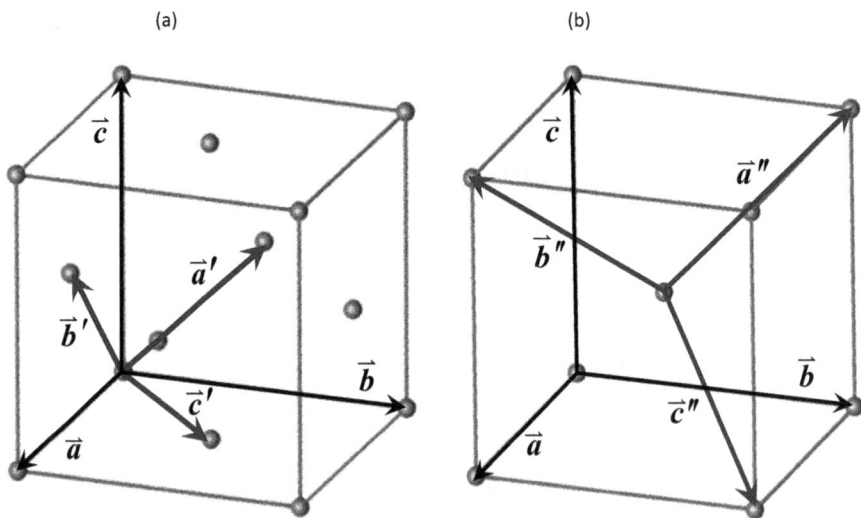

FIGURE 3.1   Basic vectors for (a) an F lattice and (b) an I lattice.

and the basic vectors of the primitive cell of a body-centered lattice are:

$$\vec{a}'' = \frac{-\vec{a} + \vec{b} + \vec{c}}{2}$$

$$\vec{b}'' = \frac{\vec{a} - \vec{b} + \vec{c}}{2}$$ (3.11)

$$\vec{c}'' = \frac{\vec{a} + \vec{b} - \vec{c}}{2}$$

The cell vectors of the reciprocal lattice of the face-centered lattice can be derived as follows:

$$\vec{a}'* = \frac{\vec{b}' \times \vec{c}'}{\vec{a}' \cdot \vec{b}' \times \vec{c}'} = \frac{\left(\dfrac{\vec{c} + \vec{a}}{2}\right) \times \left(\dfrac{\vec{a} + \vec{b}}{2}\right)}{\left(\dfrac{\vec{b} + \vec{c}}{2}\right) \cdot \left[\left(\dfrac{\vec{c} + \vec{a}}{2}\right) \times \left(\dfrac{\vec{a} + \vec{b}}{2}\right)\right]}$$

$$= \frac{\vec{c} \times \vec{a} + \vec{c} \times \vec{b} + \vec{a} \times \vec{b}}{\left(\dfrac{\vec{b} + \vec{c}}{2}\right) \cdot \left[\vec{c} \times \vec{a} + \vec{c} \times \vec{b} + \vec{a} \times \vec{b}\right]}$$

$$= 2\frac{\vec{c} \times \vec{a} + \vec{c} \times \vec{b} + \vec{a} \times \vec{b}}{\vec{b} \cdot \vec{c} \times \vec{a} + \vec{c} \cdot \vec{a} \times \vec{b}}$$

$$= \frac{-\vec{b} \times \vec{c} + \vec{c} \times \vec{a} + \vec{a} \times \vec{b}}{V}$$

$$= -\vec{a}* + \vec{b}* + \vec{c}*$$

$$= \frac{-(2\vec{a}*) + (2\vec{b}*) + (2\vec{c}*)}{2}$$

Similarly:

$$\vec{b}'* = \frac{(2\vec{a}*) - (2\vec{b}*) + (2\vec{c}*)}{2}$$

$$\vec{c}'* = \frac{(2\vec{a}*) + (2\vec{b}*) - (2\vec{c}*)}{2}$$

To compare the three equations with the expressions in Eq. (3.11), we can see clearly that the reciprocal lattice of a face-centered lattice is a body-centered lattice whose multiple cell is defined by $2\bar{a}^*$, $2\bar{b}^*$, $2\bar{c}^*$. A more detailed discussion on the reciprocal lattice can be found in Booklet No. 4, *The Reciprocal Lattice* Authier (1981).

In some crystallography textbooks, there are detailed mathematical treatments on reciprocal lattice properties, and in the following we provide some examples of derivations with similar or different methods.

Example 3.5

Use a one-dimensional infinite point row (the lattice function) to prove that the Fourier transform of the real lattice is the reciprocal lattice.

$$f(x) = \sum_{n=-\infty}^{\infty} \delta(x - na)$$

$$F(u) = \frac{1}{a} \sum_{h=-\infty}^{\infty} \delta(u - h/a)$$

Solution (Method I): For $f_n(x) = \delta(x - na)$, its Fourier transform is

$$F_n(u) = \int_{-\infty}^{+\infty} \delta(x - na) \exp(2\pi iux) dx$$

$$= \exp(2\pi iuna)$$

Therefore:

$$F(u) = \int_{-\infty}^{+\infty} \sum_{-\infty}^{+\infty} \delta(x - na) \exp(2\pi iux) dx = \sum_{-\infty}^{+\infty} \exp(2\pi iuna)$$

Since $\sum_{0}^{+\infty} x^n = \frac{1}{1-x}$

we have

$$F(u) = \sum_{-\infty}^{+\infty} \exp(2\pi iuna) = \sum_{-\infty}^{+\infty} \left[\exp(2\pi iua)\right]^n$$

$$= \sum_{0}^{\infty} \left[\exp(2\pi iua)\right]^n + \sum_{0}^{\infty} \left[\exp(-2\pi iua)\right]^n - 1$$

$$= \frac{1}{1 - \exp(2\pi iua)} + \frac{1}{1 - \exp(-2\pi iua)} - 1$$

If exp($2\pi iua$) = 1, then $F(u)$ = ∞. In this case, $2\pi ua = 2\pi h$, or $u = \dfrac{h}{a}$, where h is an integer.

If exp($2\pi iua$) ≠ 1, then $F(u)$ = 0, which is derived below.

$$\frac{1}{1-\exp(2\pi iua)} + \frac{1}{1-\exp(-2\pi iua)} - 1$$

$$= \frac{1}{\exp(\pi iua)\exp(-\pi iua) - \exp(\pi iua)\exp(\pi iua)} + \frac{1}{\exp(\pi iua)\exp(-\pi iua) - \exp(-\pi iua)\exp(-\pi iua)} - 1$$

$$= \frac{1}{\exp(\pi iua)}\frac{1}{\exp(-\pi iua) - \exp(\pi iua)} + \frac{1}{\exp(-\pi iua)}\frac{1}{\exp(\pi iua) - \exp(-\pi iua)} - 1$$

$$= \frac{1}{\exp(\pi iua) - \exp(-\pi iua)}\left[\frac{1}{\exp(-\pi iua)} - \frac{1}{\exp(\pi iua)}\right] - 1$$

$$= \frac{1}{\exp(\pi iua) - \exp(-\pi iua)}\frac{\exp(\pi iua) - \exp(-\pi iua)}{\exp(\pi iua)\exp(-\pi iua)} - 1$$

$$1 - 1$$

$$= 0$$

Therefore, $F(u)$ is a set of equally spaced delta functions, of period $\dfrac{1}{a}$, or

$$F(u) = \sum_{h=-\infty}^{\infty}\delta(u - h/a).$$

Based on Cowley's (1995) discussion, a factor $\dfrac{1}{a}$ is incorporated as the correct weight, or

$$F(u) = \frac{1}{a}\sum_{h=-\infty}^{\infty}\delta(u - h/a)$$

Solution (Method II):

Step I: Decompose $f(x) = \sum_{n=-\infty}^{\infty}\delta(x - na)$ into a Fourier series.

In Appendix A1, you can see that the Fourier series of $f(x)$ with a period λ can be written as

$$f(x) = \sum_{n=-\infty}^{\infty}c_n e^{inkx}$$

in the range of $\left[-\dfrac{\lambda}{2}, \dfrac{\lambda}{2}\right]$, where

$$C_n = \frac{1}{\lambda} \int_{-\lambda/2}^{\lambda/2} f(x)e^{-inkx}dx$$

$$\left(k = \frac{2\pi}{\lambda}, n = 0, \pm 1, \pm 2, \ldots\right)$$

Consider a function $f(x)$ with a period a, and use $\dfrac{2\pi}{a}$ to replace k, then $f(x)$ can be expressed as

$$f(x) = \sum_{n=-\infty}^{n=\infty} C_n e^{2\pi inx/a}$$

where C_n is

$$C_n = \frac{1}{a} \int_{-a/2}^{a/2} f(x)e^{-2\pi inx/a}dx.$$

When variable x is within the range of $\left(-\dfrac{a}{2}, +\dfrac{a}{2}\right)$, only one term for the function $\displaystyle\sum_{n=-\infty}^{\infty} \delta(x - na)$ is left, which is $\delta(x - 0)$ or $\delta(x)$. Therefore,

$$C_n = \frac{1}{a} \int_{-a/2}^{a/2} f(x)e^{-2\pi inx/a}dx = \frac{1}{a} \int_{-a/2}^{a/2} \delta(x)\, e^{-2\pi inx/a}dx = \frac{1}{a}$$

since

$$\int_{-a/2}^{a/2} \delta(x)e^{-2\pi inx/a}dx = \int_{-a/2}^{a/2} \delta(x)e^{0}dx = \int_{-a/2}^{a/2} \delta(x)dx = \int_{-a/2}^{a/2} \delta(x) = 1$$

Hence, $f(x) = \displaystyle\sum_{n=-\infty}^{\infty} \delta(x - na)$ can be represented in the form of a Fourier series as:

$$f(x) = \sum_{n=-\infty}^{n=\infty} C_n e^{2\pi inx/a} = \sum_{n=-\infty}^{n=\infty} \frac{1}{a} e^{2\pi inx/a}$$

Step II: Compute the Fourier transform of $f(x)$ using the obtained Fourier series $f(x) = \sum\limits_{n=-\infty}^{n=\infty} \dfrac{1}{a} e^{2\pi inx/a}$.

$$F(u) = FT\big[f(x)\big] = \int_{-\infty}^{+\infty} \left(\sum\limits_{n=-\infty}^{n=\infty} \frac{1}{a} e^{2\pi inx/a} \right) e^{-2\pi iux} dx$$

$$= \frac{1}{a} \sum\limits_{n=-\infty}^{n=\infty} \int_{-\infty}^{+\infty} e^{2\pi inx/a} e^{-2\pi iux} dx$$

$$= \frac{1}{a} \sum\limits_{n=-\infty}^{n=\infty} FT\left(e^{2\pi inx/a} \right)$$

We know that the inverse Fourier transform of $\delta\left(u - \dfrac{n}{a} \right)$ is

$$FT^{-1}\left[\delta\left(u - \frac{n}{a} \right) \right]$$

$$= \int_{-\infty}^{+\infty} \delta\left(u - \frac{n}{a} \right) e^{2\pi iux} du$$

$$= e^{2\pi i \frac{nx}{a}}$$

Therefore, $\delta\left(u - \dfrac{n}{a} \right)$ $\quad$ and $\quad$ $e^{2\pi i \frac{nx}{a}}$ are Fourier transform pairs, or

$$FT\left(e^{2\pi inx/a} \right) = \delta\left(u - \frac{n}{a} \right)$$

$$FT^{-1}\left[\delta\left(u - \frac{n}{a} \right) \right] = e^{2\pi inx/a}$$

Now, we can rewrite the Fourier transform for $f(x)$:

$$FT\big[f(x)\big] = \frac{1}{a} \sum\limits_{n=-\infty}^{n=\infty} FT\left(e^{2\pi inx/a} \right)$$

$$= \frac{1}{a} \sum\limits_{n=-\infty}^{n=\infty} \delta\left(u - \frac{n}{a} \right)$$

which means

$$F(u) = \frac{1}{a} \sum\limits_{h=-\infty}^{\infty} \delta\left(u - h/a \right)$$

So it is proved that the Fourier transform of the real lattice is the reciprocal lattice.

Example 3.6

Prove that the reciprocal lattice vector $\vec{g} = h\vec{a}^* + k\vec{b}^* + l\vec{c}^*$ is normal to the real lattice plane (hkl), and is of magnitude $\dfrac{1}{d_{hkl}}$.

Solution: We know that if a line is perpendicular to two non-parallel lines in a plane, it is perpendicular to that plane.

From Figure 3.2, we can easily find three non-parallel vectors in plane (hkl), which are

$$\overline{AB} = \frac{\vec{b}}{k} - \frac{\vec{a}}{h}$$

$$\overline{BC} = \frac{\vec{c}}{l} - \frac{\vec{b}}{k}$$

$$\overline{CA} = \frac{\vec{a}}{h} - \frac{\vec{c}}{l}$$

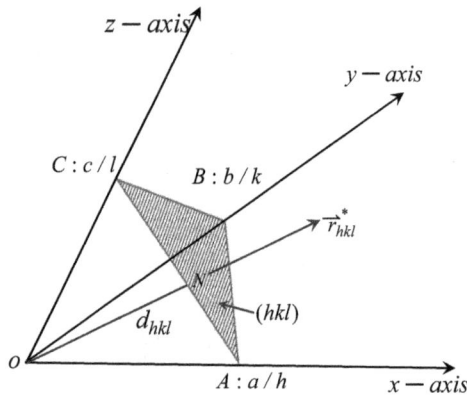

FIGURE 3.2　Relationships between a direct lattice plane and the associated reciprocal lattice vector.

Now we just use the first two vectors, and the calculations are shown below:

$$\vec{g}_{hkl} \cdot \overline{AB} = \left(h\vec{a}* + k\vec{b}* + l\vec{c}*\right) \cdot \left(\frac{\vec{b}}{k} - \frac{\vec{a}}{h}\right) = 0$$

$$\vec{g}_{hkl} \cdot \overline{BC} = \left(h\vec{a}* + k\vec{b}* + l\vec{c}*\right) \cdot \left(\frac{\vec{c}}{l} - \frac{\vec{b}}{k}\right) = 0$$

When the dot product of two vectors is zero, the angle between them is 90°. Therefore, the vector $\vec{g}_{hkl}$ in the reciprocal lattice is perpendicular to

AB and BC, or the reciprocal lattice vector $\vec{g}_{hkl}$ is perpendicular to plane (hkl) in real space.

We know that the unit vector perpendicular to plane (hkl) is $\dfrac{\vec{g}_{hkl}}{|\vec{g}_{hkl}|}$, then d_{hkl} can be obtained as:

$$d_{hkl} = ON = \frac{\vec{g}_{hkl}}{|\vec{g}_{hkl}|} \cdot \frac{\vec{a}}{h} = \frac{h\vec{a}* + k\vec{b}* + l\vec{c}*}{|\vec{g}_{hkl}|} \cdot \frac{\vec{a}}{h} = \frac{1}{h} \frac{1}{|\vec{g}_{hkl}|}$$

The above calculations tells us that the reciprocal lattice vector $\vec{g} = h\vec{a}^* + k\vec{b}^* + l\vec{c}^*$ is normal to the real lattice plane (hkl), and is of magnitude $\dfrac{1}{d_{hkl}}$.

Students are recommended to read some reference books, for example, *Elements of X-ray Diffraction* (Cullity and Stock, 2001) and *Fundamentals of Crystallography* (Giacovazzo, 2011) to better understand the derivation.

Based on Example 3.6, the methods to calculate d-spacing and the interplanar angle are given below.

$$|\vec{g}_{hkl}|^2 = \frac{1}{d_{hkl}^2} = \left(h\vec{a}^* + k\vec{b}^* + l\vec{c}^* \right) \cdot \left(h\vec{a}^* + k\vec{b}^* + l\vec{c}^* \right)$$

$$\cos\phi = \frac{\vec{g}_1 \cdot \vec{g}_2}{g_1 g_2} = \frac{1}{g_1 g_2} \left(h_1\vec{a}^* + k_1\vec{b}^* + l_1\vec{c}^* \right) \cdot \left(h_2\vec{a}^* + k_2\vec{b}^* + l_2\vec{c}^* \right)$$

Example 3.7

Prove that the d-spacing and interplanar angle equations for the orthorhombic system are

$$\frac{1}{d_{hkl}^2} = \frac{h^2}{a^2} + \frac{k^2}{b^2} + \frac{l^2}{c^2}$$

$$\cos\phi = \frac{\dfrac{h_1 h_2}{a^2} + \dfrac{k_1 k_2}{b^2} + \dfrac{l_1 l_2}{c^2}}{\sqrt{\dfrac{h_1^2}{a^2} + \dfrac{k_1^2}{b^2} + \dfrac{l_1^2}{c^2}} \sqrt{\dfrac{h_2^2}{a^2} + \dfrac{k_2^2}{b^2} + \dfrac{l_2^2}{c^2}}}$$

Solution: As $|\vec{g}_{hkl}|^2 = \dfrac{1}{d_{hkl}^2} = \left(h\vec{a}^* + k\vec{b}^* + l\vec{c}^* \right) \cdot \left(h\vec{a}^* + k\vec{b}^* + l\vec{c}^* \right)$ we   have

$$\frac{1}{d_{hkl}^2} = h^2 a^{*2} + k^2 b^{*2} + l^2 c^{*2}.$$

We know that for an orthorhombic system, $a^* = \dfrac{1}{a}$, $b^* = \dfrac{1}{b}$ and $c^* = \dfrac{1}{c}$

Therefore, $\dfrac{1}{d^2_{hkl}} = \dfrac{h^2}{a^2} + \dfrac{k^2}{b^2} + \dfrac{l^2}{c^2}$

We can calculate $\cos\phi$ using

$$\cos\phi = \frac{\vec{g}_1 \cdot \vec{g}_2}{g_1 g_2} = \frac{1}{g_1 g_2}\left(h_1 \vec{a}^* + k_1 \vec{b}^* + l_1 \vec{c}^*\right) \cdot \left(h_2 \vec{a}^* + k_2 \vec{b}^* + l_2 \vec{c}^*\right)$$

As $\left(h_1 \vec{a}^* + k_1 \vec{b}^* + l_1 \vec{c}^*\right) \cdot \left(h_2 \vec{a}^* + k_2 \vec{b}^* + l_2 \vec{c}^*\right) = h_1 h_2 a^{*2} + k_1 k_2 b^{*2} + l_1 l_2 c^{*2} = \dfrac{h_1 h_2}{a^2} + \dfrac{k_1 k_2}{b^2} + \dfrac{l_1 l_2}{c^2}$,

$g_1 = \sqrt{\dfrac{h_1^2}{a^2} + \dfrac{k_1^2}{b^2} + \dfrac{l_1^2}{c^2}}$ and $g_2 = \sqrt{\dfrac{h_2^2}{a^2} + \dfrac{k_2^2}{b^2} + \dfrac{l_2^2}{c^2}}$

then:

$$\cos\phi = \frac{\dfrac{h_1 h_2}{a^2} + \dfrac{k_1 k_2}{b^2} + \dfrac{l_1 l_2}{c^2}}{\sqrt{\dfrac{h_1^2}{a^2} + \dfrac{k_1^2}{b^2} + \dfrac{l_1^2}{c^2}} \sqrt{\dfrac{h_2^2}{a^2} + \dfrac{k_2^2}{b^2} + \dfrac{l_2^2}{c^2}}}$$

Using the above method, the equations of the d-spacings and the interplanar angles for all of the seven crystal systems can be obtained. The most general ones are for the triclinic system, which are given below:

$$\frac{1}{d^2_{hkl}} = \frac{a^2 b^2 c^2}{V^2}\left[\frac{h^2 \sin^2\alpha}{a^2} + \frac{k^2 \sin^2\beta}{b^2} + \frac{l^2 \sin^2\gamma}{c^2}\right.$$
$$+ \frac{2hk}{ab}\left(\cos\alpha\cos\beta - \cos\gamma\right)$$
$$+ \frac{2kl}{bc}\left(\cos\beta\cos\gamma - \cos\alpha\right)$$
$$\left. + \frac{2lh}{ca}\left(\cos\gamma\cos\alpha - \cos\beta\right)\right]$$

(3.12)

$$\cos\phi = \frac{d_1 d_2}{V^2}\left[h_1 h_2 (bc)^2 \sin^2\alpha + k_1 k_2 (ca)^2 \sin^2\beta + l_1 l_2 (ab)^2 \sin^2\gamma\right.$$
$$+ (h_1 k_2 + h_2 k_1)abc^2 \left(\cos\alpha\cos\beta - \cos\gamma\right)$$
$$+ (h_1 l_2 + h_2 l_1)ab^2 c \left(\cos\gamma\cos\alpha - \cos\beta\right)$$
$$\left. + (k_1 l_2 + k_2 l_1)a^2 bc \left(\cos\beta\cos\gamma - \cos\alpha\right)\right]$$

(3.13)

Now, students are encouraged to derive Eqs (3.12) and (3.13) step by step and ensure that the above equations are correct.

Table 3.2 provides a list of d-spacing and interplanar angle equations for the seven crystal systems.

TABLE 3.2
Interplanar Spacing and Interplanar Angle Equations of the Seven Crystal Systems

Crystal system	d-spacing for plane (hkl)	Interplanar angle between plane $(h_1 k_1 l_1)$ and plane $(h_2 k_2 l_2)$
Cubic	$\dfrac{1}{d_{hkl}^2} = \dfrac{h^2 + k^2 + l^2}{a^2}$ or $d = \dfrac{a}{\sqrt{h^2 + k^2 + l^2}}$	$\cos\phi = \dfrac{h_1 h_2 + k_1 k_2 + l_1 l_2}{\sqrt{h_1^2 + k_1^2 + l_1^2}\,\sqrt{h_2^2 + k_2^2 + l_2^2}}$
Tetragonal	$\dfrac{1}{d_{hkl}^2} = \dfrac{h^2 + k^2}{a^2} + \dfrac{l^2}{c^2}$	$\cos\phi = \dfrac{\dfrac{h_1 h_2 + k_1 k_2}{a^2} + \dfrac{l_1 l_2}{c^2}}{\sqrt{\dfrac{h_1^2 + k_1^2}{a^2} + \dfrac{l_1^2}{c^2}}\,\sqrt{\dfrac{h_2^2 + k_2^2}{a^2} + \dfrac{l_2^2}{c^2}}}$
Orthorhombic	$\dfrac{1}{d_{hkl}^2} = \dfrac{h^2}{a^2} + \dfrac{k^2}{b^2} + \dfrac{l^2}{c^2}$	$\cos\phi = \dfrac{\dfrac{h_1 h_2}{a^2} + \dfrac{k_1 k_2}{b^2} + \dfrac{l_1 l_2}{c^2}}{\sqrt{\dfrac{h_1^2}{a^2} + \dfrac{k_1^2}{b^2} + \dfrac{l_1^2}{c^2}}\,\sqrt{\dfrac{h_2^2}{a^2} + \dfrac{k_2^2}{b^2} + \dfrac{l_2^2}{c^2}}}$
Hexagonal	$\dfrac{1}{d_{hkl}^2} = \dfrac{4}{3}\left(\dfrac{h^2 + hk + k^2}{a^2}\right) + \dfrac{l^2}{c^2}$	$\cos\phi = \dfrac{h_1 h_2 + k_1 k_2 + \dfrac{1}{2}\left(h_1 k_2 + h_2 k_1\right) + \dfrac{3a^2}{4c^2} l_1 l_2}{\sqrt{h_1^2 + k_1^2 + h_1 k_1 + \dfrac{3a^2}{4c^2} l_1^2}\,\sqrt{h_2^2 + k_2^2 + h_2 k_2 + \dfrac{3a^2}{4c^2} l_2^2}}$
Rhombohedral	$\dfrac{1}{d_{hkl}^2} = \dfrac{1}{a^2} \cdot$ $\dfrac{\left(h^2 + k^2 + l^2\right)\sin^2\alpha + 2\left(hk + kl + lh\right)\left(\cos^2\alpha - \cos\alpha\right)}{\left(1 - 3\cos^2\alpha + 2\cos^3\alpha\right)}$	$\cos\phi = \dfrac{a^4 d_1 d_2}{V^2}\big[\sin^2\alpha\,\left(h_1 h_2 + k_1 k_2 + l_1 l_2\right)$ $+\left(\cos^2\alpha - \cos\alpha\right)\left(k_1 l_2 + k_2 l_1 + l_1 h_2 + l_2 h_1 + h_1 k_2 + h_2 k_1\right)\big]$

(Continued)

TABLE 3.2 (Continued)

Crystal system	d-spacing for plane (hkl)	Interplanar angle between plane $(h_1k_1l_1)$ and plane $(h_2k_2l_2)$

Monoclinic

$$\frac{1}{d_{hkl}^2} = \frac{h^2}{a^2 \sin^2 \beta} + \frac{k^2}{b^2} + \frac{l^2}{c^2 \sin^2 \beta} - \frac{2hl \cos \beta}{ac \sin^2 \beta}$$

$$\cos\phi = d_1 d_2 \left(\frac{h_1 h_2}{a^2 \sin^2 \beta} + \frac{k_1 k_2}{b^2} + \frac{l_1 l_2}{c^2 \sin^2 \beta} - \frac{(l_1 h_2 + l_2 h_1) \cos \beta}{ac \sin^2 \beta} \right)$$

triclinic

$$\frac{1}{d_{hkl}^2} = \frac{a^2 b^2 c^2}{V^2} \left[\frac{h^2 \sin^2 \alpha}{a^2} + \frac{k^2 \sin^2 \beta}{b^2} + \frac{l^2 \sin^2 \gamma}{c^2} \right.$$
$$+ \frac{2hk}{ab} \left(\cos\alpha \cos\beta - \cos\gamma \right)$$
$$+ \frac{2kl}{bc} \left(\cos\beta \cos\gamma - \cos\alpha \right)$$
$$\left. + \frac{2lh}{ca} \left(\cos\gamma \cos\alpha - \cos\beta \right) \right]$$

$$\cos\phi = \frac{d_1 d_2}{V^2} [h_1 h_2 (bc)^2 \sin^2 \alpha + k_1 k_2 (ca)^2 \sin^2 \beta + l_1 l_2 (ab)^2 \sin^2 \gamma$$
$$+ (k_1 l_2 + k_2 l_1) a^2 bc (\cos\beta \cos\gamma - \cos\alpha)$$
$$+ (h_1 l_2 + h_2 l_1) ab^2 c (\cos\gamma \cos\alpha - \cos\beta)$$
$$+ (h_1 k_2 + h_2 k_1) abc^2 (\cos\alpha \cos\beta - \cos\gamma)]$$

Note: $V = abc \sqrt{1 - \cos^2 \alpha - \cos^2 \beta - \cos^2 \gamma + 2\cos\alpha \cos\beta \cos\gamma}$,

and $\dfrac{a^2 b^2 c^2}{V^2} = \dfrac{1}{1 - \cos^2 \alpha - \cos^2 \beta - \cos^2 \gamma + 2\cos\alpha \cos\beta \cos\gamma}$

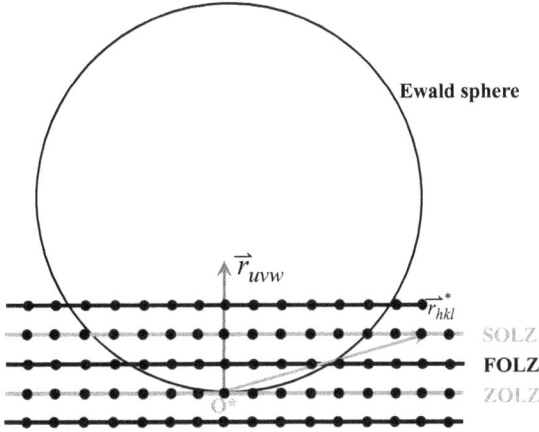

FIGURE 3.3 Ewald sphere and reciprocal lattice showing zero-order and higher-order Laue zones.

We can employ a similar concept from property (II) to derive the zone law as follows.

(II)* The direct lattice vector, $\vec{r}_{uvw} = u\vec{a} + v\vec{b} + w\vec{c}$, is normal to the reciprocal lattice plane $(uvw)^*$, and is of magnitude $1/d^*_{uvw}$.

Here, we need to calculate $\vec{r}_{uvw} \cdot \vec{r}^*_{hkl}$, where $\vec{r}^*_{hkl} = h\vec{a}^* + k\vec{b}^* + l\vec{c}^*$ is the reciprocal lattice vector. Assume that a reciprocal lattice point h, k, l lies in the nth reciprocal lattice plane from the origin as shown in Figure 3.3, from which we know that the projection of reciprocal lattice vector $\vec{r}^*_{hkl}$ on the real lattice vector $\vec{r}_{uvw}$ is $n \cdot d^*_{uvw}$. Therefore,

$$\vec{r}_{uvw} \cdot \vec{r}^*_{hkl} = \left(u\vec{a} + v\vec{b} + w\vec{c}\right) \cdot \left(h\vec{a}^* + k\vec{b}^* + l\vec{c}^*\right) = hu + kv + lw = \frac{1}{d^*_{uvw}} \cdot nd^*_{uvw} = n$$

Or

$$hu + kv + lw = n \tag{3.14}$$

The expression (3.14) is the zone law. Each of the reciprocal lattice vectors ending on a higher-order Laue zone (HOLZ) can be described by Eq. (3.14).

When we index the electron diffraction pattern from a zero-order Laue zone (ZOLZ), we just use a special case of the zone law, or the Weiss zone law. For ZOLZ, $hu + kv + lw = 0$.

Another calculation useful is the distance between two lattice points when we know the coordinates of them in the lattice. This is useful when we calculate the bond lengths and bond angles. To derive it, only the vector calculation in real space is needed (Wahab, 2009). There is no need to use the definition of the reciprocal lattice, or to use the associated properties in such a calculation. If the coordinates of a lattice point are xyz, then the vector starting from the origin and ending at the lattice point is expressed as

$$\vec{r} = x\vec{a} + y\vec{b} + z\vec{c}$$

For any two lattice points $x_1y_1z_1$ and $x_2y_2z_2$, the corresponding vectors are

$$\vec{r_1} = x_1\vec{a} + y_1\vec{b} + z_1\vec{c} \text{ and } \vec{r_2} = x_2\vec{a} + y_2\vec{b} + z_2\vec{c}$$

Therefore, the distance between the two lattice points is $d_{12} = |\vec{r_2} - \vec{r_1}|$. We know that

$$\vec{r_2} - \vec{r_1} = (x_2 - x_1)\vec{a} + (y_2 - y_1)\vec{b} + (z_2 - z_1)\vec{c} \text{ and}$$
$$(\vec{r_2} - \vec{r_1})\cdot(\vec{r_2} - \vec{r_1}) = |\vec{r_2} - \vec{r_1}|^2 = \left[(x_2 - x_1)\vec{a} + (y_2 - y_1)\vec{b} + (z_2 - z_1)\vec{c}\right]\cdot$$
$$\left[(x_2 - x_1)\vec{a} + (y_2 - y_1)\vec{b} + (z_2 - z_1)\vec{c}\right]$$

which can be simplified as:

$$|\vec{r_2} - \vec{r_1}|^2 = (x_2 - x_1)^2 a^2 + (y_2 - y_1)^2 b^2 + (z_2 - z_1)^2 c^2 + 2ab(x_2 - x_1)(y_2 - y_1)\cos\gamma$$
$$+ 2bc(y_2 - y_1)(z_2 - z_1)\cos\alpha + 2ca(z_2 - z_1)(x_2 - x_1)\cos\beta$$

So:

$$d_{12} = |\vec{r_2} - \vec{r_1}| = [(x_2 - x_1)^2 a^2 + (y_2 - y_1)^2 b^2 + (z_2 - z_1)^2 c^2 + 2ab(x_2 - x_1)(y_2 - y_1)\cos\gamma$$
$$+ 2bc(y_2 - y_1)(z_2 - z_1)\cos\alpha + 2ca(z_2 - z_1)(x_2 - x_1)\cos\beta]^{\frac{1}{2}} \tag{3.15}$$

Then the bond angle calculation becomes straightforward (Wahab, 2009). After calculating the distance between the cation and anion 1, the distance between cation and anion 2, as well as the distance between the 2 anions, the bond angle can be obtained using the law of cosines:

$$\cos\phi = \frac{d_{C-A1}^2 + d_{C-A2}^2 - d_{A1-A2}^2}{2d_{C-A1}d_{C-A2}} \tag{3.16}$$

I used a similar method to calculate the twist angles in apatite type materials after I obtained the coordinates of atoms through Rietveld refinement. In that case, I used the projected structure along the c axis of the apatite showing the metaprism twist angle (White and ZhiLi, 2003).

Students can practice further on direct and reciprocal lattice calculations. For example, if a vector $\vec{r}$ expressed by direct lattice vectors is known, how can we find the same vector $\vec{r}^*$ expressed by reciprocal lattice cell vectors through the calculation $\vec{r} = \vec{r}^*$?

The trick is to multiply each of the unit cell vectors in the direct lattice or in the reciprocal lattice by both the left sides and right sides. This kind of calculation, in general, is not well used in materials science and engineering. However, when indexing HOLZ patterns, this calculation can be helpful. We would suggest our students train their calculation skills through this example if they want to better understand the various calculations related to direct and reciprocal lattices.

SUMMARY

The reciprocal lattice is mathematically constructed based on the direct lattice for the convenience of calculations and treating diffraction problems. The reciprocal lattice does not physically exist. The orientations and dimensions of a direct lattice and its reciprocal lattice are coupled rigidly. The reciprocal lattice is used to make various crystallographic and diffraction calculations, although its utility is not obvious when working with the cubic crystal system. The reciprocal lattice calculation is a powerful tool when working with low-symmetry crystal systems, and solving electron diffraction patterns.

In quantum mechanics and solid state physics, a 2π factor is used in the definition of the reciprocal lattice. Students need to pay attention to when the 2π factor will appear or disappear while treating various problems for X-ray waves and electron waves.

The reciprocal lattice of the reciprocal lattice is the direct lattice, which can be observed from the crystallographic reciprocal lattice definition. However, when the 2π factor is included as presented in solid state physics, the relationship of reciprocality between the direct lattice and reciprocal lattice, strictly speaking, is broken. When generating the reciprocal lattice from the reciprocal lattice, the $(2\pi)^2$ factor will be multiplied to the original direct lattice vectors.

As mentioned in this chapter, many crystallography textbooks and X-ray diffraction textbooks have presented the reciprocal lattice definition and the associated calculations. This chapter has presented some selected reciprocal lattice calculations that are useful to materials scientists when they examine crystal structures and perform some relevant calculations.

REFERENCES

Authier, A. (1981). *The reciprocal lattice*. Edited by the IUCr Commission on Crystallographic Teaching. Cardiff: University College Cardiff Press.

Authier, A. (ed.) (2010) *International tables for crystallography. Volume D, Physical properties of crystals/edited by A. Authier.* 1st ed., *Physical properties of crystals.* 1st ed. Chichester, West Sussex, UK: John Wiley & Sons.

Cowley, J. M. (John M.) (1995) *Diffraction physics [electronic resource]/John M. Cowley*. 3rd rev. ed Amsterdam: Elsevier Science B.V. (North-Holland personal library).

Cullity, B. D. (Bernard D.) and Stock, S. R. (2001) *Elements of x-ray diffraction/B.D. Cullity, S.R. Stock*. 3rd ed. Upper Saddle River, NJ: Prentice Hall.

Giacovazzo, C. (2011) *Fundamentals of crystallography/G. Giacovazzo ... [et al.]*. 3rd ed. Oxford: Oxford University Press (IUCr texts on crystallography; 15).

Wahab, M. A. (Mohammad A.) (2009) *Essentials of crystallography/M.A. Wahab*. Oxford, UK: Alpha Science International.

White, T. J. and ZhiLi, D. (2003) 'Structural derivation and crystal chemistry of apatites', *Acta Crystallographica Section B: Structural Science*, 59(1), pp. 1–16. doi:10.1107/S0108768102019894.

4 Examples of Crystal Structure Representation

In this chapter, we employ apatite-type materials as examples of the representation of crystal structure characteristics. Apatite-type materials are intensively investigated due to their biomedical, optical, clean energy and environmental applications. Usually the natural apatite minerals are of the calcium phosphate type, referring to hydroxyapatite, fluorapatite and chlorapatite with $Ca_{10}(PO_4)_6(OH)_2$, $Ca_{10}(PO_4)_6F_2$ and $Ca_{10}(PO_4)_6Cl_2$ endmembers. Some museums exhibit natural apatite minerals as gems in their exhibition halls. The mineral apatite $Ca_{10}(PO_4)_6(F, Cl, OH)_2$ was also found in lunar samples (Boyce *et al.*, 2010; 2014). The apatite structure is tolerant and is able to accommodate various elements at different Wyckoff sites. Most of the apatites follow the general formula of $A^I_4A^{II}_6(BO_4)_6X_2$ with the symmetries of $P6_3/m$ (space group No. 176), where A^I and A^{II} are two cationic sites with Wyckoff symbols 4*f* and 6*h* respectively, and comprise mono-, di-, tri- or tetravalent cations, typically Cs, Ca, Sr, Ba, Cd, Pb, Ln or Th. B in the tetrahedra are metalloids, commonly P, As, V and Si. The channel anions X are halides, oxygen or hydroxyl. Smaller sized anions, such as F, occupy a 2*a* site, and larger anions, such as Cl, occupy a 2*b* site along the channel. The hexagonal lattice constants are $a \approx 10\,\text{Å}$ and $c \approx 7\,\text{Å}$.

Among the apatite family, hydroxyapatite has attracted the special attention of materials scientists as it is the main inorganic component in bones and teeth. Scientists have been studying hydroxyapatite containing composites for bone replacements or hydroxyapatite coatings for use in implants. In the Environmental Technology Institute (ETI) laboratory, we synthesized hydroxyapatite whiskers, and the analyses will be presented in the X-ray diffraction and high-resolution transmission electron microscopy parts of the book. Because of the presence of OH- ordering along the channels (Elliott *et al.*, 1973), the unit cell for hydroxyapatite is represented using a monoclinic cell consisting of two hexagonal cells with $b = 2a$; its space group is $P2_1/b$ but $P6_3/m$. As explained by Young's group (Hitmi *et al.*, 1988), in the monoclinic hydroxyapatite, all of the OH- ions are equally spaced and oriented the same way within a column. Further, all columns with the same y coordinate have similarly oriented OH- ions while those with the same x coordinate have alternately "up" and "down" orientations with increasing y.

The transformation of the low-temperature monoclinic phase with $P2_1/b$ symmetry to the high-temperature hexagonal phase with $P6/m$ symmetry for hydroxyapatite $Ca_{10}(PO_4)_6(OH)_2$ has been investigated by means of molecular dynamic simulations (Hochrein *et al.*, 2005).

The space group diagram of $P6_3/m$ (Hahn, 2011) shows that there is a three-fold rotation axis at each of the 4*f* sites. The coordinates of the 4*f* sites, or the four A^I

DOI: 10.1201/9780429351662-6

FIGURE 4.1 HRTEM images of synthetic $Nd_8Sr_2(SiO_4)_6O_2$ apatite before and after FFT processing twice along the [0001] zone axes. (a) HRTEM image, (b) enlarged HRTEM image from the selected region in (a), (c) image after FFT processing from the selected region in (a).

(Source: Reproduced from Wang *et al.* (2016), with the permission of John Wiley & Sons.)

cation sites for apatites, are 1/3 2/3 z, 2/3 1/3 z+1/2, 2/3 1/3 −z, 1/3 2/3 −z+1/2. Our studies on apatite materials show that the metaprism twist angle is a probe to monitor symmetry deviation (White and ZhiLi, 2003). The O(1)-A(1)-O(2) twist angle projected on the (0001) plane of the A(1)O6 metaprism decreases linearly as a function of the increasing average ionic radius of the formula unit.

Figure 4.1(a) is a high resolution transmission electron microscopy (HRTEM) image of the synthetic $Nd_8Sr_2(SiO_4)_6O_2$ apatite material taken from a [0001] zone axis, showing the hexagonal symmetry along the c axis (Wang *et al.*, 2016). As mentioned, A^I (or A1) and A^{II} (or A2) cations occupy 4*f* and 6*h* respectively. Si, O1 and O2, in the SiO_4 tetrahedra, all occupy 6*h*. O3 in the SiO_4 tetrahedra and O4 in the channel occupy 12*i* and 2*a*.

The image from a selected area (Figure 4.1(b)) is much clearer after FFT processing (Figure 4.1(c)). The polyhedral representations with coordination number 6 and 9 for an A1 site are shown in Figure 4.2. To highlight the twist angle, we need to show the A(1)O6 metaprism as presented in Figure 4.2(a).

It is believed that a better understanding of the apatite structure and symmetries can help in the design of new apatite materials for various applications.

SUMMARY

Many useful synthetic materials have analogues of mineral structure types. The structural characteristics of those mineral types can be well correlated with their properties. Synthetic materials with those mineral types are widely used for various applications. Among them, the apatite, spinel, olivine, garnet, perovskite, anatase and wurtzite types are well studied. Depending on the chemical compositions, those structures exhibit some unique properties. To understand these characteristics and represent them clearly enables materials scientists to well explain the properties it exhibits.

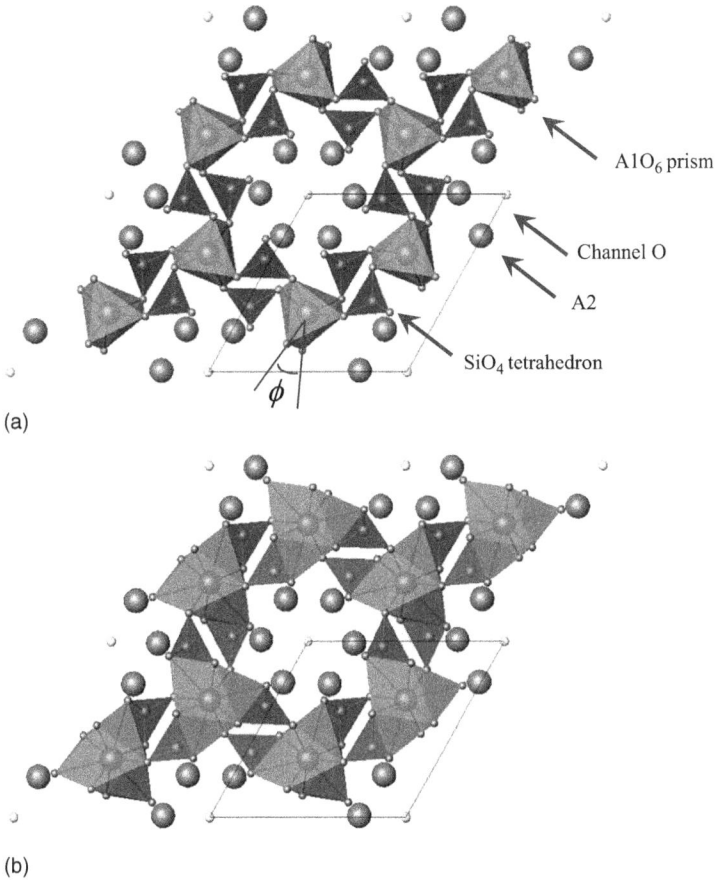

FIGURE 4.2 The crystal structure representations of $REE_8Sr_2(SiO_4)_6O_2$ along the [0001] direction from the data of XRD. (a) The polyhedral representation with coordination number 6 for the A1 site showing the twist angle, (b) the polyhedral representation with coordination number 9 for the A1 site.

This chapter has employed apatite material as a model structure for the demonstration of crystal structure characteristics. When students study different materials, the ideas presented through the apatite example may provide them with some guidance.

REFERENCES

Boyce, J. W. et al. (2010) 'Lunar apatite with terrestrial volatile abundances', *Nature*. Nature Publishing Group, 466(7305), pp. 466–469. doi:10.1038/nature09274.
Boyce, J. W. et al. (2014) 'The lunar apatite paradox', *Science*, 344(6182), pp. 400–402. doi:10.1126/science.1250398.
Elliott, J. C., Mackie, P. E. and Young, R. A. (1973) 'Monoclinic hydroxyapatite', *Science*, 180(4090), pp. 1055–1057. doi:10.1126/science.180.4090.1055.

Hahn, T. (2011) *International tables for crystallography. Volume A, Space-group symmetry/ edited by Theo Hahn.* 5th ed. re, *Space-group symmetry.* 5th ed. re. Chichester, West Sussex: Published for the International Union of Crystallography by John Wiley & Sons.

Hitmi, N., LaCabanne, C. and Young, R. A. (1988) 'Oh– reorientability in hydroxyapatites: effect of F– and Cl–', *Journal of Physics and Chemistry of Solids*, 49(5), pp. 541–550. doi:10.1016/0022-3697(88)90065-0.

Hochrein, O., Kniep, R. and Zahn, D. (2005) 'Atomistic simulation study of the order/disorder (monoclinic to hexagonal) phase transition of hydroxyapatite', *Chemistry of Materials*. American Chemical Society, 17(8), pp. 1978–1981. doi:10.1021/cm0401903.

Wang, J. et al. (2016) 'Synthesis and crystal structure characterization of oxysilicate apatites for stabilization of Sr and rare-earth elements', *Journal of the American Ceramic Society*, 99(5), pp. 1761–1768. doi:10.1111/jace.14146.

White, T. J. and ZhiLi, D. (2003) 'Structural derivation and crystal chemistry of apatites', *Acta Crystallographica Section B: Structural Science*, 59(1), pp. 1–16. doi:10.1107/S0108768102019894.

Part II

X-ray Diffraction of Materials

The aim of Part II is to explain the theories of X-ray diffraction for the analysis of polycrystalline or powder materials.

X-rays were discovered by German scientist Wilhelm Conrad Röntgen in 1895, who won the first Nobel Prize in Physics in 1901 "in recognition of the extraordinary services he has rendered by the discovery of the remarkable rays subsequently named after him". In 1912, 17 years after the discovery, an important work by Max von Laue, Walter Friedrich and Paul Knipping from Germany demonstrated that a crystal of copper sulfate diffracted X-rays, which not only suggested the wave character of X-rays, but also gave direct evidence of the underlying order of atoms in the crystal. The Laue equations proposed could be expressed in a reciprocal lattice to illustrate the X-ray diffraction conditions of a crystal lattice; Laue won the Nobel Prize in Physics 1914 "for his discovery of the diffraction of X-rays by crystals". In 1913, Lawrence Bragg and his father William Henry Bragg proposed the X-ray diffraction equation $2d\sin\theta = n\lambda$ to explain why at certain angles of incidence crystals diffract X-ray beams. They were awarded the Nobel Prize in Physics 1915 "for their services in the analysis of crystal structure by means of X-rays". Another significant contribution was made by the German physicist and crystallographer Paul Peter Ewald, who used a "Ewald sphere" to geometrically represent Bragg's law, which was widely used for the interpretation of X-ray diffraction and electron diffraction conditions in materials research. In my X-ray diffraction lectures, I introduced the Bragg equation first and then explained how to derive the Ewald sphere description for the Bragg condition. In fact, the Bragg equation is an expression in direct space, whereas the Ewald sphere construction is a reciprocal space expression for the Bragg condition. The Laue equations are also considered as expressions in reciprocal space. The equivalence of Bragg's equation with the Laue equations can be proved mathematically (Wahab, 2009; Glusker and Trueblood, 2010). After Laue and Bragg's initial

75

work on X-ray diffraction, the polycrystal diffraction or powder diffraction method was proposed independently in about 1915 by Debye and Scherrer in Germany and in 1916 by Hull in the USA. In the early X-ray diffraction experiments, powder diffraction patterns were recorded using film cameras, such as Hull/Debye–Scherrer cameras, Seemann–Bohlin cameras and Guinier cameras. In modern materials research laboratories, powder diffraction data are usually collected by powder diffractometers, which can be further processed using computers. In our discussions on the diffracted beam intensity, we focus on Bragg–Brentano geometry for the derivation of the absorption factor, as in *Elements of X-ray Diffraction* (Cullity and Stock, 2001).

X-rays are beams of electromagnetic radiation, which have high energy and short wavelengths at the angstrom level. X-ray radiation exhibits its wave–particle duality characteristics like visible light. If expressed by its wave character, X-ray radiation propagates with a sinusoidal form oscillating in electric field with a magnetic field at right angles to it.

There is a linkage between energy E and frequency v, and that between momentum $\bar{p}$ and wave vector $\bar{k}$. The wave vector's direction is the propagation direction of the wave, and its amplitude is the reciprocal of the wavelength, $1/\lambda$, called the wave number. The relationships are presented as

$$E = hv = \hbar\omega$$

$$\bar{p} = \frac{h}{\lambda}\bar{S} = \hbar\bar{k}$$

where $\bar{S}$ is the unit vector along the propagation direction, and $\frac{2\pi}{\lambda}\bar{S} = \bar{k}$; h is the Plank constant, and the values are

$$h = 6.6256 \times 10^{-34} J \cdot s$$

$$\hbar = \frac{h}{2\pi} = 1.0545 \times 10^{-34} J \cdot s$$

X-ray radiations can be generated by a synchrotron machine or a laboratory X-ray tube. The X-ray tubes are commonly used in X-ray diffractometers in materials research laboratories.

In a typical X-ray tube, the accelerated electrons bombard a metal target and eject electrons from the inner shells of the atoms of the metal target. The quick filling of those inner shell vacancies by electrons dropping down from higher levels gives rise to sharply defined characteristic X-rays. The relationship between the energy difference and the X-ray wavelength is

$$\Delta E = hv = \frac{hc}{\lambda}$$

Bremsstrahlung radiation is also emitted when electrons are decelerated or "braked" upon firing at a metal target. In classical physics, accelerated charges give off electromagnetic radiation, and when the energy of the bombarding electrons is high enough, that radiation is in the X-ray region of the electromagnetic spectrum. It is characterized by a continuous distribution of radiation which becomes more intense and shifts toward higher frequencies when the energy of the bombarding electrons is increased (Cullity and Stock, 2001). The Larmor formula is used to calculate the total power radiated by an accelerating point charge in the non-relativistic assumption.

REFERENCES

Cullity, B. D. (Bernard D.) and Stock, S. R. (2001) *Elements of x-ray diffraction/B.D. Cullity, S.R. Stock.* 3rd ed. Upper Saddle River, NJ: Prentice Hall.

Glusker, J. P. and Trueblood, K. N. (2010) *Crystal structure analysis: a primer/Jenny Pickworth Glusker, Kenneth N. Trueblood.* 3rd ed. Oxford; Oxford University Press (International Union of Crystallography texts on crystallography; no. 14).

Wahab, M. A. (Mohammad A.) (2009) *Essentials of crystallography/M.A. Wahab.* Oxford, U.K: Alpha Science International.

5 Geometry of X-ray Diffraction

5.1 BRAGG'S EQUATION

The Laue equations relate the incoming X-ray waves to the outgoing X-ray waves in the process of diffraction by a crystal lattice. The Laue condition is a vector expression in reciprocal space, which is equivalent to the Bragg condition, an expression in real space. This chapter focuses on Bragg's expression, from which the Ewald sphere construction and vector expression are further discussed.

When X-ray beams strike the surface of a material, the beams penetrate to a certain depth of the sample before they are completely absorbed. The beams will also be scattered by the atoms in the sample.

Figure 5.1 shows monochromatic incident X-ray beams striking a crystalline material. If the path difference between beam 1 and beam 2 is equal to the X-ray wavelength or is n times the X-ray wavelength, constructive interference will occur, resulting in a strongly diffracted beam.

In mathematics, the relationship is expressed as:

$SQ + QT = d_{hkl} \sin\theta + d_{hkl} \sin\theta = 2d_{hkl} \sin\theta = n\lambda$, where $n = 1,2,3, \ldots$ are the orders of reflection.

$$2d_{hkl} \sin\theta = n\lambda$$

is called Bragg's Law, which was first introduced by Sir W. H. Bragg and his son Sir W. L. Bragg in 1913.

In case the order of reflection, n, is incorporated into d-spacing, Bragg's equation becomes:

$$2d_{hkl} \sin\theta_B = \lambda \qquad (5.1)$$

For a particular plane (hkl) with an interplanar spacing, d_{hkl}, we use the first order $(n = 1)$ diffraction, the second order $(n = 2)$ diffraction and so on to calculate different diffraction angles for the same plane through the expression $2d_{hkl} \sin\theta = n\lambda$. And the indices h, k and l have no common factors. With the modified equation, $2d_{hkl} \sin\theta_B = \lambda$, the indices h, k and l are integers that can have common factors. In such a case, $(h\,'k\,'l\,')$, $(2h\,'2k\,'2l\,')$, $(3h\,'3k\,'3l\,')$, $\ldots$, and $(nh\,'nk\,'nl\,')$ are regarded as different diffraction planes, rather than a plane $(h\,'k\,'l\,')$ with different orders, 1, 2, 3, $\ldots$, and n. For example, in the original expression $2d_{hkl} \sin\theta = n\lambda$, if the (111) plane diffracts X-rays, then the first order, second order, third order, $\ldots$ and nth order are employed to correlate with different diffracted beams having different Bragg angles. However, in the modified form $2d_{hkl} \sin\theta_B = \lambda$, it is unnecessary to consider different orders of

DOI: 10.1201/9780429351662-8

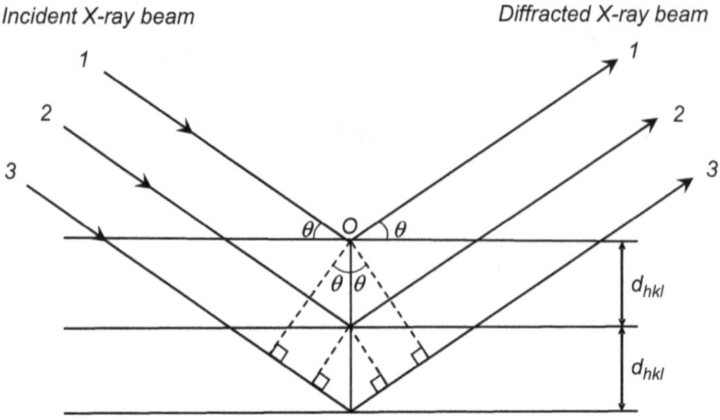

FIGURE 5.1 Constructive interference of X-ray beams occurs when $2d_{hkl} \sin\theta = n\lambda$ is satisfied, where d_{hkl} is the interplanar spacing of (hkl) and λ is the wavelength of the X-ray.

diffraction. Instead, (111), (222), (333), ... and (nnn) planes with different d-spacings are used and they diffract X-rays at different Bragg angles. The planes (222), (333), ... and (nnn) cannot be reduced to the plane (111). From the d-spacing equations obtained in Chapter 3, we know that $d_{nhnknl} = \dfrac{d_{hkl}}{n}$.

5.2 EWALD SPHERE CONSTRUCTION AND THE VECTOR FORM OF BRAGG'S LAW

In X-ray diffraction, Bragg's Law can be represented through Ewald sphere construction and the vector form. In the Ewald sphere construction, the wave number $1/\lambda$ is the radius of the sphere. The origin of vector $\vec{k}_0$ is the center of the Ewald sphere, and the tail of the vector $\vec{k}_0$ is the origin of the reciprocal lattice O^*. Based on the definition of the reciprocal lattice, once the direct lattice parameters are known, the reciprocal lattice parameters can be obtained. When we know the incident X-ray beam direction with respect to the direct lattice of the crystalline sample, the incident beam direction with respect to the reciprocal lattice is determined, as the orientations between the direct lattice and reciprocal lattice is coupled by the definition.

Supposing the incident and diffracted wave vectors are:

$$\vec{k}_0 = \frac{\vec{S}_0}{\lambda} \text{ and } \vec{k} = \frac{\vec{S}}{\lambda}.$$

$\vec{S}_0$ and $\vec{S}$ are unit vectors along the incident and diffracted beam directions respectively, as shown in Figure 5.2. Therefore, $\left|\vec{S}_0\right| = \left|\vec{S}\right| = 1$, and

$$\left|\vec{k}_0\right| = \left|\vec{k}\right| = \frac{1}{\lambda},$$

which is the radius of the Ewald sphere.

What does the vector $O*D$ represent if the Bragg condition $2d_{hkl}\,sin\theta_B = \lambda$ is satisfied?

Let

$$2d_{hkl}sin\theta_B = \lambda,$$

then

$$\frac{BD}{CD} = sin\theta_B = \frac{\lambda}{2d_{hkl}},$$

or

$$BD = \frac{\lambda}{2d_{hkl}} \cdot CD = \frac{\lambda}{2d_{hkl}} \cdot \left|\vec{k}\right| = \frac{\lambda}{2d_{hkl}} \cdot \frac{1}{\lambda} = \frac{1}{2d_{hkl}}.$$

Hence

$$O*D = 2BD = \frac{1}{d_{hkl}}.$$

We know that $\overline{O^*D}$ is perpendicular to the reflection plane, or

$$\overline{O*D}\|\vec{r}^*_{hkl}$$

From Chapter 3, we know that the reciprocal lattice vector, $\vec{g}_{hkl} = \vec{r}^*_{hkl} = h\vec{a}^* + k\vec{b}^* + l\vec{c}^*$ is normal to the real lattice plane (hkl), and is of magnitude $1/d_{hkl}$. Therefore, $\overline{O*D} = \vec{g}_{hkl}$ and

$$\vec{k} - \vec{k}_0 = \vec{g}_{hkl} \tag{5.2}$$

Equation (5.2) is the vector form of Bragg's Law, and Figure 5.2 gives the Ewald sphere representation. In X-ray and electron diffraction, it is more convenient to use the Ewald sphere construction and the vector representation to treat the complicated diffraction problems.

Bragg's Law, $2d_{hkl}\,sin\theta_B = \lambda$, provides the information on the crystal structure through d_{hkl}. For a powder X-ray diffractometer, a characteristic X-ray with a known wavelength is used. Once the diffraction angles are measured, the corresponding d-spacings can be calculated, which carry the information of the lattice parameters and Miller indices. To use the Rietveld refinement to solve the crystal structure of a powder material or a polycrystalline material, both diffraction geometry and intensity calculations are involved. The intensity of the diffracted beam will be discussed in the next chapter.

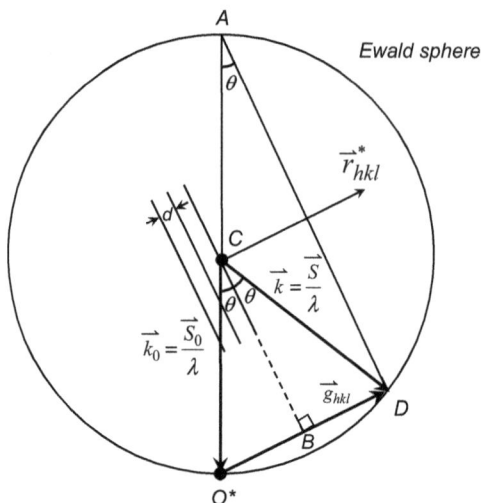

FIGURE 5.2 Illustration of Bragg's Law through Ewald sphere construction and vector expression.

In the textbook *Elements of X-ray Diffraction* (Cullity and Stock, 2001), there are some questions in the diffraction geometry part for students to answer. I have used some of those examples in my class, and we will use a similar example to explain the usage of Bragg's equation. In the early days, students were using the second edition of *Elements of X-ray Diffraction*. Since 2001, the third edition was released, which changes emphasis to diffractometry as the primary technique.

Example 5.1

Assume that we are using Cu Kα X-rays (wavelength λ = 1.54 Å) in the X-ray diffractometer. Determine, and list in order of increasing angle, the values of 2θ and (hkl) for the first three diffraction lines in the powder patterns of a substance with a simple tetragonal lattice (a = 4 Å, c = 3 Å).

Solution: From the d-spacing equation for a tetrahedral system $\dfrac{1}{d_{hkl}^2} = \dfrac{h^2 + k^2}{a^2} + \dfrac{l^2}{c^2}$,

d-spacings for different planes can be calculated.

As $2d_{hkl} \sin \theta_B = \lambda$, we have $\sin \theta_B = \dfrac{\lambda}{2d_{hkl}}$, then we can calculate θ and 2θ.

During the calculation, we need to find 2θ in order of increasing angle, which means we need to find the largest d-spacing, and the second and the third largest. The results are show below.

Plane	d (Å)	theta (°)	2 theta (°)
100	4	11.10	22.20
001	3	14.87	29.74
110	2.83	15.79	31.58

SUMMARY

Bragg's Law can be expressed through Bragg's equation $2d_{hkl} \sin \theta_B = \lambda$, the vector form $\vec{k} - \vec{k}_0 = \vec{g}_{hkl}$ and the Ewald sphere construction.

In the Bragg–Brentano powder X-ray diffraction study, the incident X-ray wavelength is known, and the 2θ values are obtained from the X-ray diffraction pattern collected. In this case the d-spacings can be obtained.

The d-spacings are calculated from the diffraction patterns. The d-spacing carries the information on the lattice parameters, and the Miller indices of diffracting planes, which allow us to retrieve the geometry information of crystal structures. In the next chapter, we will study the intensities of diffracted beams. The peak intensity can be correlated with the unit cell content, site occupancy, crystallite size, preferred orientation and so on.

REFERENCE

Cullity, B. D. (Bernard D.) and Stock, S. R. (2001) *Elements of x-ray diffraction/B.D. Cullity, S.R. Stock.* 3rd ed. Upper Saddle River, NJ: Prentice Hall.

6 The Intensity of Diffracted X-ray Beams

The intensity of the diffracted X-ray beam is highly dependent on the electron density distribution. First, we introduce the scattering of the X-ray by an electron. Then, we discuss the scattering by an atom, a unit cell and a small crystal consisting of many unit cells.

6.1 SCATTERING OF AN X-RAY BY AN ELECTRON

If a polarized, monochromatic, incident X-ray wave strikes a stationary free electron, the oscillating electric field of the incident X-ray exerts a force on the electron as it is a charged particle, which causes the electron to oscillate at the same frequency as the incident wave. According to the classical theory in physics, an accelerated charge emits electromagnetic radiation. Therefore, the oscillating electron becomes a new source of X-rays that radiates in all directions and has the same frequency as the incident X-rays. From classical theory, the intensity of X-rays scattered by a free electron was first demonstrated by the English scientist J. J. Thomson. Figure 6.1 shows the X-ray beam scattered at a 2θ angle, also called the scattering angle θ_S.

Although X-rays are scattered in all directions by an electron, the intensity of the scattered beam depends on the angle of scattering θ_S. According to J. J. Thomson, the intensity of a polarized X-ray beam scattered by an electron is expressed by:

$$I_e = I_0 \left(\frac{\mu_0}{4\pi} \right)^2 \left(\frac{e^4}{m_e^2 r^2} \right) \sin^2 \alpha \qquad (6.1)$$

where

I_O is the intensity of the incident beam, $\mu_0 = 4\pi \times 10^{-7} Tm/A$ is the permeability of free space, α is the angle between the scattering direction and acceleration direction of the electron, I_e is the scattered intensity at a distance r from the electron, m_e is the mass of an electron, and e is the charge carried by an electron. As $c = \dfrac{1}{\sqrt{\varepsilon_0 \mu_0}}$, $\dfrac{1}{4\pi\varepsilon_0 c^2}$ can be used to replace $\dfrac{\mu_0}{4\pi}$.

An unpolarized incident X-ray beam has its electric vector E in a random direction in the yz plane, which can be resolved into E_y and E_z.

For an unpolarized incident X-ray beam:

$$E^2 = E_y^2 + E_z^2$$

Therefore,

$$E_y^2 = E_z^2 = \frac{1}{2}E^2$$

DOI: 10.1201/9780429351662-9

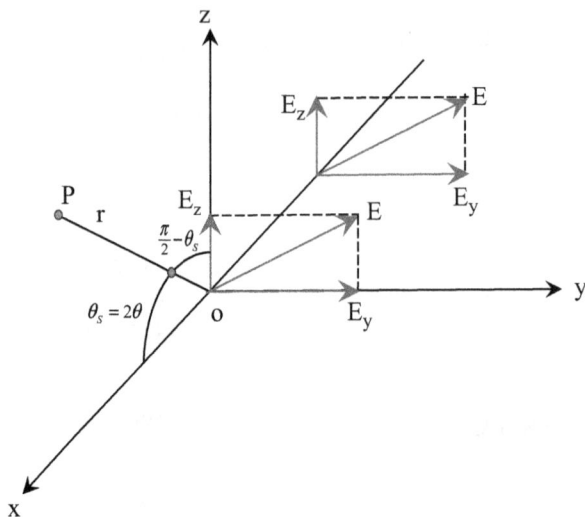

FIGURE 6.1 Coherent scattering of an X-ray by an electron.

or

$$I_{Oy} = I_{Oz} = \frac{1}{2} I_0$$

The intensity of a scattered beam at P can be expressed as

$$I_P(2\theta) = I_{Py} + I_{Pz}$$

$$= I_{Oy} \left(\frac{\mu_0}{4\pi} \right)^2 \left(\frac{e^4}{m_e^2 r^2} \right) \sin^2 90° + I_{Oz} \left(\frac{\mu_0}{4\pi} \right)^2 \left(\frac{e^4}{m_e^2 r^2} \right) \sin^2 (90° - \theta_S)$$

$$= \frac{1}{2} I_0 \left(\frac{\mu_0}{4\pi} \right)^2 \left(\frac{e^4}{m_e^2 r^2} \right) + \frac{1}{2} I_0 \left(\frac{\mu_0}{4\pi} \right)^2 \left(\frac{e^4}{m_e^2 r^2} \right) \cos^2 \theta_S$$

$$= I_0 \left(\frac{\mu_0}{4\pi} \right)^2 \left(\frac{e^4}{m_e^2 r^2} \right) \left(\frac{1 + \cos^2 2\theta}{2} \right)$$

or

$$I_e(2\theta) = \left(\frac{e^2}{4\pi\varepsilon_0 c^2 m_e r} \right)^2 \left(\frac{1 + \cos^2 2\theta}{2} \right) I_0$$

$$= \left(\frac{\mu_0}{4\pi} \right)^2 \left(\frac{e^4}{m_e^2 r^2} \right) \left(\frac{1 + \cos^2 2\theta}{2} \right) I_0$$

(6.2)

where $\dfrac{1 + \cos^2 2\theta}{2}$ is known as the polarization factor.

$I_P(2\theta)$ or $I_e(2\theta)$is the power transported by an X-ray per unit area, or the energy per unit time per unit area at a distance r from the electron at a scattering angle θ_S $(=2\theta)$.

For scattering problems, the unit we often use is the power per unit solid angle, or the energy per unit time per unit solid angle.

We know that the area at a distance r from the electron for a unit solid angle is r^2. Based on Equation (6.2), the power transported by an X-ray through the area r^2 at distance r, or through a unit solid angle, is

$$I_e\left(2\theta\right)=\left(\frac{e^2}{4\pi\varepsilon_0 c^2 m_e r}\right)^2\left(\frac{1+\cos^2 2\theta}{2}\right)I_0\cdot r^2$$

$$=\left(\frac{\mu_0}{4\pi}\right)^2\left(\frac{e^4}{m_e^2}\right)\left(\frac{1+\cos^2 2\theta}{2}\right)I_0$$

(6.3)

Therefore, the equation above is the X-ray energy transported per unit time per unit solid angle. When we calculate the integrated intensity contributed by one crystallite, we use this equation.

6.2   SCATTERING BY AN ATOM

To better understand the scattering of X-rays by an atom, in theory we need to consider the contributions from both the positively charged nucleus and the negatively charged electron clouds. When we use the same principle to calculate the contribution by a positively charged proton inside the nucleus, we notice that $\frac{1}{m_p^2}$ is much smaller than $\frac{1}{m_e^2}$. Therefore, the contribution to the intensity from a proton can be ignored.

In X-ray diffraction the atomic scattering factor of X-rays by an atom is defined by

$$f_X = \frac{amplitude\ of\ the\ wave\ scattered\ by\ an\ atom}{amplitude\quad of\quad the\quad wave\quad scattered\quad by\quad one\quad electron} = \frac{A_a}{A_e}$$

(6.4)

Therefore, we have

$$f_X^2 = \frac{A_a^2}{A_e^2} = \frac{I_a}{I_e}$$

or

$$I_a = f_X^2 I_e$$

The calculation of f_X involves integration over the whole range of electron clouds around the atom, and the table of f_X can be found from different sources (Doyle and Turner, 1968). In fact, the effect of anomalous dispersion is strong only when the incoming X-ray radiation has a wavelength near one of the absorption edges of the

FIGURE 6.2 Path difference between two beams scattered by the electrons associated with an atom.

scattering atom. When the frequency of X-ray radiation is very large compared with the frequency of every absorption edge of the scattering atom, the atomic scattering factor is independent of the X-ray frequency or wavelength (Barrett and Massalski, 1980). In Rietveld refinement, the valence state needs to be specified for the atomic scattering factor as the number of electrons associated with an atom affects the value of the atomic scattering factor.

Before calculating the atomic structure factor, we need to find the phase difference between the waves from two scattering centers (Figure 6.2) when the scattering angle θ_S (=2θ) is given.

As shown in Figure 6.2, the beam path difference can be expressed by:

$$\delta_j = \vec{r}_j \cdot \vec{S}_0 - \vec{r}_j \cdot \vec{S} = -\vec{r}_j \cdot \left(\vec{S} - \vec{S}_0\right) = -\vec{r}_j \cdot \left(\vec{k} - \vec{k}_o\right)\lambda = -\vec{r}_j \cdot \vec{q}\lambda \tag{6.5}$$

And the phase difference is given by:

$$\phi_j = -\vec{r} \cdot \left(\vec{S} - \vec{S}_0\right)\left(\frac{2\pi}{\lambda}\right) = -2\pi\left(\vec{k} - \vec{k}_0\right) \cdot \vec{r}_j = -2\pi\vec{q} \cdot \vec{r}_j \tag{6.6}$$

where $\vec{r}_j = x_j\vec{a} + y_j\vec{b} + z_j\vec{c}$ is the position in direct space, $\vec{S}_0$ and $\vec{S}$ are the unit vectors of the incident and the scattered beams, $\vec{k}_0 = \dfrac{\vec{S}_0}{\lambda}$ and $\vec{k} = \dfrac{\vec{S}}{\lambda}$ are the wave vectors of the incident wave and scattered wave, $\left|\vec{k}_o\right| = \dfrac{1}{\lambda}$ and $\left|\vec{k}\right| = \dfrac{1}{\lambda}$ are the wave numbers of the incident and scattered waves, and $\vec{q} = \vec{k} - \vec{k}_0 = u\vec{a}^* + v\vec{b}^* + w\vec{c}^*$ is the scattering vector represented in reciprocal space. For a chemical element, the values of the atomic scattering factor of X-rays vary with s, where $s = \dfrac{\sin\theta}{\lambda}$ and $2s = \dfrac{2\sin\theta}{\lambda} = |\vec{q}| = \left|\vec{k} - \vec{k}_o\right|$.

In Figure 6.2, O is the origin and A is the scattering center at position $\vec{r}$; $2\theta(=\theta_S)$ is the scattering angle.

Materials science students should notice that many physics textbooks use $\left|\vec{k}_o\right| = \left|\vec{k}\right| = \dfrac{2\pi}{\lambda}$ to represent wave numbers, in which case $\left|\vec{k} - \vec{k}_o\right| = \left|\vec{q}\right| = \dfrac{4\pi \sin\theta}{\lambda}$.

Given $\rho(\vec{r})$ as the electron density distribution of the electron cloud for an atom, the atomic scattering factor f_X is presented by the integral

$$f_X = \int e^{-i\phi} \rho \, dV \tag{6.7}$$

or

$$
\begin{aligned}
f_X &= \int e^{(2\pi i/\lambda)\left(\vec{S} - \vec{S}_0\right)\cdot\vec{r}} \rho \, dV \\
&= \int e^{2\pi i \left(\vec{k} - \vec{k}_0\right)\cdot\vec{r}} \rho \, dV \\
&= \int e^{2\pi i \vec{q}\cdot\vec{r}} \rho \, dV
\end{aligned}
\tag{6.8}
$$

To calculate the integral, we assume that the electron cloud of an atom is spherically symmetric and has the radial density distribution $\rho(r)$. When a spherical coordinate system is introduced as shown in Figure 6.3, we fix the $\left(\vec{S} - \vec{S}_0\right)$ or $\left(\vec{k} - \vec{k}_0\right)$ direction along the z direction.

We can use $dV = r^2 \sin\alpha \cdot d\alpha \cdot d\varphi \cdot dr$, $\left(\vec{S} - \vec{S}_0\right)\cdot\vec{r} = (2\sin\theta)\cdot r \cdot \cos\alpha$ and $s = \dfrac{\sin\theta}{\lambda}$, then:

$$f_X = \int_{r=0}^{\infty} \int_{\varphi=0}^{2\pi} \int_{\alpha=0}^{\pi} e^{i(4\pi s)r\cos\alpha} \rho(r) r^2 \sin\alpha \cdot d\alpha \cdot d\varphi \cdot dr \tag{6.9}$$

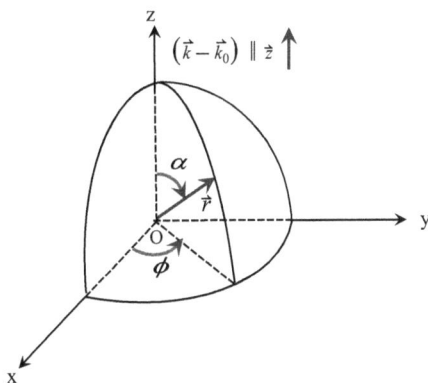

FIGURE 6.3 The spherical coordinate system for the calculation of the atomic scattering factor based on the electron density distribution $\rho(r)$ associated with an atom. The $\left(\vec{k} - \vec{k}_0\right)$ direction is set along the z direction, and angle α instead of θ is used in the spherical coordinate system so that it will not be confused with the semi-scattering angle. The volume element is $dV = r^2 \sin\alpha \cdot d\alpha \cdot d\varphi \cdot dr$.

If integrations with respect to α and φ are performed first, we have

$$f_X = \int_{r=0}^{\infty} \int_{\varphi=0}^{2\pi} \int_{\alpha=0}^{\pi} e^{i(4\pi s)r\cos\alpha} \rho(r) r^2 \sin\alpha \cdot d\alpha \cdot d\varphi \cdot dr$$

$$= \int_{r=0}^{\infty} \int_{\varphi=0}^{2\pi} \rho(r) r^2 \left[\frac{1}{-i(4\pi s)r} \int_{\alpha=0}^{\pi} e^{i(4\pi s)r\cos\alpha} d(i4\pi sr\cos\alpha) \right] \cdot d\varphi \cdot dr$$

$$= \int_{r=0}^{\infty} \int_{\varphi=0}^{2\pi} \rho(r) r^2 \left[\frac{1}{-i(4\pi s)r} \left(e^{-i4\pi sr} - e^{i4\pi sr} \right) \right] \cdot d\varphi \cdot dr$$

$$= \int_{r=0}^{\infty} \int_{\varphi=0}^{2\pi} \rho(r) r^2 \left[\frac{2\sin 4\pi sr}{4\pi sr} \right] \cdot d\varphi \cdot dr$$

$$= 4\pi \int_{r=0}^{\infty} \rho(r) r^2 \frac{\sin 4\pi sr}{4\pi sr} \cdot dr$$

In different X-ray diffraction textbooks, we can find the expression of

$$f_X = \int_0^{\infty} 4\pi r^2 \rho(r) \frac{\sin 4\pi sr}{4\pi sr} \cdot dr \tag{6.10}$$

or

$$f_X = \int_0^{\infty} 4\pi r^2 \rho(r) \frac{\sin 2\pi qr}{2\pi qr} \cdot dr \tag{6.11}$$

where $q = 2s$.

In some physics textbooks, the expression is written as

$$f_X = \int_0^{\infty} 4\pi r^2 \rho(r) \frac{\sin qr}{qr} \cdot dr \tag{6.12}$$

where $q = (2\pi)(2s) = \dfrac{4\pi \sin\theta}{\lambda}$

For instance, in the textbook *X-ray Diffraction* (Warren, 1990), you can find the expression

$$f_x = \sum_n \int_0^{\infty} 4\pi r^2 \rho(r) \frac{\sin kr}{kr} \cdot dr$$

In the case of forward scattering, $\theta = 0$ and $q = 0$, so we have

$$f_X = \int_0^\infty 4\pi r^2 \rho(r) \cdot dr = Z$$

We would like to mention here that in some cases of small angle X-ray scattering similar mathematical treatments are employed, and that the electron density contribution is not just from an atom as shown above. Instead, the electron density contribution is from a relatively large sized particle; the form factor is related to the scattering resulting from a particle.

In the structure factor discussion (Section 6.3), the integration is performed for the electron density inside a unit cell of a crystal. As only the scattering that satisfies the Bragg condition is considered, $\vec{q} = \vec{k} - \vec{k}_0 = u\vec{a}^* + v\vec{b}^* + w\vec{c}^*$ is just the reciprocal lattice vector $\vec{g} = h\vec{a}^* + k\vec{b}^* + l\vec{c}^*$, representing a crystallographic plane in the real lattice. This mathematical treatment allows us to correlate the electron density map with the structure factor:

$$\int_{\substack{unit \\ cell}} \rho(x,y,z) \exp\left[2\pi i \vec{q} \cdot \vec{r}\right] dv \xrightarrow{\vec{q}=\vec{g}} \int_{\substack{unit \\ cell}} \rho(x,y,z) \exp\left[2\pi i \vec{g} \cdot \vec{r}\right] dv = F_{hkl}$$

The following is a brief summary of the above discussion on the integration. (i) For atomic scattering factor calculation, we only consider the distribution of the electron cloud in an atom. (ii) For structure factor calculation in single crystal X-ray diffraction analysis, we consider the electron density contribution inside a unit cell in the crystal. The scattering occurs under the Bragg condition and the electron density map can be correlated with the structure factor. (iii) In small angle X-ray scattering, we consider the electron density contribution inside a particle, and the form factor in small angle X-ray scattering is associated with the shape and size of an individual particle. In mathematics, the formats of the integrals in the above three cases look similar.

Furthermore, in small angle X-ray scattering, both the form factor and structure factor determine the X-ray scattering intensity, and the structure factor in small angle X-ray scattering employs different expressions in comparison with that in X-ray diffraction. In X-ray diffraction, the structure factor is associated with a unit cell in a crystal. In my lectures provided to materials science students, I have focused on wide-angle powder X-ray diffraction but not small angle X-ray scattering. Students who are interested in the latter techniques can use the third edition of *Elements of X-ray Diffraction*, in which a chapter for small angle scattering has been included (Cullity and Stock, 2001).

When the incoming wave has a frequency near any of the absorption edges of the scattering atom, the atomic scattering factor must be modified. In fact, the atomic scattering factor f_X is independent of the frequency of the diffracted radiation only if the frequency is very large compared with the frequency of every absorption edge of

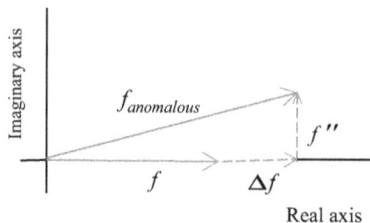

FIGURE 6.4 Schematical representation of the influence of an anomalous phase shift on the atomic scattering factor.

the scattering atom. The atomic scattering factor changes dramatically if the incoming radiation has a wavelength near one of the absorption edges of the scattering atom (Barrett and Massalski, 1980), in which case the scattering process shows unusual behavior and there is an anomalous phase shift of the scattered wave. Figure 6.4 shows that the correction includes a real part and an imaginary part.

The corrected expression is in the following form (Barrett and Massalski, 1980):

$$f_{corr} = f_0 + \Delta f' + i\Delta f''$$
$$|f_{corr}| = \sqrt{\left(f_0 + \Delta f'\right)^2 + \left(\Delta f''\right)^2} \tag{6.13}$$

With analytical approximation (Doyle and Turner, 1968), f_X can be expressed in the form of

$$f_X(s) = f_X\left(\frac{\sin\theta}{\lambda}\right) = \sum_{i=1}^{N} a_i \exp\left(-b_i s^2\right) + c$$

where s has been defined before, which is given by $s = \dfrac{\sin\theta}{\lambda}$. When the Bragg condition is satisfied, $s = \dfrac{\sin\theta_B}{\lambda} = \dfrac{1}{2d_{hkl}}$. A similar expression was presented by an even earlier study (Vand et al., 1957).

6.3 SCATTERING BY A UNIT CELL

In order to calculate the X-ray intensity after scattering by a unit cell, the path difference and phase difference after scattering by different atoms within a unit cell need to be obtained.

From Figure 6.5, it can be seen that the path difference between the two beams after scattering from the atoms at points O and A is $\delta = CA - OB$, and

$$\delta = \vec{r} \cdot \vec{S}_0 - \vec{r} \cdot \vec{S} = \vec{r} \cdot \left(\lambda \vec{k}_0\right) - \vec{r} \cdot \left(\lambda \vec{k}\right) = -\lambda \vec{r} \cdot \vec{q}$$

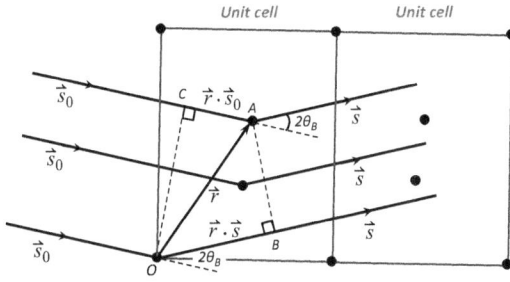

FIGURE 6.5 Path difference between beams scattered by atoms in a unit cell. The Bragg condition is satisfied.

So the phase difference is:

$$\phi = -2\pi \left(\vec{k} - \vec{k}_0 \right) \cdot \vec{r}_j = -2\pi \vec{q} \cdot \vec{r}_j$$

where $\vec{r}_j = x_j \vec{a} + y_j \vec{b} + z_j \vec{c}$ is the position of the jth atom, and $\vec{q} = \vec{k} - \vec{k}_0 = u\vec{a}^* + v\vec{b}^* + w\vec{c}^*$ is the scattering vector in reciprocal space, and u, v and w can be non-integers or integers. The mathematical treatment of the above phase difference calculation is the same as that shown in the atomic scattering factor part.

When $\theta = \theta_B$, or $\vec{k} - \vec{k}_0 = \vec{g}_{hkl} = h\vec{a}^* + k\vec{b}^* + l\vec{c}^*$, which means the Bragg condition is satisfied for plane (hkl), the phase difference is:

$$\phi = -2\pi \left(\vec{k} - \vec{k}_0 \right) \cdot \vec{r}_j = -2\pi \vec{g}_{hkl} \cdot \vec{r}_j = -2\pi \left(hx_j + ky_j + lz_j \right) \qquad (6.14)$$

It is known that there are different conventions in quantum mechanics and crystallography for structure factor calculation (Spence, 2013). In X-ray diffraction, we use the crystallographic convention, and in electron diffraction we use the quantum mechanical convention. In crystallographic convention, the free space wave is expressed as:

$$\exp\left\{ -2\pi i \left(\vec{k} \cdot \vec{r} - vt \right) \right\}$$

and the structure factor is expressed as:

$$\sum_j f_j \exp\left\{ +2\pi i \left(\vec{g} \cdot \vec{r}_j \right) \right\}$$

Whereas in quantum mechanical convention, the free space wave is expressed as:

$$\exp\left\{ +2\pi i \left(\vec{k} \cdot \vec{r} - vt \right) \right\}$$

and the structure factor is expressed as:

$$\sum_j f_j \exp\left\{-2\pi i\left(\vec{g}\cdot\vec{r}_j\right)\right\}$$

In quantum mechanical convention, some textbooks use $\exp\left\{i\left(\vec{k}\cdot\vec{r}-\omega t\right)\right\}$ to represent a wave instead of $\exp\left\{2\pi i\left(\vec{k}\cdot\vec{r}-vt\right)\right\}$.

If $\exp\left\{i\left(\vec{k}\cdot\vec{r}-\omega t\right)\right\}$ is used, it implies that $\left|\vec{k}_0\right|=\left|\vec{k}\right|=\dfrac{2\pi}{\lambda}$, $\left|\vec{k}-\vec{k}_0\right|=\left|\vec{q}\right|=\dfrac{(2\pi)(2\sin\theta)}{\lambda}=4\pi s$, where $\theta=\dfrac{\theta_s}{2}$.

If $\exp\left\{2\pi i\left(\vec{k}\cdot\vec{r}-vt\right)\right\}$ is used, it means $\left|\vec{k}_0\right|=\left|\vec{k}\right|=\dfrac{1}{\lambda}$ and $\left|\vec{k}-\vec{k}_0\right|=\left|\vec{q}\right|=\dfrac{2\sin\theta}{\lambda}=2s$.

Therefore, students should be very careful when reading textbooks in the areas of crystallography, X-ray diffraction and physics. Our students need to check what convention is used in each textbook.

Following the crystallographic convention, the amplitude of the X-ray after scattering by a unit cell can be derived.

In X-ray diffraction, the structure factor is defined as the ratio of the amplitude of X-rays scattered by a unit cell to that by an electron in the Bragg condition, or

$$F = \frac{A_{cell}}{A_e}$$

The amplitude of the scattered X-ray contributed by a unit cell under the Bragg condition is:

$$A_{cell} = A_e \sum_{\substack{all\ atoms\\per\ cell}} f_j \exp\left(-i\phi_j\right)$$

$$= A_e \sum_{\substack{all\ atoms\\per\ cell}} f_j \exp\left[2\pi i \vec{g}_{hkl}\cdot\vec{r}_j\right]$$

$$= A_e \sum_{\substack{all\ atoms\\per\ cell}} f_j \exp\left[2\pi i\left(hx_j+ky_j+lz_j\right)\right]$$

Therefore, the structure factor $F(=A_{cell}/A_e)$ can be expressed as

$$F_{hkl} = \sum_{\substack{all\ atoms\\per\ cell}} f_j \exp\left[2\pi i \vec{g}_{hkl}\cdot\vec{r}_j\right] \tag{6.15}$$

or

$$F_{hkl} = \sum_{\substack{all \ atoms \\ per \ cell}} f_j \exp\left[2\pi i\left(hx_j + ky_j + lz_j\right)\right] \tag{6.16}$$

The diffracted beam intensity contributed by a unit cell is:

$$I_{cell} = \left|F_{hkl}\right|^2 \cdot I_e \tag{6.17}$$

For a plane (hkl) that satisfies the Bragg condition, $\vec{k} - \vec{k}_0 = \vec{g}_{hkl}$, the diffraction will be forbidden if $F_{hkl} = 0$. The "allowed" reflecting planes are those with $F_{hkl} \neq 0$.

In the *International Table for Crystallography, Volume A*, we can find the allowed or reflection conditions for each space group. It is known that the reflection condition for the face-centered lattice (not only for the cubic, but also for the non-cubic) is when h, k, l are all odd or all even, whereas for the body-centered lattice, $h+k+l$ must be even.

There is a method to correlate the electron density map with the structure factor as presented in many X-ray diffraction textbooks (Warren, 1990; Stout and Jensen, 1989). First, the three-dimensional periodic electron density in a crystal is presented by a three-dimensional Fourier series

$$\rho\left(x,y,z\right) = \sum_{h'=-\infty}^{+\infty}\sum_{k'=-\infty}^{+\infty}\sum_{l'=-\infty}^{+\infty} C_{h'k'l'} \exp\left[2\pi i\left(h'x + k'y + l'z\right)\right],$$

where h', k', l' are integers between $-\infty$ and $+\infty$.

Then, integration throughout the unit cell is performed to correlate the electron density with the structure factor. In this case, the intermediate step of integrating around an atom and obtaining the atomic scattering factor is skipped.

The structure factor is:

$$
\begin{aligned}
F_{hkl} &= \int_{\substack{unit \\ cell}} \rho\left(x,y,z\right)\exp\left[2\pi i\left(hx + ky + lz\right)\right]dv \\
&= \int_{\substack{unit \\ cell}} \sum_{h'}\sum_{k'}\sum_{l'} C_{h'k'l'} \exp\left[2\pi i\left(h'x + k'y + l'z\right)\right]\exp\left[2\pi i\left(hx + ky + lz\right)\right]dv \\
&= \int_{\substack{unit \\ cell}} \sum_{h'}\sum_{k'}\sum_{l'} C_{h'k'l'} \exp\left\{2\pi i\left[\left(h+h'\right)x + \left(k+k'\right)y + \left(l+l'\right)z\right]\right\}dv
\end{aligned}
$$

The integral over one period (unit cell) is zero for all terms, except when $h' = -h$, $k' = -k$, $l' = -l$. Thus,

$$F_{hkl} = \int_{\substack{unit \\ cell}} C_{\overline{hkl}}\, dv = V_c C_{\overline{hkl}}$$

$$C_{\overline{hkl}} = (1/V_c) F_{hkl}$$

Using the relationship $C_{\overline{hkl}} = (1/V_c) F_{hkl}$, or $C_{hkl} = (1/V_c) F_{\overline{hkl}}$, and removing' in

the electron density expression $\rho(x,y,z) = \sum_{h'}\sum_{k'}\sum_{l'} C_{h'k'l'} \exp\left[2\pi i (h'x + k'y + l'z)\right]$,

the electron density in a unit cell in direct space is represented in terms of the structure factors in reciprocal space as shown in Equation (6.18) (Luger, 2014), which is an import expression in single crystal structure analysis.

$$\rho(x,y,z) = \frac{1}{V_c} \sum_h \sum_k \sum_l F_{\overline{hkl}} \exp\left[2\pi i (hx + ky + lz)\right]$$

$$= \frac{1}{V_c} \sum_h \sum_k \sum_l F_{hkl} \exp\left[-2\pi i (hx + ky + lz)\right] \tag{6.18}$$

The structure factor, F_{hkl}, is a complex number ($F_{hkl} = |F_{hkl}|e^{i\alpha_{hkl}}$), and the phase problem is the central problem in single crystal analysis (Luger, 2014). There are different methods for the solution of the phase problem.

As discussed previously, $I_{cell} = |F_{hkl}|^2 \cdot I_e$. For planes (hkl) that satisfy the Bragg condition $\vec{k} - \vec{k}_0 = \vec{g}_{hkl}$, there are "allowed" and "forbidden" diffracting planes. If $F_{hkl} = 0$, then $I \propto |F_{hkl}|^2 = 0$. If $F_{hkl} \neq 0$, then $I \neq 0$. Thus the "allowed" diffraction planes are those for which $F_{hkl} \neq 0$.

The structure factor tells us which diffraction spots/peaks are missing due to the position of the atoms in the unit cell at the Bragg diffraction conditions. For the four lattice types, the forbidden reflection conditions are:

(I) Simple unit cell: no systematic absence, or $F_{hkl} \neq 0$ for any h, k, l.
(II) Base-face-centered cell: h, k: mix of odd and even numbers, $F_{hkl} = 0$.
(III) Face-centered cell: h, k, l: mix of odd and even numbers, $F_{hkl} = 0$.
(IV) Body-centered cell: $h + k + l$ = odd number, $F_{hkl} = 0$.

A more general explanation is that only symmetry operations with translational elements lead to classes of reflections with zero intensity, which can be summarized as follows (Ladd and Palmer, 2003).

(I) Screw symmetry operations may affect axial reflections: h00, 0k0, and 00ℓ.
(II) Glide symmetry operations may affect zonal reflections: hk0, h0ℓ, and 0kℓ.
(III) Cell centering may affect all reflections: all hkℓ.

As mentioned previously, the *International Table for Crystallography, Volume A* provides the allowed or forbidden reflection conditions for each space group.

In the following, we will discuss how to calculate F_{hkl} and $|F_{hkl}|^2$.

As

$$F_{hkl} = \sum_{\substack{all \quad atoms \\ per \quad cell}} f_j \exp\left[2\pi i \bar{g}_{hkl} \cdot \bar{r}_j\right]$$

$$= \sum_{\substack{all \quad atoms \\ per \quad cell}} f_j \exp\left[2\pi i\left(hx_j + ky_j + lz_j\right)\right]$$

the amplitude and phase satisfy the following relations

$$\left|F\left(hkl\right)\right| = \left|F\left(\overline{hkl}\right)\right|$$

$$\alpha\left(hkl\right) = -\alpha\left(\overline{hkl}\right)$$

The diffracted intensity for (*hkl*) plane is

$$I\left(hkl\right) = C\left|F\left(hkl\right)\right|^2$$

where C depends upon various physical factors that influence the intensities measured by X-ray experiments. Therefore,

$$I\left(hkl\right) = I\left(\overline{hkl}\right)$$

no matter whether the crystal structure is centrosymmetric or not. This is Friedel's Law, which states that the diffraction intensities from plane (*hkl*) and plane $\left(\overline{hkl}\right)$ are equal.

Example 6.1

In the face-centered structure as presented in Figure 6.6, one unit cell contains four atoms, located in the following positions:

$$000, \ ½ \ ½ \ 0, \ ½ \ 0 \ ½ \ \text{and} \ 0 \ ½ \ ½.$$

Derive the selection rules.

Solution: From the coordinates provided, we can write the structure factor for reflection (hkl) as

$$F_{hkl} = f\left[e^0 + e^{\pi i(h+k)} + e^{\pi i(k+l)} + e^{\pi i(l+h)}\right].$$

We know that $e^{m\pi i} = (-1)^m = \begin{cases} +1, & m \quad even \\ -1, & m \quad odd \end{cases}$, so we can find the selection

rules for the face-centered structure, which are

$$F_{hkl} = \begin{cases} 4f, & h,k,l \quad all \quad even \quad or \quad all \quad odd \\ 0, & h,k,l \quad mix \quad of \quad odd \quad and \quad even \end{cases}$$

Some metals have the face-centered structure, for instance, γ-Fe, Al, Ni, Cu, Ag and Au, and the allowed reflections are those with h, k, l being all odd or all even.

From the example, we notice that the type of crystal system is not included in the calculation. Therefore, the above selection rules are true for a face-centered cubic (FCC) structure and a face-centered orthorhombic structure. During the calculation, the atom coordinates, but not the lattice constants, in the unit cell are needed.

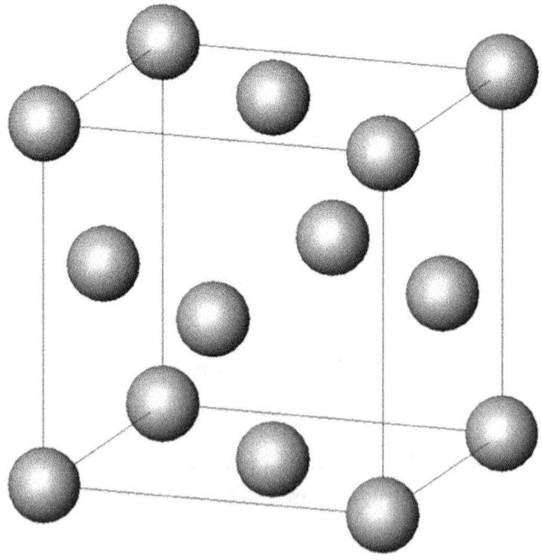

FIGURE 6.6 Faced-centered crystal structure showing atom positions.

Example 6.2

Derive a simplified expression for the structure factor of NaCl and give the selection rules.

NaCl has a cubic lattice with four Na and four Cl atoms per unit cell (Figure 6.7), located as follows:

Na: 0 0 0   ½ ½ 0   ½ 0 ½   0 ½ ½
Cl: ½ ½ ½ 0 0 ½   0 ½ 0   ½ 0 0

Solution: By substituting x_j, y_j, z_j values into the structure factor equation

$$F_{hkl} = \sum_{\substack{all \ atoms \\ per \ cell}} f_j \exp\left[2\pi i\left(hx_j + ky_j + lz_j\right)\right]$$

we get

$$F_{hkl} = f_{Na}\left\{1 + \exp\left[\pi i(h+k)\right] + \exp\left[\pi i(h+l)\right] + \exp\left[\pi i(k+l)\right]\right\}$$
$$+ f_{Cl}\left\{\exp\left[\pi ih\right] + \exp\left[\pi ik\right] + \exp\left[\pi il\right] + \exp\left[\pi i(h+k+l)\right]\right\}$$
$$= f_{Na}\left\{1 + \exp\left[\pi i(h+k)\right] + \exp\left[\pi i(h+l)\right] + \exp\left[\pi i(k+l)\right]\right\} +$$
$$f_{Cl}\exp\left[\pi i(h+k+l)\right]\left\{1 + \exp\left[\pi i(-h-k)\right] + \exp\left[\pi i(-h-l)\right] + \exp\left[\pi i(-k-l)\right]\right\}$$
$$= \left\{f_{Na} + f_{Cl}\exp\left[\pi i(h+k+l)\right]\right\}\left\{1 + \exp\left[\pi i(h+k)\right] + \exp\left[\pi i(h+l)\right] + \exp\left[\pi i(k+l)\right]\right\}$$

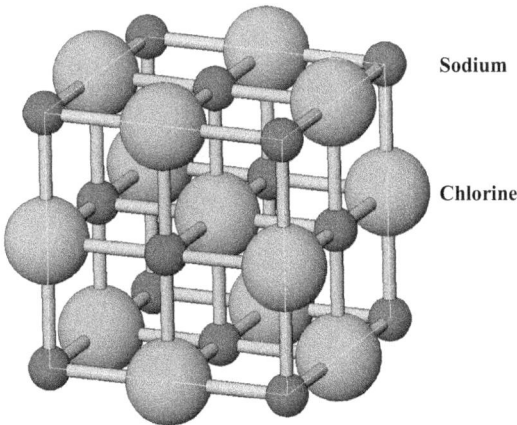

Sodium

Chlorine

FIGURE 6.7 Sodium chloride structure showing positions of Na and Cl atoms. Na is at the origin of the unit cell.

By considering different combinations for h, k and l, we can obtain the selection rules, which are:

$$F_{hkl} = \begin{cases} 4(f_{Na} + f_{Cl}), & h,k,l \quad all \quad even \\ 4(f_{Na} - f_{Cl}), & h,k,l \quad all \quad odd \\ 0, & for \quad mixed \quad indices \end{cases}$$

In fact, if we use the concept of *crystal = lattice + basis*, we realize that the NaCl lattice has the face-centered type, and the basis contains one Na atom and one Cl atom. As the lattice is face-centered, there are reflections when h, k and l are all odd or all even. For mixed indices, $F_{hkl} = 0$. The basis contains more than one atom with different coordinates; the F_{hkl} expressions for all odd indices and all even indices are different. For this particular case, {1 + exp [$\pi i(h + k)$] + exp [$\pi i(h + l)$] + exp [$\pi i(k + l)$]} reflects the effect of the lattice type, and {$f_{Na} + f_{Cl}$ exp [$\pi i(h + k + l)$]} reflects the effect of the basis. Therefore, the selection rules for NaCl are more restrictive than the simple face-centered structure, such as Au, Ag and Cu.

For sodium chloride, there are two commonly used crystal structure representations, one is to locate Na in the origin of the unit cell (Figure 6.7), and the other is to have Cl in the origin (Figure 6.8). If the origin is changed, what will happen?

Using a similar calculation as shown above, the selection rules for NaCl, with Cl at the origin of the unit cell, are

$$F_{hkl} = \begin{cases} 4(f_{Cl} + f_{Na}), & h,k,l \quad all \quad even \\ 4(f_{Cl} - f_{Na}), & h,k,l \quad all \quad odd \\ 0, & for \quad mixed \quad indices \end{cases}$$

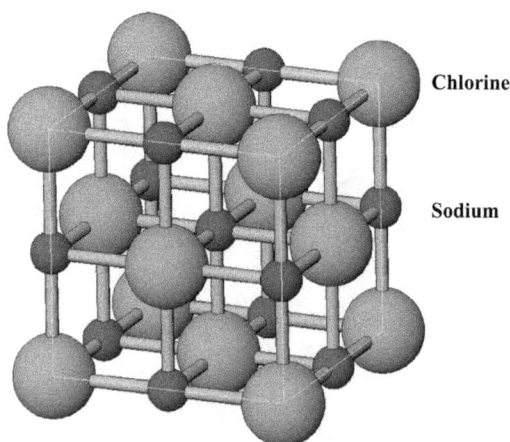

FIGURE 6.8 Sodium chloride structure showing positions of Na and Cl atoms. Cl is at origin of the unit cell.

Example 6.3

Diamond has high symmetries with the space group $Fd\bar{3}m$. The 000, ¾ ¼ ¾ positions of the unit cell are occupied with carbon atoms. There are face-centered structure transitions: (0, 0, 0)+, (0, ½, ½)+, (½, 0, ½)+ and (½, ½, 0)+, which give the eight carbon atom positions: 000, 0 ½ ½, ½ 0 ½, ½ ½ 0, ¾ ¼ ¾, ¾ ¾ ¼, ¼ ¼ ¼ and ¼ ¾ ¾ (Figure 6.9). Derive a simplified structure factor expression for diamond and give the selection rules.

Solution: Substitute x_j, y_j, z_j values into the structure factor equation:

$$F_{hkl} = \sum_{\substack{all \ atoms \\ per \ cell}} f_j \exp\left[2\pi i\left(hx_j + ky_j + lz_j\right)\right]$$

$$F_{hkl} = f_C\left[e^0 + e^{\pi i(k+l)} + e^{\pi i(h+l)} + e^{\pi i(h+k)} + e^{\frac{\pi i}{2}(3h+k+3l)} + e^{\frac{\pi i}{2}(3h+3k+l)} + e^{\frac{\pi i}{2}(h+k+l)} + e^{\frac{\pi i}{2}(h+3k+3l)}\right]$$

$$= F_{FCC} + f_C e^{\frac{\pi i}{2}(h+k+l)}\left[1 + e^{\pi i(h+l)} + e^{\pi i(h+k)} + e^{\pi i(k+l)}\right]$$

$$= F_{FCC} + F_{FCC}e^{\frac{\pi i}{2}(h+k+l)}$$

$$= F_{FCC}\left[1 + e^{\frac{\pi i}{2}(h+k+l)}\right]$$

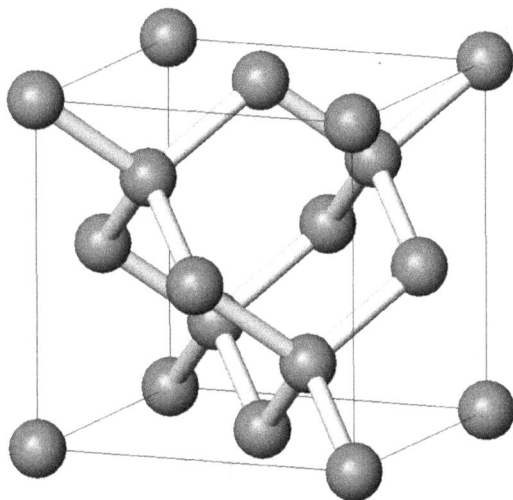

FIGURE 6.9 Diamond crystal structure showing positions of C atoms.

By considering different combinations for h, k and l, we can obtain the selection rules, which are:

h,k,l are mixed odd and even

$F_{hkl} = 0$

h,k,l are   all   odd,   and $h+k+l = 4n+1$,

$F_{hkl} = F_{FCC}\left(1+i\right) = 4\left(1+i\right)f_C$

h,k,l are   all   odd,   and $h+k+l = 4n+3$,

$F_{hkl} = F_{FCC}\left(1-i\right) = 4\left(1-i\right)f_C$

h,k,l are   all   even,   and $h+k+l = 4n$,

$F_{hkl} = F_{FCC}\left(1+1\right) = 8f_C$

h,k,l are   all   even,   and $h+k+l = 4n+2$,

$F_{hkl} = F_{FCC}\left(1-1\right) = 0$

We can also use F_{hkl}^2 values to represent the selection rules. And we have

$$F_{hkl}^2 = F_{FCC}{}^2\left[1+e^{\frac{\pi i}{2}(h+k+l)}\right]\left[1+e^{-\frac{\pi i}{2}(h+k+l)}\right] = F_{FCC}{}^2\left[2+e^{\frac{\pi i}{2}(h+k+l)}+e^{-\frac{\pi i}{2}(h+k+l)}\right]$$

$$= F_{FCC}{}^2[2+2\cos\frac{\pi}{2}(h+k+l)] = 2F_{FCC}{}^2\left[1+\cos\frac{\pi}{2}(h+k+l)\right]$$

Then, the selection rules are

h,k,l are mixed odd and even,

$F_{hkl} = 0$

h,k,l are   all   odd ,

$F_{hkl}^2 = 2F_{FCC}^2 = 32f^2$

h,k,l are   all   even,   and $h+k+l = 4n$,

$F_{hkl}^2 = 2F_{FCC}^2\left(1+1\right) = 64f^2$

h,k,l are   all   even,   and $h+k+l \neq 4n$,

$F_{hkl}^2 = 2F_{FCC}^2\left(1-1\right) = 0$

If we use the concept of *crystal* = *lattice* + *basis*, the Bravais lattice for diamond is FCC, and the basis contains two carbon atoms. The two carbon atoms are separated by ¼ of the body diagonal length along the body diagonal direction. Because the basis has two carbon atoms, it modifies the structure factor. The selection rules for a diamond cubic are more restrictive than the simple FCC structure. In the case where h, k, l are all even numbers, $F_{hkl} \neq 0$ for a simple FCC structure. However, for a diamond-type structure, even though h, k, l are all even numbers, F_{hkl} is 0 when $h + k + l \neq 4n$.

Students can practice the selection rules for a Cu_3Au intermetallic compound in comparison with a simple FCC structure. The key is that when an order-disorder transformation occurs, both lattice type and basis are changed. The disorder structure has an FCC lattice with a basis of $1/4Au + 3/4Cu$. The ordered structure has a primitive lattice with a basis of a four-atom cluster.

Example 6.4

ZrO_2 has a cubic crystal structure (space group $Fm\overline{3}m$), and contains four zirconium and eight oxygen atoms per unit cell (Figure 6.10), located in the following positions:

Zr: 0 0 0, ½ ½ 0, ½ 0 ½, 0 ½ ½;
O: ¼ ¼ ¼, ¾ ¾ ¼, ¾ ¼ ¾, ¼ ¾ ¾, ¼ ¼ ¾, ¾ ¾ ¾, ¾ ¼ ¼ and ¼ ¾ ¼.

Calculate the value of $|F|^2$ for a 002 reflection in terms of the atomic scattering factors (f_{Zr} and f_O), where F is the structure factor.

Solution: From the coordinates of oxygen and zirconium atoms, the structure factor for zirconia can be written as

$$F_{hkl} = f_{Zr}\left[e^0 + e^{\pi i(h+k)} + e^{\pi i(h+l)} + e^{\pi i(k+l)}\right]$$

$$+ f_O\left[e^{\frac{\pi i}{2}(h+k+l)} + e^{\frac{\pi i}{2}(3h+3k+l)} + e^{\frac{\pi i}{2}(3h+k+3l)} + e^{\frac{\pi i}{2}(h+3k+3l)}\right.$$

$$\left. + e^{\frac{\pi i}{2}(h+k+3l)} + e^{\frac{\pi i}{2}(3h+3k+3l)} + e^{\frac{\pi i}{2}(3h+k+l)} + e^{\frac{\pi i}{2}(h+3k+l)}\right]$$

For the 002 reflection

$$F_{002} = 4f_{Zr} - 8f_O$$

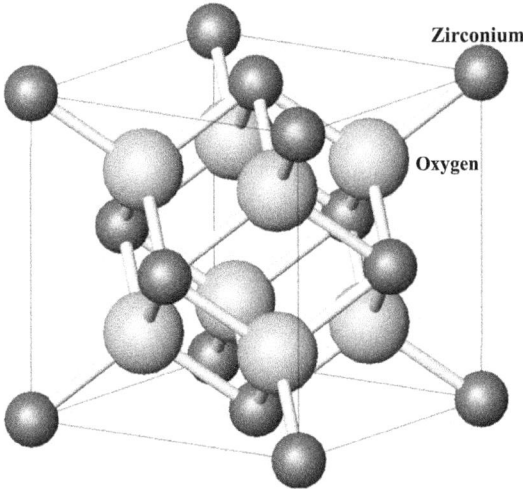

FIGURE 6.10 ZrO_2 crystal structure showing the positions of the Zr and O atoms.

Therefore,

$$\left| F_{002} \right|^2 = 16\left(f_{Zr} - 2f_O \right)^2$$

In summary, when calculating F_{hkl} use

$$F_{hkl} = \sum_{\substack{all \ atoms \\ per \ cell}} f_j \exp\left[2\pi i \vec{g}_{hkl} \cdot \vec{r}_j \right] = \sum_{\substack{all \ atoms \\ per \ cell}} f_j \exp\left[2\pi i \left(hx_j + ky_j + lz_j \right) \right]$$

We may notice that F_{hkl} can be the complex number

$$F_{hkl} = A_{hkl} + iB_{hkl}$$

where

$$A_{hkl} = \sum_{\substack{all \ atoms \\ per \ cell}} f_j \cos\left[2\pi i \left(hx_j + ky_j + lz_j \right) \right]$$

$$B_{hkl} = \sum_{\substack{all \ atoms \\ per \ cell}} f_j \sin\left[2\pi i \left(hx_j + ky_j + lz_j \right) \right]$$

$$tg\alpha_{hkl} = \frac{B_{hkl}}{A_{hkl}}$$

(I) If there is a center of symmetry in the crystal, the sine terms will cancel out.

(II) If we shift the origin of the unit cell, then the phase α of the structure factor F_{hkl} will be changed. In different crystal structure databases or textbooks, the choice for the origin of the unit cell for the same crystal structure may be different. The effect is that the origin shift results in changes to the coordinates. In the case where $\vec{r}_{dis}$ is the change of coordinates of all the atoms in the unit cell due to a different choice of the unit cell origin, F_{hkl} becomes

$$F_{hkl} = \sum_{\substack{all \ atoms \\ per \ cell}} f_j \exp\left[2\pi i \vec{g}_{hkl} \cdot \left(\vec{r}_j + \vec{r}_{dis} \right) \right]$$

$$= \sum_{\substack{all \ atoms \\ per \ cell}} f_j \exp\left[2\pi i \vec{g}_{hkl} \cdot \vec{r}_j \right] \exp\left[2\pi i \vec{g}_{hkl} \cdot \vec{r}_{disp} \right]$$

$$= \exp\left[2\pi i \vec{g}_{hkl} \cdot \vec{r}_{disp} \right] \sum_{\substack{all \ atoms \\ per \ cell}} f_j \exp\left[2\pi i \vec{g}_{hkl} \cdot \vec{r}_j \right]$$

Which means $\Delta\alpha = 2\pi \vec{g}_{hkl} \cdot \vec{r}_{disp}$. In the NaCl example, $\Delta\alpha$ is π if h, k and l are all odd.

In some situations, calculations for $|F_{hkl}|^2$ are needed, as the intensity of the diffracted beam is proportional to $|F_{hkl}|^2$.

$$|F_{hkl}|^2 = A_{hkl}^2 + B_{hkl}^2$$

$$= \left\{ \sum_{\substack{all \ atoms \\ per \ cell}} f_j \cos\left[2\pi i\left(hx_j + ky_j + lz_j\right)\right] \right\}^2 + \left\{ \sum_{\substack{all \ atoms \\ per \ cell}} f_j \sin\left[2\pi i\left(hx_j + ky_j + lz_j\right)\right] \right\}^2$$

$$= \sum_{i=1}^{N}\sum_{j=1}^{N} f_i f_j \left[\cos\left(2\pi \vec{g}_{hkl}\cdot\vec{r}_i\right)\cos\left(2\pi \vec{g}_{hkl}\cdot\vec{r}_j\right)\right] + \sum_{i=1}^{N}\sum_{j=1}^{N} f_i f_j \left[\sin\left(2\pi \vec{g}_{hkl}\cdot\vec{r}_i\right)\sin\left(2\pi \vec{g}_{hkl}\cdot\vec{r}_j\right)\right]$$

$$= \sum_{i=1}^{N}\sum_{j=1}^{N} f_i f_j \left[\cos\left(2\pi \vec{g}_{hkl}\cdot\vec{r}_i\right)\cos\left(2\pi \vec{g}_{hkl}\cdot\vec{r}_j\right) + \sin\left(2\pi \vec{g}_{hkl}\cdot\vec{r}_i\right)\sin\left(2\pi \vec{g}_{hkl}\cdot\vec{r}_j\right)\right]$$

$$= \sum_{i=1}^{N}\sum_{j=1}^{N} f_i f_j \cos\left[2\pi \vec{g}_{hkl}\cdot\left(\vec{r}_i - \vec{r}_j\right)\right]$$

$$= \sum_{i=1}^{N}\sum_{j=1}^{N} f_i f_j \cos\left\{2\pi\left[h\left(x_i - x_j\right) + k\left(y_i - y_j\right) + l\left(z_i - z_j\right)\right]\right\}$$

(6.19)

Again, this result tells us that:

(I) If the origin of the unit cell is changed, the relative positions between atoms will not change, or $\vec{r}_i - \vec{r}_j$ will not change. The modulus of the structure factor and $|F_{hkl}|^2$ remain the same, as suggested by Equation (6.19).

(II) As the cosine is an even function, this suggests that $\left|F_{hkl}\right|^2 = \left|F_{\overline{hkl}}\right|^2$, or $I_{hkl} = I_{\overline{hkl}}$, which is Friedel's Law.

Example 6.5

The $A_5B_2O_{12}$ structural family of compounds has a particularly interesting low-dimensional structure due to the presence of bioctahedral B_2O_{10} dimers arranged in one-dimensional edge-sharing chains along the direction of the metal–metal bonds. These compounds offer the potential to generate technologically relevant physical properties. For example, f subshell electrons may exhibit unusual behavior, including heavy fermion transport and superconductivity associated with those f electrons (Colabello et al., 2017). Scientists from the United States have successfully synthesized the $La_5Mo_2O_{12}$ compound and analyzed the crystal structure. This compound possesses a body-centered orthorhombic lattice with lattice parameters $a = 12.3$ Å, $b = 5.9$ Å and $c = 4.0$ Å. If Cu Kα radiation, with a wavelength λ of 1.5418 Å, is used in the X-ray diffractometer to investigate this powder sample, determine and list in order of increasing angle the values of 2θ and (hkl) for the first five diffraction peaks in the diffraction pattern.

Solution: For a body-centered cell, when $h+k+l$ = odd numbers, $F_{hkl} = 0$, diffraction is forbidden. When $h+k+l$ = even numbers, $F_{hkl} \neq 0$. Students need to choose those planes such that $h+k+l$ = even numbers.

Given a plane (*hkl*) with non-zero F_{hkl}, then d-spacing and 2θ can be calculated using the following equations:

$$\frac{1}{d_{hkl}^2} = \frac{h^2}{a^2} + \frac{k^2}{b^2} + \frac{l^2}{c^2}$$

and

$$2d \cdot sin\theta = \lambda$$

As $sin\theta \leq 1$, only $d \geq \dfrac{\lambda}{2}$ is considered. The results are shown below:

No.	Plane	d (Å)	2θ(°)
1	200	6.15	14.40
2	110	5.32	16.66
3	101	3.80	23.38
4	310	3.37	26.47
5	011	3.31	26.93

6.4   FURTHER DISCUSSION ON STRUCTURE FACTOR

The structure factor, F_{hkl}, when the site occupancy and temperature factor are included, is expressed as:

$$F_{hkl} = \sum_{\substack{all \ atoms \\ per \ cell}} f_j Occ_j T_j \exp\left[2\pi i \left(hx_j + ky_j + lz_j\right)\right]$$

where f_j and Occ_j are the atomic scattering factor and the site occupancy. T_j is the temperature factor, or the Debye–Waller factor (Trueblood *et al.*, 1996), which reflects the effect of thermal vibrations on the intensity of the scattered X-ray.

For the isotropic case (Stout and Jensen, 1989), the Debye–Waller factor is:

$$T_j = \exp\left[-B_j\left(\frac{sin\theta}{\lambda}\right)^2\right] = \exp\left[-\frac{B_j}{4}\left(\frac{2sin\theta}{\lambda}\right)^2\right] = \exp\left[-\frac{B_j}{4}\left(\frac{1}{d_{hkl}}\right)^2\right] \quad (6.20)$$

where B_j is of dimension (length)^2 and is related to the mean-square displacement $\overline{u_j^2}$, or expressed by $\langle u_j^2 \rangle$, through the relationship of

$$B_j = 8\pi^2 \overline{u_j^2} \quad (6.21)$$

Castellano reported the classical interpretation of thermal probability ellipsoids and the Debye–Waller factor (Castellano and Main, 1985). In the discussion, the Boltzmann distribution function is employed to explain density probability as a function of the amplitude of atomic displacement.

If we assume that the vibration of each atom in a crystal is a classical harmonic oscillator, which means $E = \frac{1}{2}ku^2$, where u is the amplitude of atomic displacement and k is a constant related to the bonding behavior of the material. Based on the Boltzmann distribution of energy $f(E) \propto \exp\left(-\frac{E}{k_B T}\right) dE = \exp\left(-\frac{k}{2}\frac{u^2}{k_B T}\right) dE$, the probability density of atomic displacement is a Gaussian function that can be normalized to give (Castellano and Main, 1985):

$$p(u) = \frac{1}{\sqrt{2\pi\overline{u^2}}} \exp\left(\frac{-u^2}{2\overline{u^2}}\right) \tag{6.22}$$

where $\overline{u^2}$ is the average of u^2 for each atom over the whole range of energies. $\overline{u^2}$ values depend on the binding behaviors and the absolute temperatures of the associated material. The higher the temperature T, the large the amplitude of atomic displacement u, the mean-square displacement u^2 and the B parameter.

Based on the discussions by Castellano, a simplified structure, or a one-dimensional structure, is employed to represent the concepts. Therefore, instead of using

$$F_{hkl} = \sum_{\substack{all\ atoms \\ per\ cell}} f_j \exp\left\{2\pi i\left[h(x_j + u_j / a) + k(y_j + v_j / b) + l(z_j + w_j / c)\right]\right\} \tag{6.23}$$

to show thermal motion, which modifies each atomic position by displacements of u, v and w along the three edge axes from the equilibrium position, here we use

$$F(h) = \sum_{j=1}^{N} f_j \exp\left[2\pi i h(x_j + u_j / a)\right] \tag{6.24}$$

With $\frac{h}{a} = \frac{1}{d} = \frac{2\sin\theta}{\lambda}$ for the one-dimensional structure, (6.24) becomes

$$F(h) = \sum_{j=1}^{N} f_j \exp\left[4\pi i\left(\frac{\sin\theta}{\lambda}\right) u_j\right] \exp(2\pi i h x_j) \tag{6.25}$$

Using the equation of the probability density of atomic displacement obtained from the simple harmonic oscillation model, we can calculate the average of the exponential factor $\exp\left[4\pi i\left(\dfrac{\sin\theta}{\lambda}\right)u_j\right]$ over all u_j (Castellano and Main, 1985), which is shown below:

$$\left\langle\exp\left[4\pi i\left(\frac{\sin\theta}{\lambda}\right)u_j\right]\right\rangle$$

$$=\int_{-\infty}^{\infty}\exp\left[4\pi i\left(\frac{\sin\theta}{\lambda}\right)u_j\right]\frac{1}{\sqrt{2\pi u^2}}\exp\left(\frac{-u_j^{\,2}}{2u^2}\right)du_j$$

$$=\frac{1}{\sqrt{2\pi u^2}}\int_{-\infty}^{\infty}\exp\left[4\pi i\left(\frac{\sin\theta}{\lambda}\right)u_j\right]\exp\left(\frac{-u_j^{\,2}}{2u^2}\right)du_j$$

$$=\exp\left[-8\pi^2\left(\frac{\sin^2\theta}{\lambda^2}\right)\overline{u_j^{\,2}}\right]$$

(6.26)

$\langle\,\rangle$ indicates the mean value or expected value. In the above calculation, we have used the result of the following integration:

Let $a=\dfrac{1}{2u^2}$ and $b=4\pi\left(\dfrac{\sin\theta}{\lambda}\right)$, then

$$\int_{-\infty}^{\infty}\exp\left[4\pi i\left(\frac{\sin\theta}{\lambda}\right)u_j\right]\exp\left(\frac{-u_j^{\,2}}{2u^2}\right)du_j$$

$$=\int_{-\infty}^{\infty}\exp\left[ibu_j\right]\exp\left(-au_j^{\,2}\right)du_j$$

$$=\int_{-\infty}^{\infty}\left[\cos\left(bu_j\right)+i\sin\left(bu_j\right)\right]\exp\left(-au_j^{\,2}\right)du_j$$

As $\sin(bu_j)$ is an odd function and $\exp(-au_j^2)$ is an even function, the integral is

$$\int_{-\infty}^{\infty}\left[\cos\left(bu_j\right)+i\sin\left(bu_j\right)\right]\exp\left(-au_j^{\,2}\right)du_j$$

$$=\int_{-\infty}^{\infty}\cos\left(bu_j\right)\exp\left(-au_j^{\,2}\right)du_j$$

From the power-series expansions, we have

$$\cos x = 1 - \frac{x^2}{2!} + \frac{x^4}{4!} - \ldots$$

When bu_j is small, we have

$$\int_{-\infty}^{\infty} \cos(bu_j) \exp(-au_j^2) du_j$$

$$\approx \int_{-\infty}^{\infty} \left(1 - \frac{b^2}{2} u_j^2\right) \exp(-au_j^2) du_j$$

$$= \int_{-\infty}^{\infty} \exp(-au_j^2) du_j - \frac{b^2}{2} \int_{-\infty}^{\infty} u_j^2 \exp(-au_j^2) du_j$$

$$= \sqrt{\frac{\pi}{a}} - \frac{b^2}{2} \frac{1}{2a} \sqrt{\frac{\pi}{a}}$$

$$= \sqrt{2\pi \overline{u^2}} \left(1 - 8\pi^2 \left(\frac{\sin^2 \theta}{\lambda^2}\right) \overline{u_j^2}\right)$$

As $8\pi^2 \left(\dfrac{\sin^2 \theta}{\lambda^2}\right) \overline{u_j^2}$ is small,

$$\sqrt{2\pi \overline{u^2}} \left(1 - 8\pi^2 \left(\frac{\sin^2 \theta}{\lambda^2}\right) \overline{u_j^2}\right)$$

$$\approx \sqrt{2\pi \overline{u^2}} \exp\left(-8\pi^2 \left(\frac{\sin^2 \theta}{\lambda^2}\right) \overline{u_j^2}\right)$$

Then we have the results as presented in Equation (6.26). Students can also calculate the following integral, which seems easier:

$$\int_{-\infty}^{\infty} \exp(-ax^2 + bx) dx$$

$$= \int_{-\infty}^{\infty} \exp\left[-a\left(x - \frac{b}{2a}\right)^2 + \frac{b^2}{4a}\right] dx$$

$$= \exp\left(\frac{b^2}{4a}\right) \int_{-\infty}^{\infty} \exp\left[-a\left(x - \frac{b}{2a}\right)^2\right] dx$$

By comparing (6.20) with (6.26), we get $B_j = 8\pi^2 \overline{u_j^2}$ as shown in Equation (6.21),

and the Debye–Waller factor is represented as $\exp\left[-\dfrac{B_j}{4}\left(\dfrac{2\sin\theta}{\lambda}\right)^2\right]$. Although the

above discussion is simply based on a one-dimensional structure, the method employed can be applied to the calculations for three-dimensional crystal structures.

In the textbook *X-ray Diffraction* (Warren, 1990), there is a discussion about the effect of temperature vibration, indicating that the contribution comes from the component of displacement normal to the diffracting planes. In fact, for a value of deviation $\vec{\delta}_j$ from its equilibrium position $\vec{r}_j$

$$F_{hkl} = \sum_{\substack{all\ atoms \\ per\ cell}} f_j \exp\left[2\pi i \vec{g}_{hkl} \cdot \left(\vec{r}_j + \vec{\delta}_j\right)\right] = \sum_{\substack{all\ atoms \\ per\ cell}} f_j \exp\left[2\pi i \vec{g}_{hkl} \cdot \vec{r}_j\right]\exp\left[2\pi i \vec{g}_{hkl} \cdot \vec{\delta}_j\right]$$

Therefore, only the component of $\vec{\delta}_j$ along $\vec{g}_{hkl}$, or perpendicular to (hkl), affects the structure factor F_{hkl}.

The above discussion on parameter B is simply based on the isotropic displacement. More discussion on anisotropic displacement associated parameters, including B^{ij}, U^{ij} and β^{ij}, can be found in the textbooks *X-ray Structure Determination: A Practical Guide* (Stout and Jensen, 1989) and *Modern X-ray Analysis on Single Crystals* (Luger, 2014). In X-ray single crystal structure analysis, a description of the anisotropic vibrational behavior of an atom by an ellipsoid is employed.

6.5   DIFFRACTION BY A SMALL CRYSTAL

In powder X-ray diffraction, an X-ray beam strikes a sample that contains many powders or many grains. When we analyze the intensity of an X-ray beam diffracted by a crystal, we mean that the crystal size or the grain size is very small, often in a range from a few nanometers to a few micrometers. In a calculation in Cullity and Stock (2001), we know that, for Cu Ka radiation, the diffracted beam intensity from a thin layer of depth 132 um of a nickel powder sample is only about 1/1000 that from a thin layer on the sample surface, which suggests that X-ray beams can generally penetrate to a depth ranging from a few tens of micrometers to about the hundred micrometer level. When the grain size is small, the sample exposed to the X-ray beam contains sufficient crystallites or grains; statistically the peak shape can be fitted using a Gaussian, or Lorentzian, or another function.

Now we will discuss diffraction by a small crystal. Based on the definition of the structure factor in X-ray diffraction, $F_{hkl} = \sum_{\substack{all\ atoms \\ per\ cell}} f_j \exp\left[2\pi i \vec{g}_{hkl} \cdot \vec{r}_j\right] =$

$\sum_{\substack{all\ atoms \\ per\ cell}} f_j \exp\left[2\pi i\left(hx_j + ky_j + lz_j\right)\right]$ is only used when the Bragg condition is satisfied for plane (hkl). If there is a slight deviation from the Bragg condition, or the scattering vector $\vec{q} = u\vec{a}^* + v\vec{b}^* + w\vec{c}^*$ is slightly different from the reciprocal lattice point $\vec{g}_{hkl} = h\vec{a}^* + k\vec{b}^* + l\vec{c}^*$, $F_{\vec{q}}$ is almost the same as $F_{\vec{g}}$. Therefore, as an approximation, we can still use $F_{\vec{g}}$ instead of $F_{\vec{q}}$ when the scattering is slightly deviated from the Bragg condition. The following expression shows the amplitude of a diffracted beam from a small crystal that is obtained by summing over all the unit cells in the crystal at the Bragg condition or very close to it. In such a condition, $F_{\vec{q}}$ is replaced by $F_{\vec{g}}$:

$$A_{crystal} = \sum_{\substack{all \\ unit \\ cell}} \left(A_e F_{hkl} \right) \exp\left[2\pi i \left(\vec{k} - \vec{k}_0 \right) \cdot \vec{r}_{mnp} \right]$$

$$= A_e F_{hkl} \sum_{\substack{all \\ unit \\ cell}} \exp\left[2\pi i \left(mu + nv + pw \right) \right] \tag{6.27}$$

$$= A_e F_{hkl} G$$

Thus

$$I_{crystal} \propto A_e^2 F_{hkl}^2 G^2 \tag{6.28}$$

where $\vec{r}_{mnp} = m\vec{a} + n\vec{b} + p\vec{c}$ is the location of a unit cell in direct space, and m, n and p are integers; u, v and w are the coordinates of the scattering vector in reciprocal space. When u, v and w are all integers, they are replaced by the Miller indices h, k and l to represent the crystallographic planes. If u, v and w are very close to integers h, k and l, the deviation from the Bragg condition is very limited. In this case, the $\displaystyle\sum_{\substack{all \quad atoms \\ per \quad cell}} f_j \exp\left[2\pi i\vec{q} \cdot \vec{r}_j \right]$ values are almost the same as that of

$$F_{hkl} = \sum_{\substack{all \quad atoms \\ per \quad cell}} f_j \exp\left[2\pi i\vec{g} \cdot \vec{r}_j \right].$$

In comparison with $\left| F_{\vec{q}} \right|$, the $|G(u, v, w)|$ values are more sensitive to u, v and w, and drop very fast when the latter deviate from h, k and l, especially when the crystallite size is large, or N_1, N_2 and N_3 are large. N_1, N_2 and N_3 are the numbers of the unit cells along the three cell-edge directions for a crystallite.

The factor $G = \displaystyle\sum_{\substack{all \\ unit \\ cell}} \exp\left[2\pi i \left(mu + nv + pw \right) \right]$ reflects the crystal shape effect, and

it can be calculated when the external shape of the crystallite is not complicated. Very similar to the discussions on electron diffraction (Williams, 1996), an idealized crystallite modeled as a parallelepiped, which is made up of triclinic parallelepiped unit cells having the same interaxial angles α, β, and γ, is employed in the following calculation. There are N_1 cells in the $\bar{a}_1$ direction, N_2 cells in the $\bar{a}_2$ direction and N_3 cells in the $\bar{a}_3$ direction.

$$G = \sum_{\substack{all \\ unit \\ cell}} \exp\left[2\pi i\left(mu + nv + pw\right)\right]$$

$$= \sum_{m=1}^{N_1-1} \exp\left[2\pi i\left(mu\right)\right] \sum_{n=1}^{N_2-1} \exp\left[2\pi i\left(nv\right)\right] \sum_{p=1}^{N_3-1} \exp\left[2\pi i\left(pw\right)\right] \qquad (6.29)$$

$$= G_1 G_2 G_3$$

We know that

$$S = \sum_{n=0}^{n=N-1} a_1 r^n = a_1 + a_1 r + a_1 r^2 + \ldots + a_1 r^{N-1} = \frac{a_1\left(1-r^N\right)}{1-r}$$

Therefore,

$$G_1 = \sum_{0}^{N_1-1} e^{2\pi i m u} = \frac{1-\left(e^{2\pi i u}\right)^{N_1}}{1-\left(e^{2\pi i u}\right)} = \frac{1-e^{2\pi i N_1 u}}{1-e^{2\pi i u}}$$

Simplifying:

$$G_1 = \frac{e^{\pi i N_1 u} e^{-\pi i N_1 u} - e^{\pi i N_1 u} e^{\pi i N_1 u}}{e^{\pi i u} e^{-\pi i u} - e^{\pi i u} e^{\pi i u}}$$

$$= \frac{e^{\pi i N_1 u}\left(e^{-\pi i N_1 u} - e^{\pi i N_1 u}\right)}{e^{\pi i u}\left(e^{-\pi i u} - e^{\pi i u}\right)} \qquad (6.30)$$

$$= e^{\pi i (N_1-1)u} \frac{\sin \pi N_1 u}{\sin \pi u}$$

Therefore,

$$|G_1| = \frac{\sin \pi N_1 u}{\sin \pi u} \qquad (6.31)$$

$$|G_1|^2 = \frac{\sin^2 \pi N_1 u}{\sin^2 \pi u}$$

Another way to calculate $|G_1|^2$ is as:

$$G_1 = \sum_{0}^{N_1-1} e^{2\pi i m u} = \frac{1-e^{2\pi i N_1 u}}{1-e^{2\pi i u}}$$

Then

$$|G_1|^2 = G_1 G_1^*$$

$$= \frac{1 - e^{2\pi i N_1 u}}{1 - e^{2\pi i u}} \cdot \frac{1 - e^{-2\pi i N_1 u}}{1 - e^{-2\pi i u}}$$

$$= \frac{2 - \left(e^{2\pi i N_1 u} + e^{-2\pi i N_1 u}\right)}{2 - \left(e^{2\pi i u} + e^{-2\pi i u}\right)}$$

$$= \frac{2 - 2\cos 2\pi N_1 u}{2 - 2\cos 2\pi u}$$

$$= \frac{\sin^2 \pi N_1 u}{\sin^2 \pi u}$$

And we obtain:

$$|G|^2 = \frac{\sin^2 \pi N_1 u}{\sin^2 \pi u} \cdot \frac{\sin^2 \pi N_2 v}{\sin^2 \pi v} \cdot \frac{\sin^2 \pi N_3 w}{\sin^2 \pi w} \tag{6.32}$$

The $|G|$ or $|G|^2$ function of the central peak has values within the range of:

$$u = h \pm \frac{1}{N_1}, v = k \pm \frac{1}{N_2}, w = l \pm \frac{1}{N_3},$$

where N_1, N_2 and N_3 are the numbers of the unit cells along the three cell-edge directions.

As N_1, N_2 and N_3 are large, $1/N_1$, $1/N_2$ and $1/N_3$ are very small. Therefore, the u, v and w values are close to integers h, k and l for the central peak.

The diffracted beam intensity from a particular diffracting plane (hkl) of the crystal can be calculated using:

$$I_{crystal} = I_e\left(\theta_S\right) \cdot \left|F_{hkl}\right|^2 \cdot |G|^2 \tag{6.33}$$

Notice that the diffracted beams are concentrated around the reciprocal lattice points only. Far away from the lattice points, the intensity is reduced dramatically and will be difficult to detect. In some experimental conditions, the calculation of $|G|^2$ is required in order to know the diffracted beam intensity, especially the maximum value and its integral. The results are:

$$|G|^2_{max} = N_1^2 N_2^2 N_3^2 = N^2$$

$$\iiint\limits_{crystal} |G|^2 \, du dv dw = N_1 N_2 N_2 = N \tag{6.34}$$

where $N(=N_1 N_2 N_2)$ is the total number of unit cells in the small crystal.

In some X-ray diffraction textbooks, there are detailed mathematical treatments of the effect of crystal size, or $|G|^2$, and some examples of derivations are presented in this section.

When discussing diffraction by a very small crystal, Guinier explained clearly how identical crystal domains can produce increasingly broad lines as the diffraction angle approaches $\frac{\pi}{2}$ (Guinier, 1994). This concept will also be discussed further through Debye–Sherrer formula in this section.

Example 6.6   Calculate the maximum value for $|G|^2$.

Solution: It is known that $|G|^2 = \dfrac{\sin^2 \pi N_1 u}{\sin^2 \pi u} \cdot \dfrac{\sin^2 \pi N_2 v}{\sin^2 \pi v} \cdot \dfrac{\sin^2 \pi N_3 w}{\sin^2 \pi w}$. If we draw the curve of $|G_1|^2$ as a function of u, we find that $|G_1|^2$ has a highest value at the reciprocal lattice points only, or when u is an integer. Far away from the lattice points, $|G_1|^2$ is reduced dramatically. If we calculate the maximum value for $|G_1|^2$ when u is approaching zero, or any other intêgers, we need to use L'Hôpital's Rule.

This tells us that if we have an indeterminate form $0/0$ or ∞/∞, all we need to do is differentiate the numerator and denominator and then take the limit. Using L'Hôpital's Rule, the value of $\underset{u \to 0}{Lim}|G_1|^2 = |G_1|^2_{max}$ can be calculated:

$$
\begin{aligned}
|G_1|^2_{max} = Lim_{u \to 0}|G_1|^2 &= Lim_{u \to 0}\frac{\sin^2 \pi N_1 u}{\sin^2 \pi u} \\[2mm]
&= Lim_{u \to 0}\frac{\dfrac{d}{du}\left(\sin^2 \pi N_1 u\right)}{\dfrac{d}{du}\left(\sin^2 \pi u\right)} = Lim_{u \to 0}\frac{\left(2\sin \pi N_1 u\right)\left(\cos \pi N_1 u\right)\left(\pi N_1\right)}{\left(2\sin \pi u\right)\left(\cos \pi u\right)\left(\pi\right)} \\[2mm]
&= N_1 \cdot Lim_{u \to 0}\frac{\sin \pi N_1 u}{\sin \pi u} \\[2mm]
&= N_1 \cdot Lim_{u \to 0}\frac{\dfrac{d}{du}\left(\sin \pi N_1 u\right)}{\dfrac{d}{du}\left(\sin \pi u\right)} = N_1 \cdot Lim_{u \to 0}\frac{\left(\cos \pi N_1 u\right)\pi N_1}{\left(\cos \pi u\right)\pi} \\[2mm]
&= N_1^2
\end{aligned}
\tag{6.35}
$$

Similarly, we have

$$|G_2|^2_{max} = Lim_{v \to 0}|G_2|^2 = N_2^2$$

$$|G_3|^2_{max} = Lim_{w \to 0}|G_3|^2 = N_3^2$$

Therefore,

$$|G|^2_{max} = N_1^2 N_2^2 N_3^2 = N^2 \tag{6.36}$$

The curve of $|G_1|^2$ as a function of u is shown in Figure 6.11.

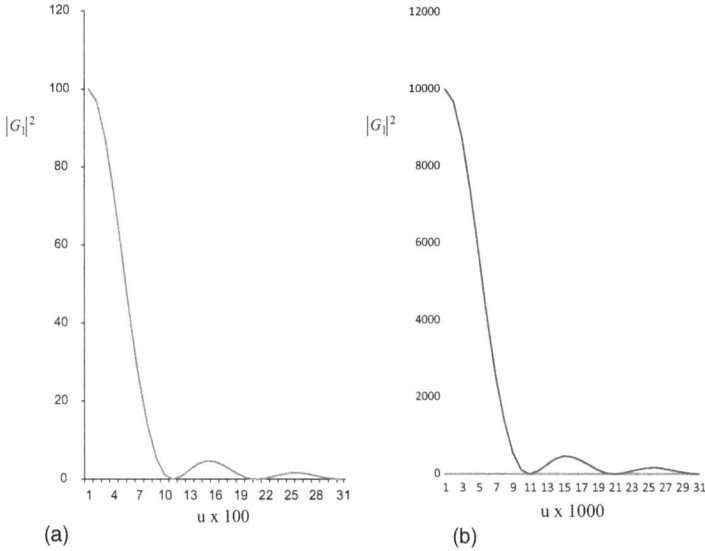

FIGURE 6.11 The curve of $|G_1|^2$ as a function of u through the relationship $|G_1|^2 = \dfrac{\sin^2 \pi N_1 u}{\sin^2 \pi u}$. (a) The number of the unit cell $N_1 = 10$, (b) the number of the unit cell $N_1 = 100$.

Example 6.7

What is the value of the following integral $\displaystyle\iiint\limits_{crystal} |G|^2 \, du\,dv\,dw$?

Solution: Around any reciprocal lattice point, the distribution of $|G|^2$ is the same. Therefore, we can perform integration around reciprocal lattice origin O*, and choose the ranges of $-1/2 \le u \le 1/2$, $-1/2 \le v \le 1/2$, $-1/2 \le w \le 1/2$ for integration. In fact, the central peak around each reciprocal lattice point is within the range

$$(h-1/2) \le u \le (h+1/2), \quad (k-1/2) \le v \le (k+1/2), \quad (l-1/2) \le w \le (l+1/2).$$

Before calculating $\displaystyle\int_{-\frac{1}{2}}^{\frac{1}{2}}\int_{-\frac{1}{2}}^{\frac{1}{2}}\int_{-\frac{1}{2}}^{\frac{1}{2}} |G|^2 \, du\,dv\,dw$

let us rewrite $|G|^2$ again:

$$G = \sum_{mnp} \exp\left[2\pi i(mu + nv + pw)\right]$$

$$= \sum_{m=1}^{N_1-1}\sum_{n=1}^{N_2-1}\sum_{p=1}^{N_3-1} \exp\left[2\pi i(mu + nv + pw)\right]$$

(6.37)

Thus

$$|G|^2 = G \cdot G*$$

$$= \left\{ \sum_{m=1}^{N_1-1} \sum_{n=1}^{N_2-1} \sum_{p=1}^{N_3-1} \exp\left[2\pi i \left(mu + nv + pw\right)\right] \right\} \left\{ \sum_{m=1}^{N_1-1} \sum_{n=1}^{N_2-1} \sum_{p=1}^{N_3-1} \exp\left[-2\pi i \left(mu + nv + pw\right)\right] \right\}$$

$$= \left\{ \sum_{m=1}^{N_1-1} \sum_{n=1}^{N_2-1} \sum_{p=1}^{N_3-1} \exp\left[2\pi i \left(mu + nv + pw\right)\right] \right\} \left\{ \sum_{m'=1}^{N_1-1} \sum_{n'=1}^{N_2-1} \sum_{p'=1}^{N_3-1} \exp\left[-2\pi i \left(m'u + n'v + p'w\right)\right] \right\}$$

$$= \sum_{m=1}^{N_1-1} \sum_{n=1}^{N_2-1} \sum_{p=1}^{N_3-1} \left\{ \sum_{m'=1}^{N_1-1} \sum_{n'=1}^{N_2-1} \sum_{p'=1}^{N_3-1} \exp\left[-2\pi i \left(m'u + n'v + p'w\right)\right] \right\} \exp\left[2\pi i \left(mu + nv + pw\right)\right]$$

$$= \sum_{m=1}^{N_1-1} \sum_{n=1}^{N_2-1} \sum_{p=1}^{N_3-1} \left\{ \sum_{m'=1}^{N_1-1} \sum_{n'=1}^{N_2-1} \sum_{p'=1}^{N_3-1} \exp 2\pi i \left[(m - m')u + (n - n')v + (p - p')w\right] \right\}$$

Or just simply write

$$|G|^2 = G \cdot G* = \sum_{mnp} \sum_{m'n'p'} \exp\left\{2\pi i \left[(m - m')u + (n - n')v + (p - p')w\right]\right\}$$

Now we can calculate the integral:

$$\int_{w=-\frac{1}{2}}^{\frac{1}{2}} \int_{v=-\frac{1}{2}}^{\frac{1}{2}} \int_{u=-\frac{1}{2}}^{\frac{1}{2}} |G|^2 \, dudvdw$$

$$= \int_{w=-\frac{1}{2}}^{\frac{1}{2}} \int_{v=-\frac{1}{2}}^{\frac{1}{2}} \int_{u=-\frac{1}{2}}^{\frac{1}{2}} = \sum_{m=1}^{N_1-1} \sum_{n=1}^{N_2-1} \sum_{p=1}^{N_3-1} \left\{ \sum_{m'=1}^{N_1-1} \sum_{n'=1}^{N_2-1} \sum_{p'=1}^{N_3-1} \exp 2\pi i \left[(m - m')u + (n - n')v + (p - p')w\right] \right\} dudvdw$$

$$= \int_{w=-\frac{1}{2}}^{\frac{1}{2}} \int_{v=-\frac{1}{2}}^{\frac{1}{2}} \int_{u=-\frac{1}{2}}^{\frac{1}{2}} = \sum_{m=1}^{N_1-1} \sum_{n=1}^{N_2-1} \sum_{p=1}^{N_3-1} \left\{ \sum_{m'=1}^{N_1-1} \sum_{n'=1}^{N_2-1} \sum_{p'=1}^{N_3-1} \exp 2\pi i \left(m - m'\right)u \cdot \exp 2\pi i \left(n - n'\right)v \cdot \exp 2\pi i \left(p - p'\right)w \right\} dudvdw$$

$$= \int_{w=-\frac{1}{2}}^{\frac{1}{2}} \int_{v=-\frac{1}{2}}^{\frac{1}{2}} \int_{u=-\frac{1}{2}}^{\frac{1}{2}} \left[\sum_{m=0}^{N_1-1} \sum_{m'=0}^{N_1-1} \exp 2\pi i \left(m - m'\right)u \right] \cdot \left[\sum_{n=0}^{N_2-1} \sum_{n'=0}^{N_2-1} \exp 2\pi i \left(n - n'\right)v \right] \cdot \left[\sum_{p=0}^{N_3-1} \sum_{p'=0}^{N_3-1} \exp 2\pi i \left(p - p'\right)w \right] \cdot dudvdw$$

$$= \sum_{m=0}^{N_1-1} \sum_{m'=0}^{N_1-1} \int_{u=-1/2}^{1/2} \exp 2\pi i \left[(m - m')u\right] du \cdot \sum_{n=0}^{N_2-1} \sum_{n'=0}^{N_2-1} \int_{v=-1/2}^{1/2} \exp 2\pi i \left[(n - n')v\right] dv \cdot \sum_{p=0}^{N_3-1} \sum_{p'=0}^{N_3-1} \int_{w=-1/2}^{1/2} \exp 2\pi i \left[(p - p')w\right] dw$$

We know that when $m \neq m'$

$$\int\limits_{u=-1/2}^{1/2} \exp 2\pi i \big[(m-m')u\big]du = \frac{\exp 2\pi i \left[(m-m')\left(\frac{1}{2}\right)\right] - \exp 2\pi i \left[(m-m')\left(-\frac{1}{2}\right)\right]}{2\pi i (m-m')}$$

$$= \frac{\sin \pi (m-m')}{\pi (m-m')}$$

$$= 0$$

We also know that when $m = m'$

$$\int\limits_{u=-1/2}^{1/2} \exp 2\pi i \big[(m-m')u\big]du$$

$$= \int\limits_{u=-1/2}^{1/2} 1du = 1$$

Therefore,

$$\sum_{m=0}^{N_1-1}\sum_{m'=0}^{N_1-1}\int\limits_{u=-1/2}^{1/2} \exp 2\pi i \big[(m-m')u\big]du = \sum_{m=0}^{N_1-1}1 = N_1 \tag{6.38}$$

Similarly, we have

$$\sum_{n=0}^{N_2-1}\sum_{n'=0}^{N_2-1}\int\limits_{v=-1/2}^{1/2} \exp 2\pi i \big[(n-n')v\big]dv = N_2$$

and

$$\sum_{p=0}^{N_3-1}\sum_{p'=0}^{N_3-1}\int\limits_{w=-1/2}^{1/2} \exp 2\pi i \big[(p-p')w\big]dw = N_3$$

Thus, the integral is

$$\int_{w=-\frac{1}{2}}^{\frac{1}{2}} \int_{v=-\frac{1}{2}}^{\frac{1}{2}} \int_{u=-\frac{1}{2}}^{\frac{1}{2}} |G|^2 \, du\,dv\,dw$$

$$= \sum_{m=0}^{N_1-1} \sum_{m'=0}^{N_1-1} \int_{u=-1/2}^{1/2} \exp 2\pi i [(m-m')u]\,du \cdot \sum_{n=0}^{N_2-1} \sum_{n'=0}^{N_2-1} \int_{v=-1/2}^{1/2} \exp 2\pi i [(n-n')v]\,dv \quad (6.39)$$

$$\cdot \sum_{p=0}^{N_3-1} \sum_{p'=0}^{N_3-1} \int_{w=-1/2}^{1/2} \exp 2\pi i [(p-p')w]\,dw$$

$$= N_1 N_2 N_3$$

$$= N$$

In the above calculation, the integral includes the central peak and other peaks around each reciprocal lattice point. As the central peak is much higher than others, the integration over the central peak alone is very close to value N.

Figure 6.12 is example showing the effect of spherical crystal size and shape on the reciprocal lattice nodes. There is an extension of the reciprocal lattice point to a node of finite size, which is the reciprocal of the diffracting crystal dimension perpendicular to the diffracting plane. This means when the grain size is small, each reciprocal lattice point becomes a domain with a volume.

The above discussion shows that crystallite size has an effect on the diffraction peak width. Measuring peak broadening allows the quantification of grain size for materials with grain size at about the hundred nanometer level or smaller. The Debye–Scherrer formula (or Scherrer formula) is presented as follows, which can be used to calculate the crystallite size:

$$t = \frac{k\lambda}{B \cos\theta_B} \quad (6.40)$$

(a)   (b)   (c)

Spherical crystal size in direct space

Domain shape and volume associated with reciprocal lattice point

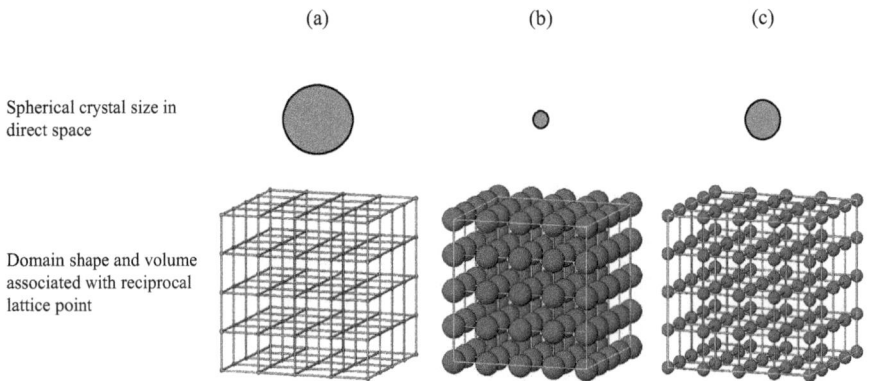

FIGURE 6.12 Schematic demonstration of effects of the size and shape of the diffracting crystal on the domain shape and volume associated with the reciprocal lattice points. (a) Large spherical crystal, (b) small spherical crystal and (c) crystal size in between.

where t is the crystallite size, constant $k = 0.8$–1.39 (usually close to unity, e.g. 0.9) and B is full width at half maximum (FWHM).

Cullity and Stock presented the derivation of the formula $t = \dfrac{\lambda}{B \cos \theta_B}$ in their textbook (Cullity and Stock, 2001), and Guinier provided a schematic illustration of peak broadening using Ewald sphere construction (Guinier, 1994), showing us that the broadening is higher for higher theta angles as it is proportional to $\dfrac{1}{\cos \theta}$.

SUMMARY

The intensity of the diffracted X-ray beam is highly dependent on the electron density distribution. This chapter has discussed the scattering of the X-ray by an electron, by an atom, by a unit cell and by a small grain having many unit cells.

For planes (hkl) that satisfy the Bragg condition, there are "allowed" and "forbidden" diffracting planes. The structure factor F_{hkl} tells us which diffraction spots/peaks are missing due to the position of the atoms in the unit cell. If $F_{hkl} = 0$, diffraction is forbidden. If $F_{hkl} \neq 0$, diffraction is allowed. Only symmetry operations with translational elements lead to classes of reflections with zero intensity. In the *International Table for Crystallography, Volume A*, students can find the allowed or forbidden reflection conditions for each space group.

Structure factor depends on the Miller indices of the reflecting plane, the occupied atoms and vacancies in the unit cell, the coordinates of each atom in the unit cell, and lattice vibrations.

The shape and size of grains affect the reflection domain of each reciprocal lattice point, and the peak width and intensity of the diffracted beam. Grain size can be calculated using the Scherrer formula.

REFERENCES

Barrett, C. S. (Charles S.) and Massalski, T. B. (1980) *Structure of metals: crystallographic methods, principles and data/C.S. Barrett, T.B. Massalski*. 3rd rev. e. Oxford: Pergamon (International series on materials science and technology: v. 35).

Castellano, E. E. and Main, P. (1985) 'On the classical interpretation of thermal probability ellipsoids and the Debye{--}Waller factor', *Acta Crystallographica Section A*, 41(2), pp. 156–157. doi:10.1107/S0108767385000307.

Colabello, D. M. et al. (2017) 'Observation of vacancies, faults, and superstructures in $Ln_5Mo_2O_{12}$ (Ln = La, Y, and Lu) compounds with direct Mo–Mo bonding', *Inorganic Chemistry*. American Chemical Society, 56(21), pp. 12866–12880. doi:10.1021/acs.inorgchem.6b02531.

Cullity, B. D. (Bernard D.) and Stock, S. R. (2001) *Elements of x-ray diffraction/B.D. Cullity, S.R. Stock*. 3rd ed. Upper Saddle River, NJ: Prentice Hall.

Doyle, P. A. and Turner, P. S. (1968) 'Relativistic Hartree–Fock X-ray and electron scattering factors', *Acta Crystallographica Section A*. International Union of Crystallography (IUCr), 24(3), pp. 390–397. doi:10.1107/S0567739468000756.

Guinier, A. (1994) *X-ray diffraction in crystals, imperfect crystals, and amorphous bodies/A. Guinier; translated by Paul Lorrain and Dorothée Sainte-Marie Lorrain.* New York: Dover.

Ladd, M. F. C. (Marcus F. C.) and Palmer, R. A. (Rex A.) (2003) *Structure determination by X-ray crystallography/Mark Ladd and Rex Palmer.* 4th ed. New York: Kluwer Academic/Plenum Publishers.

Li, S. T. (1990) *Fundamentals of x-ray diffraction of crystals/Li, S. T.* Beijing: Chinese Metallurgical Industry Publisher.

Luger, P. (2014) *Modern X-ray analysis on single crystals: a practical guide/by Peter Luger.* 2nd ed. Berlin: Walter de Gruyter GmbH & Co. KG.

Spence, J. C. H. (2013) *High-resolution electron microscopy/John C.H. Spence, Department of Physics and Astronomy, Arizona State University/LBNL California.* 4th ed. Oxford: Oxford University Press.

Stout, G. H. and Jensen, L. H. (1989) *X-ray structure determination: a practical guide/George H. Stout, Lyle H. Jensen.* 2nd ed. New York: Wiley.

Trueblood, K. N. et al. (1996) 'Atomic displacement parameter nomenclature report of a sub-committee on atomic displacement parameter nomenclature', *Acta Crystallographica Section A: Foundations of Crystallography*, 52(5), pp. 770–781. doi:10.1107/S0108767396005697.

Vand, V., Eiland, P. F. and Pepinsky, R. (1957) 'Analytical representation of atomic scattering factors', *Acta Crystallographica.* International Union of Crystallography (IUCr), 10(4), pp. 303–306. doi:10.1107/S0365110X57000882.

Warren, B. E. (Bertram E.) (1990) *X-ray diffraction.* Dover ed. New York: Dover Publications.

Williams, D. B. (1996) *Transmission Electron Microscopy [electronic resource]: A Textbook for Materials Science/by David B. Williams, C. Barry Carter.* Edited by C. B. Carter. Boston, MA: Springer US. doi:10.1007/978-1-4757-2519-3.

7 Experimental Methods and the Powder X-Ray Diffractometer

From the geometry of X-ray diffraction, we have noticed that, if a monochromatic X-ray source is used to irradiate a single crystal, the probability of a reciprocal lattice point intersecting the Ewald sphere is very low, which means there may be no diffraction occurring. In order to have more planes to satisfy the Bragg conditions, or to allow more reciprocal lattice points to touch the Ewald spheres, there are at least three possible methods that can be used for the examination of powder samples, polycrystalline samples and single crystals (Cullity and Stock, 2001) (Figure 7.1).

(I) The first method is to use a single wavelength X-ray striking a rotating single crystal. Some of the rotating reciprocal lattice points will cut through the Ewald sphere, giving rise to diffraction, as shown in Figure 7.1, where it can be seen that if the length of a reciprocal lattice vector is larger than the diameter of the Ewald sphere, it will never touch the sphere, and there is no possibility for diffraction to occur. Therefore, $\left| \vec{r}_{hkl}^{*} \right| \leq 2\dfrac{1}{\lambda}$, or $\dfrac{1}{d_{hkl}} \leq 2\dfrac{1}{\lambda}$ is required. For planes with $d_{hkl} < \dfrac{\lambda}{2}$, they cannot satisfy the Bragg condition, no matter how the sample is oriented.

(II) The second method is the Laue method that uses white (polychromatic) X-rays to irradiate a single crystal. In this case, the Ewald spheres have continuous radii within a range, and each sphere has a different center. All of the Ewald spheres pass through the origin of the reciprocal lattice. From Figure 7.2 it can be seen that many reciprocal lattice points may locate on those Ewald spheres, and the Bragg conditions are satisfied. For planes with $d_{hkl} < \dfrac{\lambda_{min}}{2}$, the Bragg condition cannot be satisfied.

(III) The third method uses a single wavelength X-ray and polycrystalline samples, or powder samples. There are so many small crystallites oriented differently, some of which satisfy $\vec{k} - \vec{k}_0 = \vec{g}$ as shown in Figure 7.3. For planes with $d_{hkl} < \dfrac{\lambda}{2}$, the Bragg condition cannot be satisfied.

The third method is employed in powder diffractometers with various designs, and materials science and engineering students are required to learn the techniques in their studies. However, when the sample is highly textured, the Bragg condition may not be satisfied for certain planes.

DOI: 10.1201/9780429351662-10

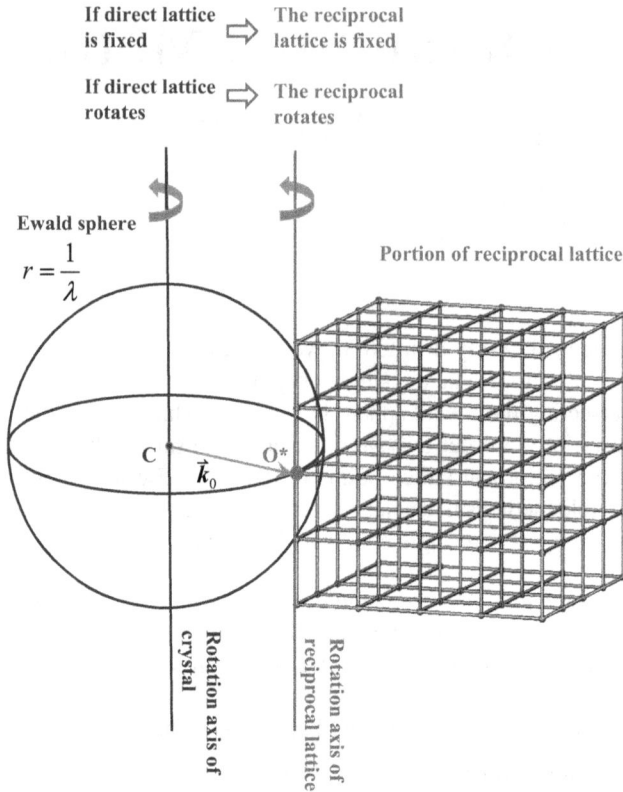

FIGURE 7.1 Rotating crystal method with single wavelength X-ray striking a rotating single crystal, showing that some of the rotating reciprocal lattice points can cut through the Ewald sphere, and diffraction can occur. If the crystal is fixed in an arbitrary orientation with respect to the X-ray beam direction, the reciprocal lattice is also fixed as it is coupled with the direct lattice, in which case the diffraction can only occur by chance.

In powder X-ray diffraction experiments based on the third method, round or rectangular shaped sample holders made of glass, aluminum, polymers or other materials are used; for example, one of the typical sample holders is a 2 mm thick aluminum plate with a 20 mm square hole in the center. In general, X-ray diffraction does not require any complex sample preparation. When polycrystalline bulk materials are studied, a small piece which can fit into the sample holder is enough. The sample piece should be fixed properly with tapes or plasticine clay, which should not be exposed to the X-ray beam. Sometimes, the sample itself can be directly mounted on the sample holder position without using any sample holder. The surface of the sample should be flat and have the same height as that of the sample holder.

In the case where powder samples are studied, peak broadening can occur when the particle size falls into the nano range. Measuring peak broadening allows the quantification of grain size for materials with grains less than about 0.1 μm. Some experiments allow mechanical grinding to be employed, others do not; this ensures that the results are from exactly the received sample forms.

(a)

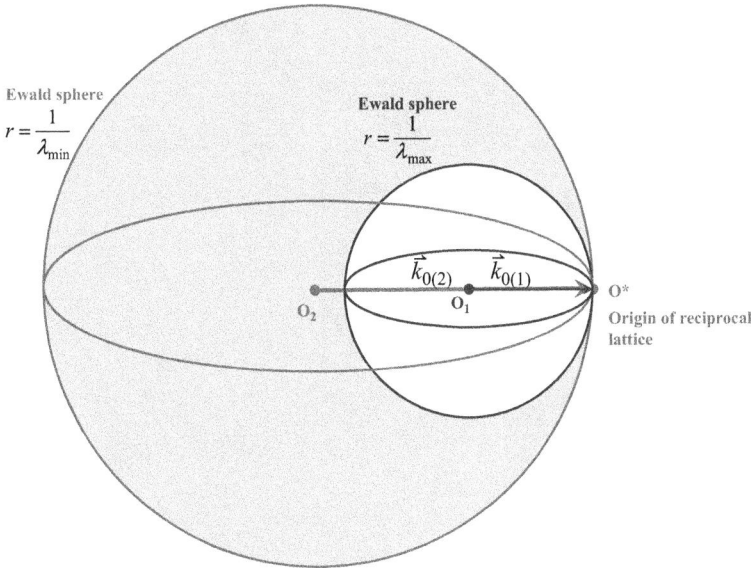

$$\vec{k}_{0(2)} \;=\; \overrightarrow{O_2 O}^* = \frac{\vec{S}_0}{\lambda_{min}} \qquad \vec{k}_{0(1)} \;=\; \overrightarrow{O_1 O}^* = \frac{\vec{S}_0}{\lambda_{max}}$$

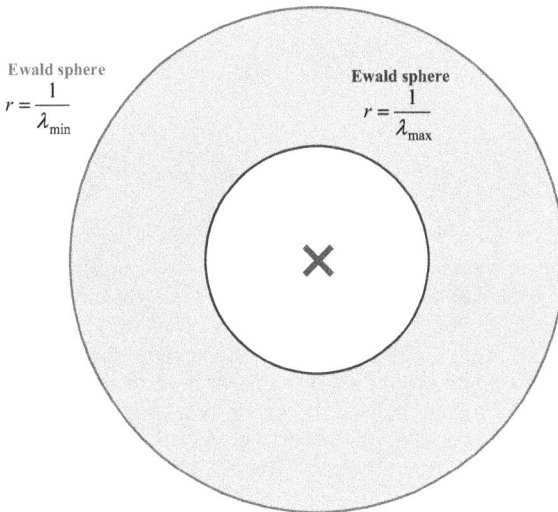

(b)

FIGURE 7.2 (a) Laue method with polychromatic X-rays irradiating a fixed single crystal, in which case many reciprocal lattice points can touch the Ewald spheres, and the Bragg conditions are satisfied. Only the largest and smallest Ewald spheres are shown. Other Ewald spheres are in between with a continual variation in wave numbers $1/\lambda$. All of the incident beam vectors end at the origin O* of the fixed reciprocal lattice. (b) View along the directions of incident X-ray beams.

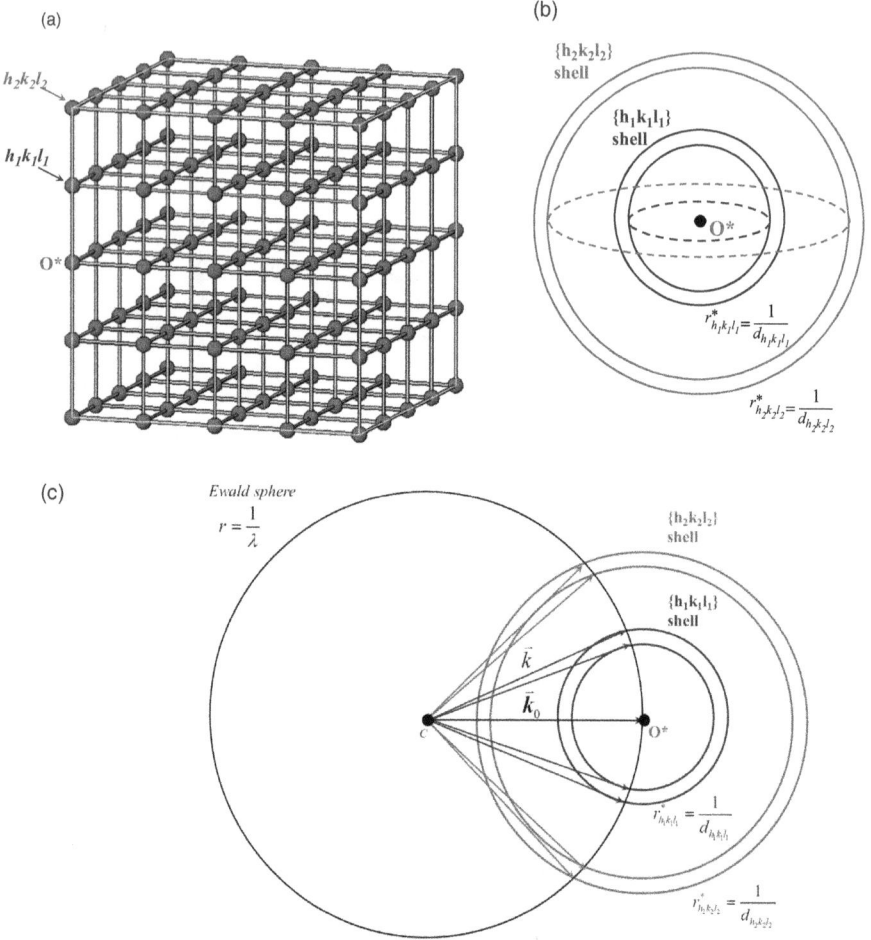

FIGURE 7.3 Powder method with single wavelength X-ray irradiating a polycrystalline sample, or a powder sample. (a) Schematic representation of the reciprocal lattice of a small crystallite in the sample. Domains of the reciprocal lattice nodes formed are due to the small size of the crystallite. (b) Shells formed in the reciprocal space, which are due to the large number of randomly oriented crystallites. Only two shells in the reciprocal space are shown although other allowed reflections form shells in the reciprocal space as well. (c) Section of the diffraction cones showing the directions of diffracted beams passing though the intersections of the Ewald sphere and the two reciprocal lattice shells in (b).

For quantitative phase analysis such as Rietveld refinement analysis, randomly oriented fine powders are preferred. To manually grind the powder in an agate mortar is good practice. When the powders are packed into a sample holder, it is recommended that they are pressed very gently using a slide glass. If the powder samples are pressed with too much force, a preferred orientation can occur, especially for the powder particles with plate and fiber shapes, which will cause the unfavorable variation of peak intensities and, in turn, will affect the quantitative phase analysis.

An ideal sample contains small polycrystals, but not so small as to cause unfavorable peak broadening for randomly oriented crystals. During data collection, the sample can spin about the axis perpendicular to the sample holder.

In one of our research projects with the conventional setting for the X-ray diffractometer, the powders were ground to fine granules and filled to a round polymer sample holder. Reliable data were collected by scanning over the angles 10–140° 2θ with a step size of 0.01° and a counting time of 5 seconds per step. During data collection, samples were spun around the holder axis at 30 rpm. Under these conditions the gross intensity of the strongest peak was 2000–3000 counts (Dong and White, 2004).

For the powder diffractometer, the intensity of a peak can be derived, which will be discussed further on. In some X-ray diffraction textbooks, there are different mathematical treatments of the peak intensity calculations, and in this chapter I include my own explanations. Some calculations may be similar to others, and other calculations are based on a summary of my lectures.

In my lectures, the diffracted peak intensity calculations are divided into a few steps.

Step I: Consider one grain and the (hkl) plane. Calculate the power of the diffracted X-ray beam.

Based on Equation (6.33), the diffracted intensity contributed by a crystallite along a direction with $F_{hkl} \neq 0$ and $G(u, v, w) \neq 0$ is

$$I_{crystal} = F_{hkl}^{2} \cdot G(u,v,w)^{2} \cdot I_{e}(2\theta)$$

Here we use $F_{\vec{g}}(= F_{hkl})$ to replace $F_{\vec{q}}$ even if the scattering vector $\vec{q}(= u\vec{a}^{*} + v\vec{b}^{*} + w\vec{c}^{*})$ is slightly deviated from the reciprocal vector $\vec{g}(= h\vec{a}^{*} + k\vec{b}^{*} + l\vec{c}^{*})$. The difference between $F_{\vec{g}}$ and $F_{\vec{q}}$ is very small when $\vec{q}$ is slightly deviated from the reciprocal vector $\vec{g}$. This approximation was discussed in Chapter 6.

The reciprocal lattice point becomes a domain due to the crystal size effect (Figure 7.4). As $G(u, v, w)$ has values within the solid angle Ω, the integrated diffracted X-ray beam intensity within the small solid angle range Ω is

$$P_{\Omega} = \int_{\Omega}\left[F_{hkl}^{2} \cdot G(u,v,w)^{2} \cdot I_{e}(2\theta)\right]d\Omega$$
$$= I_{e}(2\theta) \cdot F_{hkl}^{2} \cdot \int_{\Omega}\left[G(u,v,w)^{2}\right]d\Omega$$

(7.1)

where $d\Omega$ is the solid angle element. In Figure 7.4, the sectional view of $d\Omega$ is displayed.

(a)

(b)

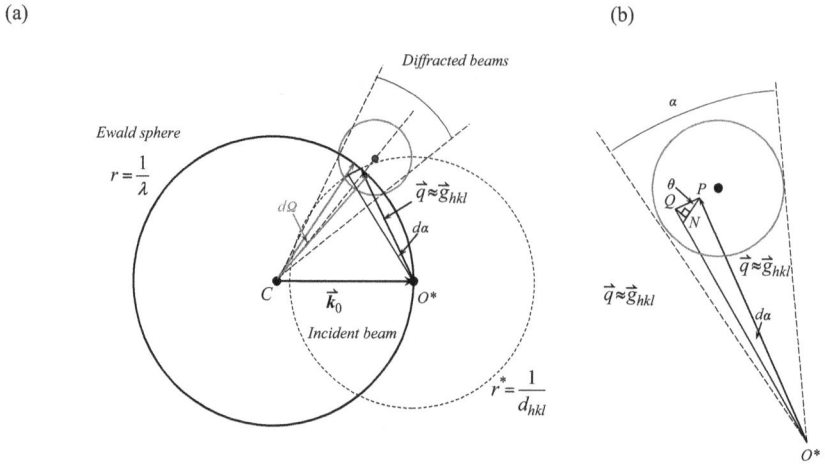

FIGURE 7.4 (a) Domain of the reciprocal lattice node and Ewald sphere passing through it. (b) Domain of the reciprocal lattice node in (a); PQ is perpendicular to the scattering vector and PN is along the diffracted beam direction.

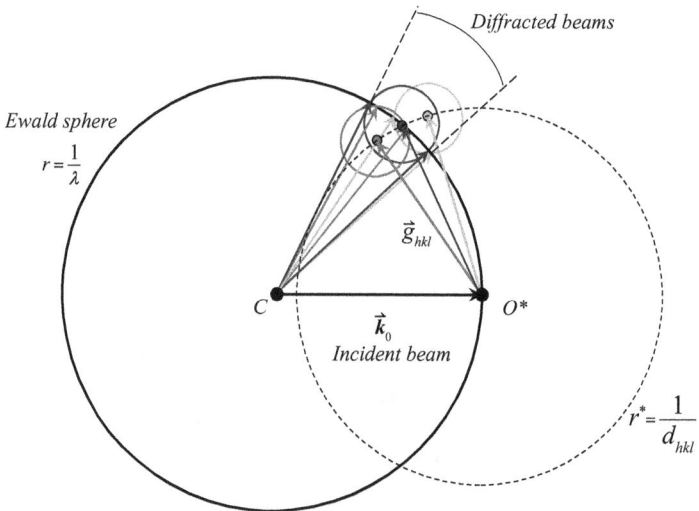

FIGURE 7.5 Schematic of diffracted beam broadening. The reciprocal lattice points extend to spheres for small spherical crystallites, some of which intersect the Ewald sphere.

Step II: Calculate the number of grains contributing to diffraction from plane (hkl) and family $\{hkl\}$.

In a polycrystalline material, or for a powder sample, there is a large number of crystallites. In order to let more crystallites contribute to the diffraction, and also to let each crystallite that satisfies the Bragg condition contribute more to the diffraction, the experimental arrangement allows the sample to rotate about an axis perpendicular to the incident beam as shown in Figure 7.5.

In fact, no matter whether the sample rotates faster or slower, at any moment, those grains with their reciprocal lattice points ($\vec{g}(hkl)$) locating within the angle range $\alpha/2$ inside or outside the Ewald sphere, will contribute to diffraction, as the Ewald sphere will pass the domain G associated with each reciprocal lattice (Figure 7.5). Therefore, only grains within such orientations will be considered.

We further assume that the crystallites in the sample are randomly oriented, and q is the number of crystallites exposed to the X-ray beam, in which case the number of crystallites contributing to the diffraction, dq, can be obtained from the area element dS on the surface of the sphere with surface area S (Figure 7.6). dS can be correlated with the element of angle $d\alpha$ (Figure 7.4), which is within the angle range α (Figure 7.5). That means the reciprocal lattice points ($\vec{g}(hkl)$) locating within the angle range $\alpha/2$ inside and outside the Ewald sphere are considered. Integration will be through the whole angle range α.

We know that

$$\frac{dq}{q} = \frac{dS}{S}$$

where

$$dS \approx 2\pi \left[r^* \sin\left(\pi/2 - \theta \right) \right] \left(r * d\alpha \right)$$

$$S = 4\pi r^{*2}$$

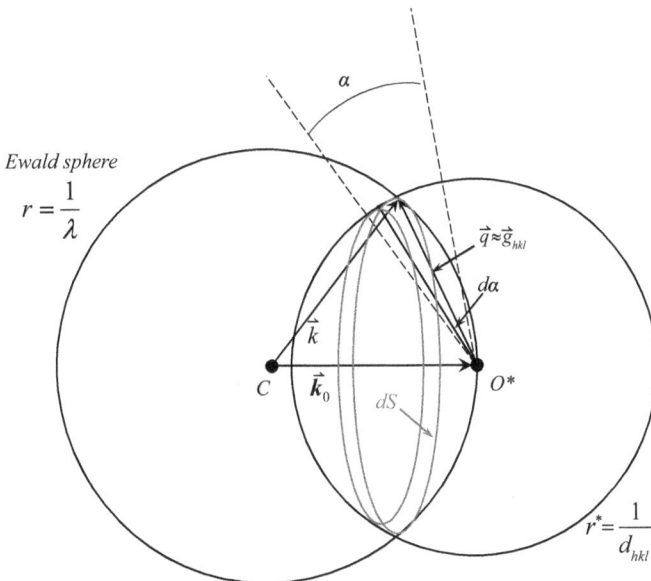

FIGURE 7.6 Number of crystals contributing to the diffraction from a plane in the family {hkl} that are oriented within ring belt dS, or angle $d\alpha$.

and

$$r^* = |g_{hkl}| = \frac{1}{d_{hkl}}$$

Therefore,

$$dq = q\frac{dS}{S} \approx q\frac{\cos\theta}{2} \cdot d\alpha \qquad (7.2)$$

The above discussion is for a particular plane (hkl). As we need to consider all of the crystallographically equivalent planes, or the planes of the $\{hkl\}$ family with the same d-spacing, the multiplicity factor M of the $\{hkl\}$ family in the corresponding crystal system should be included. Therefore, the number of grains contributing to diffraction from a plane in the $\{hkl\}$ family is

$$dQ = Mdq = Mq\frac{\cos\theta}{2} \cdot d\alpha \qquad (7.3)$$

where q is the number of crystallites exposed to the X-ray beam in the polycrystalline sample.

We have already noticed that the intensity by one plane from the $\{hkl\}$ family from any of the crystallites, which satisfy the Bragg condition, is:

$$P_\Omega = I_e(2\theta) \cdot F_{hkl}^2 \cdot \int_\Omega \left[G(u,v,w)^2 \right] d\Omega$$

To multiply P_Ω by the number dQ, which is the number of grains contributing to the diffraction by a plane from the $\{hkl\}$ family within the range of $d\alpha$, the diffracted power of the $\{hkl\}$ family can be expressed as

$$P_{ring\ (within\ d\alpha)} = I_e(2\theta) \cdot F_{hkl}^2 \cdot \left(\frac{\cos\theta}{2} Mqd\alpha \right) \cdot \int_\Omega \left[G(u,v,w)^2 \right] d\Omega \qquad (7.4)$$

Then the integrated power over whole ranges of Ω and α is

$$P_{ring} = \int_\alpha P_{ring(within\ d\alpha)} d\alpha = I_e(2\theta) \cdot F_{hkl}^2 \cdot \frac{\cos\theta}{2} Mq \int_\alpha \int_\Omega \left[G(u,v,w)^2 \right] d\Omega d\alpha \qquad (7.5)$$

Step III: Obtain the correlation between $d\Omega d\alpha$ and $dudvdw$ in reciprocal space, and calculate the integral.

Due to the sample rotation about the axis perpendicular to the incident and diffracted beams, the reciprocal lattice rotates about the reciprocal lattice origin O*, and the

domain defined by G around the reciprocal lattice point will scan over the Ewald sphere. Therefore, the integrated powder over whole ranges of Ω and α is
$$P_{ring} = I_e(2\theta)\cdot F_{hkl}{}^2 \cdot \frac{\cos\theta}{2} Mq \iint \left[G(u,v,w)^2\right] d\Omega d\alpha \text{ as shown by Equation (7.5).}$$
As presented by Li (1990), we can corelate $d\Omega d\alpha$ with $dudvdw$. From the Ewald sphere construction shown in Figures 7.4 and 7.5, it is known that the area element on the Ewald sphere surface associated with $d\Omega$ is

$$dS_{Ewald} = k^2 d\Omega = \frac{d\Omega}{\lambda^2}$$

From the reciprocal lattice point region (NP = small), it is known that

$$NP = PQ\cos\theta \approx \left(O^* P d\alpha\right)\cos\theta = \frac{2\sin\theta}{\lambda} da \cdot \cos\theta$$

Therefore, the element of volume in the reciprocal space is

$$dV^* = NP \cdot dS = \frac{2\sin\theta\cos\theta}{\lambda} d\alpha dS = \frac{\sin 2\theta}{\lambda^3} d\alpha d\Omega \qquad (7.6)$$

From the definition of a reciprocal lattice, we know that

$$dV^* = \left(\vec{a}^* du\right)\cdot\left(\vec{b}^* dv\right)\times\left(\vec{c}^* dw\right) = V_{cell}{}^* dudvdw = \frac{1}{V_{cell}} dudvdw \qquad (7.7)$$

Thus we have

$$d\alpha d\Omega = \frac{\lambda^3}{V_{cell}\sin 2\theta} dudvdw \qquad (7.8)$$

Substituting Equation (7.8) into (7.5), we have

$$P_{ring} = I_e(2\theta)\cdot F_{hkl}{}^2 \cdot \frac{\cos\theta}{2} Mq \iint_{\alpha\ \Omega}\left[G(u,v,w)^2\right]d\Omega d\alpha$$

$$= I_e(2\theta)\cdot F_{hkl}{}^2 \cdot \frac{\cos\theta}{2} Mq \cdot \frac{\lambda^3}{V_{cell}\sin 2\theta}\iiint_{u\ v\ w}\left[G(u,v,w)^2\right]dudvdw \qquad (7.9)$$

Suppose that the sample is rotating at an angular velocity ω, the time costed to cross the angle range α is Δt, and $\Delta t = \alpha/\omega$. During the above time interval, the energy emitted through diffraction from the hkl family is $E = P_{ring}\Delta t$. Therefore, the total power is $E/\Delta t = P_{ring}$. This suggests that the angular velocity does not affect the power of the diffracted beam. However, it changes the collection time as well as the overall photon counts (or integrated peak area) from that diffracting family of planes.

We also know that the intensity of the X-ray beam scattered by one electron is

$$I_e(2\theta) = \left(\frac{\mu_0}{4\pi}\right)^2 \left(\frac{e^4}{m_e^2}\right) \left(\frac{1+\cos^2 2\theta}{2}\right) I_0 = \left(\frac{1}{4\pi\varepsilon_0}\right)^2 \frac{1}{c^4} \left(\frac{e^4}{m_e^2}\right) \left(\frac{1+\cos^2 2\theta}{2}\right) I_0$$

and the integral

$$\iiint\limits_{crystal} |G|^2 \, du\,dv\,dw = N$$

where N is the number of cells in a crystallite. Therefore, we have

$$P_{ring} = I_e(2\theta) \cdot F_{hkl}^2 \cdot \frac{\cos\theta}{2} Mq \cdot \frac{\lambda^3}{V_{cell} \sin 2\theta} \iiint\limits_{u\ v\ w} \left[G(u,v,w)^2 \right] du\,dv\,dw$$

$$= I_0 \left(\frac{\mu_0}{4\pi}\right)^2 \left(\frac{e^4}{m_e^2}\right) \left(\frac{1+\cos^2 2\theta}{2}\right) \cdot F_{hkl}^2 \cdot \frac{\cos\theta}{2} Mq \cdot \frac{\lambda^3}{V_{cell} \sin 2\theta} N$$

(7.10)

In a polycrystalline sample with uniform grain size, if the volume V_{grain} of a grain is known, then the number of cells in a crystallite is $N = \dfrac{V_{grain}}{V_{cell}}$.

We also know that the number of crystallites exposed to the X-ray beam is q, so the volume of the polycrystalline sample exposed to the X-ray beam is $V_{exposure} = qV_{grain}$, and

$$N = \frac{V_{grain}}{V_{cell}} = \frac{V_{exposure}}{qV_{cell}}$$

(7.11)

Therefore,

$$P_{ring} = I_0 \left(\frac{\mu_0}{4\pi}\right)^2 \left(\frac{e^4}{m_e^2}\right) \left(\frac{1+\cos^2 2\theta}{2}\right) \cdot F_{hkl}^2 \cdot \frac{\cos\theta}{2} Mq \cdot \frac{\lambda^3}{V_{cell} \sin 2\theta} N$$

$$= I_0 \left(\frac{\mu_0}{4\pi}\right)^2 \left(\frac{e^4}{m_e^2}\right) \left(\frac{1+\cos^2 2\theta}{2}\right) \cdot F_{hkl}^2 \cdot \frac{\cos\theta}{2} M \cdot \frac{\lambda^3}{V_{cell}^2 \sin 2\theta} V_{exposure}$$

(7.12)

$$= I_0 \lambda^3 \left(\frac{\mu_0}{4\pi}\right)^2 \left(\frac{e^4}{m_e^2}\right) \left(\frac{1+\cos^2 2\theta}{8\sin\theta}\right) \cdot \left(\frac{F_{hkl}}{V_{cell}}\right)^2 \cdot M \cdot V_{exposure}$$

Step IV: Calculate the intensity based on the power per unit length on the cone of diffracted rays at the diffractometer circle.

If the radius of the diffractometer circle is R, then the diffraction ring circumference at the distance of the receiving slit of the diffractometer is $2\pi(R\sin 2\theta)$ (see Figure 7.7).

Therefore, the intensity, or the power per unit length, can be obtained as:

$$\frac{P_{ring}}{2\pi\left(R\sin 2\theta\right)}$$

$$=\frac{I_0}{2\pi\left(R\sin 2\theta\right)}\lambda^3\left(\frac{\mu_0}{4\pi}\right)^2\left(\frac{e^4}{m_e^2}\right)\left(\frac{1+\cos^2 2\theta}{8\sin\theta}\right)\left(\frac{F_{hkl}}{V_{cell}}\right)^2\cdot M\cdot V_{exposure} \qquad (7.13)$$

$$\left(\frac{I_0\lambda^3}{32\pi R}\right)\left(\frac{\mu_0}{4\pi}\right)^2\left(\frac{e^4}{m_e^2}\right)\cdot\left(\frac{F_{hkl}}{V_{cell}}\right)^2\cdot M\cdot\left(\frac{1+\cos^2 2\theta}{\sin^2\theta\cos\theta}\right)\cdot V_{exposure}$$

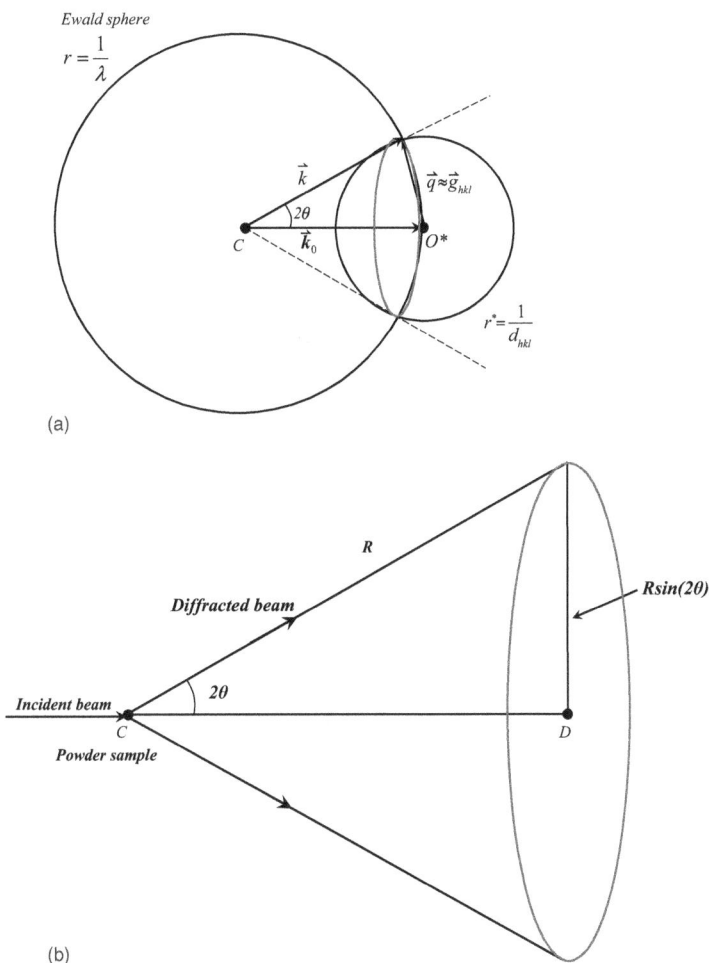

(a)

(b)

FIGURE 7.7   (a) Geometry of diffraction represented through Ewald sphere construction. (b) Cone of diffraction rays passing through the diffractometer circle.

Step V: Include the absorption effect.

Now we consider the absorption effect using Bragg–Brentano geometry, or the θ–2θ scanning mode. In Bragg–Brentano geometry, the absorption term $e^{-2\mu z/\sin\theta}$ is added during calculation. If $V_{exposure}$ is replaced with the element of exposure volume $dV_{exposure} = \dfrac{A_0 dz}{\sin\theta}$ in Equation (7.13), where dz is the element of thickness as shown in Figure 7.8, then the intensity can be obtained through integration over the sample depth (Cullity and Stock, 2001; Warren, 1990).

$$
\begin{aligned}
I_{hkl} &= \int_{z=0}^{\infty} \left(\frac{I_0\lambda^3}{32\pi R}\right)\left(\frac{\mu_0}{4\pi}\right)^2\left(\frac{e^4}{m_e^2}\right)\cdot\left(\frac{F_{hkl}}{V_{cell}}\right)^2\cdot M\cdot\left(\frac{1+\cos^2 2\theta}{\sin^2\theta\cos\theta}\right)\cdot e^{-2\mu z/\sin\theta}\cdot\frac{A_0 dz}{\sin\theta} \\
&= \left(\frac{I_0\lambda^3}{32\pi R}\right)\left(\frac{\mu_0}{4\pi}\right)^2\left(\frac{e^4}{m_e^2}\right)\cdot\left(\frac{F_{hkl}}{V_{cell}}\right)^2\cdot M\cdot\left(\frac{1+\cos^2 2\theta}{\sin^2\theta\cos\theta}\right)\cdot\left(\frac{A_0}{-2\mu}\right)\int_{z=0}^{\infty} e^{-2\mu z/\sin\theta}d(-2\mu z/\sin\theta) \\
&= \left(\frac{I_0\lambda^3}{32\pi R}\right)\left(\frac{\mu_0}{4\pi}\right)^2\left(\frac{e^4}{m_e^2}\right)\cdot\left(\frac{F_{hkl}}{V_{cell}}\right)^2\cdot M\cdot\left(\frac{1+\cos^2 2\theta}{\sin^2\theta\cos\theta}\right)\cdot\left(\frac{A_0}{-2\mu}\right)\cdot(-1) \\
&= \left(\frac{I_0 A_0\lambda^3}{32\pi R}\right)\left(\frac{\mu_0}{4\pi}\right)^2\left(\frac{e^4}{m_e^2}\right)\cdot\left(\frac{F_{hkl}}{V_{cell}}\right)^2\cdot M\cdot\left(\frac{1+\cos^2 2\theta}{\sin^2\theta\cos\theta}\right)\cdot\left(\frac{1}{2\mu}\right)
\end{aligned}
$$

The constant in the equation can be expressed in two different ways:

$$
\begin{aligned}
I_{hkl} &= \left(\frac{I_0 A_0\lambda^3}{32\pi R}\right)\left(\frac{\mu_0}{4\pi}\right)^2\left(\frac{e^4}{m_e^2}\right)\cdot\left(\frac{F_{hkl}}{V_{cell}}\right)^2\cdot M\cdot\left(\frac{1+\cos^2 2\theta}{\sin^2\theta\cos\theta}\right)\cdot\left(\frac{1}{2\mu}\right) \\
&= \left(\frac{I_0 A_0\lambda^3}{32\pi R}\right)\left(\frac{1}{4\pi\varepsilon_0 c^2}\right)^2\left(\frac{e^4}{m_e^2}\right)\cdot\left(\frac{F_{hkl}}{V_{cell}}\right)^2\cdot M\cdot\left(\frac{1+\cos^2 2\theta}{\sin^2\theta\cos\theta}\right)\cdot\left(\frac{1}{2\mu}\right)
\end{aligned}
\tag{7.14}
$$

The above absorption coefficient term $(1/2\mu)$ in the equation is derived from the geometry of the Bragg–Brentano setting. For other geometries, the absorption factor $A(\theta)$ is not $(1/2\mu)$.

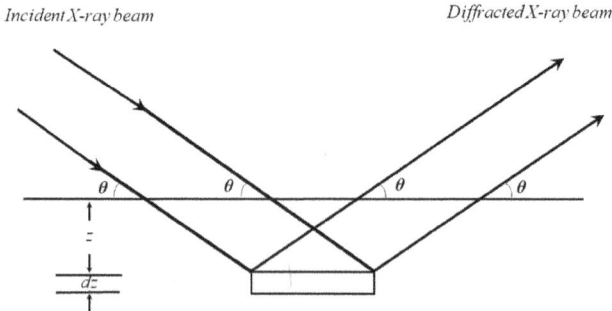

FIGURE 7.8 Absorption effect for a flat sample in θ–2θ scanning mode.

When we analyze the effects of different factors on intensity from Equation (7.14), we can simply use

$$I_{hkl} = \left(\frac{I_0 A_0 \lambda^3}{32\pi R}\right)\left(\frac{\mu_0}{4\pi}\right)^2 \left(\frac{e^4}{m_e^2}\right) \cdot \left(\frac{F_{hkl}}{V_{cell}}\right)^2 \cdot M \cdot \left(\frac{1+\cos^2 2\theta}{\sin^2 \theta \cos\theta}\right) \cdot A(\theta)$$

$$\propto \left(\frac{F_{hkl}}{V_{cell}}\right)^2 \cdot M \cdot \left(\frac{1+\cos^2 2\theta}{\sin^2 \theta \cos\theta}\right) \cdot A(\theta)$$

(7.15)

where I_{hkl} is the diffracted beam power per unit length of the diffraction circle at a distance R; I_0 is the intensity of the incident X-ray; A_0 is the cross-sectional area of the incident beam; $I_0 A_0 = P_0$ is the incident beam power striking on the sample surface; λ is the X-ray wavelength; R is the radius of the diffractometer circle; $\frac{\mu_0}{4\pi} = \frac{1}{4\pi\varepsilon_0 c^2}$; ε_0 is electric permittivity; c is the speed of light; μ_0 is magnetic permeability; m_0 is the mass of an electron; e is the charge of an electron; V_{cell} is the unit cell volume; M is the multiplicity factor; $\frac{1+\cos^2 2\theta}{\sin^2 \theta \cos\theta} = L$ (the Lorentz–polarization factor); and $A(\theta)$ is the absorption factor.

For thick flat samples analyzed using Bragg–Brentano geometry, $A(\theta) = 1/2\mu$, and μ is the linear absorption coefficient. If another powder diffraction geometry, and not the Bragg–Brentano set-up, is employed, the absorption term $A(\theta)$ will be expressed differently. A schematic diagram of the Bragg–Brentano diffractometer is shown in Figure 7.9.

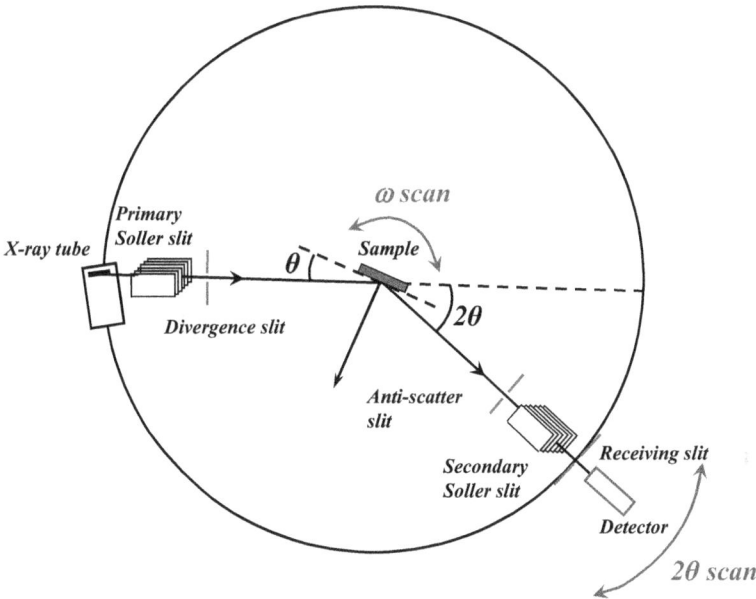

FIGURE 7.9 Schematic diagram of the Bragg–Brentano diffractometer. The sample rotates through an angle θ while the detector rotates through an angle 2θ. The detector rotates at twice the angular rate of the sample.

Equation (7.14) is only true for single phase samples. In the case where there are two or more crystalline phases in the sample, the intensity from the α phase can be calculated by multiplying the equation by the volume fraction c_α, and replacing the linear absorption coefficient μ with $\mu_{mixture}$. For example, for the sample containing two phases α and β,

$$I_\alpha = \left(\frac{I_0 A_0 \lambda^3}{32\pi R}\right)\left(\frac{\mu_0}{4\pi}\right)^2\left(\frac{e^4}{m_e^2}\right)\cdot\left(\frac{F_{hkl}}{V_{cell}}\right)^2 \cdot M \cdot \left(\frac{1+\cos^2 2\theta}{\sin^2\theta\cos\theta}\right)\cdot\left(\frac{1}{2\mu_{mixture}}\right)\cdot c_\alpha \quad (7.16)$$

or

$$I_\alpha = \frac{K_1 c_\alpha}{\mu_{mixture}}$$

where K_1 is defined as $\left(\frac{I_0 A_0 \lambda^3}{32\pi R}\right)\left(\frac{\mu_0}{4\pi}\right)^2\left(\frac{e^4}{m_e^2}\right)\cdot\left(\frac{F_{hkl}}{V_{cell}}\right)^2 \cdot M \cdot\left(\frac{1+\cos^2 2\theta}{\sin^2\theta\cos\theta}\right)\cdot\frac{1}{2}$

Following the presentation in Cullity and Stock (2001), we will use a material containing two phases α and β as an example to explain the diffracted beam intensity calculation. We know that

$$\frac{\mu_{mixture}}{\rho_{mixture}} = w_\alpha\left(\frac{\mu_\alpha}{\rho_\alpha}\right)+w_\beta\left(\frac{\mu_\beta}{\rho_\beta}\right) \quad (7.17)$$

where w denotes the weight fraction and ρ the density. Thus

$$\mu_{mixture} = w_\alpha\left(\frac{\mu_\alpha}{\rho_\alpha}\right)\rho_{mixture}+w_\beta\left(\frac{\mu_\beta}{\rho_\beta}\right)\rho_{mixture}$$

Consider a unit volume of the mixture; its weight is $\rho_{mixture}$ and the weight of the α phase is $w_\alpha\rho_{mixture}$; therefore the volume of the α phase is $w_\alpha\rho_{mixture}/\rho_\alpha$. As this calculation is based on a unit volume, $w_\alpha\rho_{mixture}/\rho_\alpha$ means c_α. Therefore,

$$\mu_{mixture} = w_\alpha\left(\frac{\mu_\alpha}{\rho_\alpha}\right)\rho_{mixture}+w_\beta\left(\frac{\mu_\beta}{\rho_\beta}\right)\rho_{mixture}$$
$$= c_\alpha\mu_\alpha + c_\beta\mu_\beta \quad (7.18)$$
$$= c_\alpha\mu_\alpha + (1-c_\alpha)\mu_\beta$$
$$= c_\alpha(\mu_\alpha-\mu_\beta)+\mu_\beta$$

Then the intensity of the α phase is

$$I_\alpha = K_1 \cdot \frac{c_\alpha}{c_\alpha(\mu_\alpha-\mu_\beta)+\mu_\beta} \quad (7.19)$$

The volume fraction c_α can be correlated with weight fraction w_α (Cullity and Stock, 2001):

$$c_\alpha = \frac{\dfrac{w_\alpha}{\rho_\alpha}}{\dfrac{w_\alpha}{\rho_\alpha} + \dfrac{w_\beta}{\rho_\beta}} = \frac{\dfrac{w_\alpha}{\rho_\alpha}}{w_\alpha \left(\dfrac{1}{\rho_\alpha} - \dfrac{1}{\rho_\beta} \right) + \dfrac{1}{\rho_\beta}} \qquad (7.20)$$

Then the intensity of the α phase can be presented using either a volume fraction or weight fraction as shown in Equation (7.21). The relationships are very useful for quantitative phase analysis.

$$\begin{aligned} I_\alpha &= K_1 \cdot \frac{c_\alpha}{c_\alpha \left(\mu_\alpha - \mu_\beta \right) + \mu_\beta} \\ &= K_1 \cdot \frac{w_\alpha}{\rho_\alpha \left\{ w_\alpha \left[\left(\mu / \rho \right)_\alpha - \left(\mu / \rho_\beta \right) \right] + \left(\mu / \rho_\beta \right) \right\}} \end{aligned} \qquad (7.21)$$

In summary, the intensity based on the Bragg–Brentano powder X-ray diffractometer has been calculated, which corresponds to a particular reciprocal lattice point (or a real lattice plane) with the non-zero structure factor F_{hkl}. If the Ewald sphere is extremely close to a reciprocal lattice point when G is non-zero, there will be a diffracted beam. Figure 7.5 shows three different grains that satisfy the diffraction condition, showing the diffracted beams from a particular reciprocal lattice vector with slight variations in directions that cause peak broadening. The Scherrer formula is often used to calculate the average crystal size when peak width is measured for a strain-free sample. The diffracted beam intensity from a single-phase powder specimen in a Bragg–Brentano X-ray diffractometer is presented in detail in Cullity and Stock (2001).

Bragg–Brentano geometry is not effective for studying polycrystalline thin film at angle ranges higher than 2θ, as the peak intensities are low. In order to increase the peak intensity of a polycrystalline thin film, an incident beam direction with a very low glancing angle can be set up, in which case the beam path along the surface layer is increased. However, for some experimental settings, the diffracted beams may not be focused, and the resolutions for angles higher than 2θ can be low. To improve the resolution, focused geometry such as Seemann–Bohlin can be employed (Tao et al., 1985).

In comparison with the powder X-ray diffraction, thin epitaxial film analysis often involves reciprocal space mapping. Three basic scanning modes are presented in Figure 7.10.

The three basic operation modes include an $\omega - 2\theta$ scan, an ω scan and a 2θ scan which can be well understood through Ewald sphere construction. In my lectures, I have not focused on thin film study. Therefore, students are encouraged to study further by themselves if their projects include thin film analysis by X-ray diffraction.

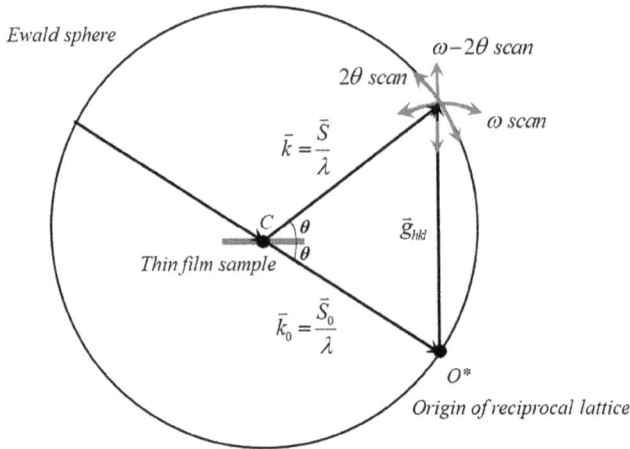

FIGURE 7.10   Three typical scanning modes in thin film X-ray diffraction analysis.

SUMMARY

Based on Bragg's Law, Bragg diffraction can only occur when a reciprocal lattice point lies on the Ewald sphere. For powders and polycrystalline materials, the grain size is small, and each reciprocal lattice point becomes a domain with a volume. Due to such an effect, if a reciprocal lattice point is close to the Ewald sphere, and the domain touches the sphere, diffraction can occur. Ewald sphere construction enables us to better understand the Bragg condition in reciprocal space and the different diffraction methods proposed. In this chapter, three X-ray diffraction methods have been proposed.

(I)   By using "white" radiation to examine a single crystal sample, the spotty X-ray diffraction pattern can be collected, which is the Laue method.

(II)   By using monochromatic radiation to examine a rotating single crystal sample the rotating reciprocal lattice points have a chance to pass through the Ewald sphere and give rise to diffraction. The diffraction pattern collected depends on the rotation axis arrangement.

(III) Use of monochromatic radiation and a polycrystal sample or powder sample containing crystallites with various orientations is called the powder method.

In the powder method, there is no fixed position for each reciprocal lattice point $\bar{g}_{hkl}$, but there are many sets of points at different orientations. If the number of crystallites or grains is large enough, the reciprocal lattice points from the same family $\{hkl\}$ form a sphere with $r = 1/d_{hkl}$. If we consider the shape factor G, then it is seen to be a spherical shell with a very small thickness. The Bragg–Brentano powder method is often used in materials analysis.

If there are preferred orientations, the intensity of each peak is different from that of a randomly oriented sample.

Unlike the Bragg–Brentano method, polycrystalline thin film analysis can fix the incident beam direction with a very low glancing angle to enhance the diffraction peak intensities contributed by the surface layer.

For epitaxially grown thin films, X-ray diffraction analysis often involves reciprocal space mapping using $\omega - 2\theta$, ω and 2θ scan modes.

The key is to wisely apply the geometry based on Ewald sphere construction to the experimental setting, so that diffraction can occur and crystal structure information can be obtained.

REFERENCES

Cullity, B. D. (Bernard D.) and Stock, S. R. (2001) *Elements of x-ray diffraction/B.D. Cullity, S.R. Stock*. 3rd ed. Upper Saddle River, NJ: Prentice Hall.

Dong, Z. L. and White, T. J. (2004) 'Calcium-lead fluoro-vanadinite apatites. I. Disequilibrium structures', *Acta Crystallographica Section B: Structural Science*, 60(2), pp. 138–145. doi:10.1107/S0108768104001831.

Li, S. T. (1990) *Fundamentals of x-ray diffraction of crystals/Li, S. T.* Beijing: Chinese Metallurgical Industry Publisher.

Tao, K. et al. (1985) 'An X-ray diffraction attachment tor thin filam analysis of high sensitivity', *MRS Proceedings*. 2011/02/26. Cambridge University Press, 54, p. 687. doi:10.1557/PROC-54-687.

Warren, B. E. (Bertram E.) (1990) *X-ray diffraction*. Dover ed. New York: Dover Publications.

8 Rietveld Refinement of Powder X-Ray Diffraction Patterns

It is necessary to remind our materials science and engineering students that if the compositions of the sample for X-ray diffraction analysis are unknown, it is suggested that Energy Dispersive X-ray Spectroscopy (EDX) analysis be carried out to obtain the approximate composition information before the X-ray diffraction analysis is conducted. This step is truly effective when doing a phase search during qualitative analysis. In the following quantitative phase analysis, we will imply that composition analysis and qualitative phases analysis are completed and that the phases present in the materials are known.

If more than one phase is present in a polycrystalline sample or a powder sample, the diffracted beam intensity from the $\{hkl\}$ family for any of the crystalline phases, for instance from phase alpha, can be calculated as:

$$I_\alpha = \left(\frac{I_0 A_0 \lambda^3}{32\pi R}\right)\left(\frac{\mu_0}{4\pi}\right)^2\left(\frac{e^4}{m_e^2}\right)\cdot\left(\frac{F_{hkl}}{V_{cell}}\right)^2\cdot M\cdot\left(\frac{1+\cos^2 2\theta}{\sin^2\theta\cos\theta}\right)\cdot A(\theta)\cdot c_\alpha$$

If there are only two phases in the sample, namely the alpha and beta phases, then

$$I_\alpha = K_1\cdot\frac{c_\alpha}{c_\alpha\left(\mu_\alpha-\mu_\beta\right)+\mu_\beta} = K_1\cdot\frac{w_\alpha}{\rho_\alpha\left\{w_\alpha\left[\left(\mu/\rho\right)_\alpha-\left(\mu/\rho_\beta\right)\right]+\left(\mu/\rho_\beta\right)\right\}}$$

where K_1 is defined as $\left(\dfrac{I_0 A_0 \lambda^3}{32\pi R}\right)\left(\dfrac{\mu_0}{4\pi}\right)^2\left(\dfrac{e^4}{m_e^2}\right)\cdot\left(\dfrac{F_{hkl}}{V_{cell}}\right)^2\cdot M\cdot\left(\dfrac{1+\cos^2 2\theta}{\sin^2\theta\cos\theta}\right)\cdot\dfrac{1}{2}$

In the above expressions, it is assumed that there is no preferred orientation in the sample. In the case where there is an amorphous phase present, we can mix a known crystalline standard phase, for instance 10wt% silica or 10wt% alumina, into the powder sample for calibration. In some versions of the Rietveld refinement software, an amorphous fraction can be calculated as well.

There is a detailed discussion on quantitative X-ray diffraction analysis in Zevin et al. (1995). In this chapter, we briefly describe the Rietveld refinement method

DOI: 10.1201/9780429351662-11

(Rietveld, 1969) for quantitative phase analysis. In this refinement of an X-ray diffraction pattern, both diffraction geometry and beam intensity are considered. R. A. Young (1995) summarized the calculated intensities with background contribution from Equation (8.1):

$$y_{ci} = s \sum L_K \left| F_K \right|^2 \varphi\left(2\theta_i - 2\theta_K\right) P_K A + y_{bi} \tag{8.1}$$

where s is the scale factor; K represents the Miller indices h, k, l, for a Bragg reflection; L_K contains the Lorentz, polarization and multiplicity factors; ϕ is the reflection profile function; P_K is the preferred orientation function; A is the absorption factor; F_K is the structure factor for the Kth Bragg reflection; and y_{bi} is the background intensity at the ith step.

For a particular X-ray diffraction pattern, different profile functions can be chosen to fit the peaks. I will not discuss the profile functions and preferred orientation for Rietveld refinement. Instead, we are more interested in the "fundamental parameter approach" (Cheary and Coelho, 1992). In this method, the observed peak profile shape is considered as a convolution of: (i) the emission profile of an X-ray, (ii) the instrument component and (iii) the specimen contribution. The instrument component includes a primary Soller slit, a divergence slit, a secondary Soller slit and a receiving slit. The specimen contribution includes the parameters given in Equation (7.16), as well as crystal size broadening and strain broadening. During refinement, the background contribution and the amorphous component, if present, need to be considered.

From the peak positions and reflection indices, the lattice parameters, a, b, c and α, β, γ, are calculated. Once the intensity is considered, more information, such as atom type, position, site occupancy and displacement parameters, can be retrieved. Those are key factors that influence the structure factor F_{hkl}, as indicated in:

$$F_{hkl} = \sum_{\substack{all \ atoms \\ per \ cell}} f_j Occ_j T_j \exp\left[2\pi i \left(hx_j + ky_j + lz_j\right)\right].$$

During whole pattern fitting, the calculated patterns are compared with the experimental one, and the R-Bragg factor is calculated, which reflects the difference between the observed and the calculated intensities for the Bragg reflections. The R-Bragg factor is defined by

$$R_B = \frac{\sum_i \left| I_{K(obs)} - I_{K(calc)} \right|}{\sum_i I_{K(obs)}} \tag{8.2}$$

where I_K is the intensity assigned to the Kth Bragg reflection at the end of the refinement cycles (Young, 1995).

Besides R-Bragg, a few other factors, for instance the expected R-factor (R_{exp}) and the weighted profile R-factor (R_{wp}), are often used during the Rietveld refinement (Toby, 2006).

The weighted profile R-factor is expressed by:

$$R_{wp} = \sqrt{\frac{\sum_i w_i \left(y_{C,i} - y_{O,i}\right)^2}{\sum_i w_i \left(y_{O,i}\right)^2}} \qquad (8.3)$$

where $y_{O,i}$ is the observed intensity at the ith step; $y_{C,i}$ is the calculated intensity at the ith step; and $w_i = \dfrac{1}{\sigma^2 \lfloor y_{O,i} \rfloor} = \dfrac{1}{\left\langle \left(y_{O,i} - \left\langle y_{O,i}\right\rangle\right)^2 \right\rangle}$. $\langle \rangle$ indicates the expected value.

The expected R-factor is expressed by

$$R_{exp} = \sqrt{\frac{N}{\sum_i w_i \left(y_{O,i}\right)^2}} \qquad (8.4)$$

In this formula, N (or $N - P$) is the number of data points, less the number of varied parameters (Toby, 2006). Some researchers prefer using $(N - P)$ in Equation (8.4).

During refinement, different crystal structure parameters can be released and their values changed in a way to reduce the difference between the calculated and observed patterns until it converges to a minimum. Parameters should be refined step by step, instead of releasing them all together at the same time.

During the refinement, the goodness of fit indicator S and Chi square χ^2 are calculated from the expected R-factor (R_{exp}) and the weighted profile R-factor (R_{wp}). The goodness of fit indicator $S = R_{wp}/R_{exp}$ and Chi square $\chi^2 = [R_{wp}/R_{exp}]^2$. It should be noted that χ^2 should never drop below one, or equivalently the smallest that R_{wp} should ever be is R_{exp} (Toby, 2006).

R_B and χ^2 can be used to evaluate the reliability of the obtained results. In *Fundamentals of Powder Diffraction and Structural Characterization of Materials* (Pecharsky and Zavalij, 2003), many examples are given to explain the method of powder X-ray diffraction pattern refinement. Using the example of a synthetic apatite materials research project that was completed at the Environmental Technology Institute (ETI) labs, we demonstrated how to apply the Rietveld refinement to crystal structure analysis.

CASE STUDY: APPLICATION OF RIETVELD REFINEMENT TO THE ANALYSIS OF SYNTHETIC HYDROXYAPATITE MATERIALS

Hydroxyapatite can be synthesized in many different ways, such as by the sol-gel approach and the hydrothermal method (Liu et al., 1997). In our experiment, a new precursor was prepared using AR grade $CaCO_3$ and H_3PO_4 as starting materials, with the $CaCO_3$ decomposed at 900 °C for two hours before the mixing. During the precursor preparation, 0.30 moles of H_3PO_4 acid was diluted slowly into 100 ml using distilled water accompanied by a magnetic string. Then, 0.15 moles and 0.35 moles of the decomposed $CaCO_3$ were added to the solution successively. Before the 0.35 moles of the decomposed $CaCO_3$ was introduced to the solution, distilled water was added to keep the solution at the 150 ml level. The solution was dried in an oven overnight and the powders obtained were properly kept in sealed bottles. The precursor and distilled water were placed into Teflon liners with a weight ratio of 1:10, which were then encapsulated by stainless steel autoclaves. The autoclaves were kept in furnaces at temperatures ranging from 175 to 250 °C for different durations ranging from 1 to 96 hours. Each of the reacted products was washed three times, using distilled water to remove residual ions, and then dried in the oven at about 95 °C. The powders collected were examined using XRD and HRTEM for crystal structure determination.

Powder X-ray diffraction patterns were collected using a Siemens D5005 X-ray diffractometer fitted with a Cu-tube that was operated at 40 kV and 40 mA. The divergence, anti-scatter, receiving and detector slits were fixed at 0.5°, 0.5°, 0.1 mm and 0.6 mm respectively. The primary Soller slit in the incident path and the secondary Soller slit in the diffracted path were used to limit beam divergence along the longitude of the above slits within 2.3°. Scans were collected from 10–140° 2θ with a step size 0.01°, and counting time 5 seconds per step. The XRD patterns were refined using the Rietveld method (with Topas software) with fundamental parameters. The four-coefficient background polynomial and peak shift due to zero error and sample displacement were refined. The space group was set at $P6_3/m$, instead of $P2_1/b$, although it is believed that $P2_1/b$ reflects the crystal structure more precisely due to OH ordering (Elliott et al., 1973; Hitmi et al., 1988).

The morphologies of the hydrothermally synthesized products were examined using electron microscopy: a JEM-3010 (Cs = 1.4 mm) operating at 300 kV. The TEM specimen was prepared by dispersing the powders with ethanol in an ultrasonic bath and a drop of suspension deposited onto a holey-carbon coated copper grid. The hydroxyapatite whiskers formed after longer duration were longer, as expected. Figure shows the materials obtained when hydrothermal synthesis was conducted at 225 °C for 3 hours and 96 hours respectively.

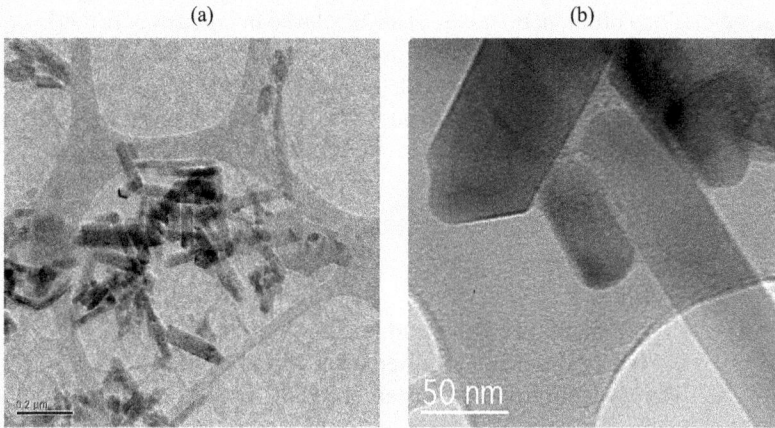

FIGURE 8.1   TEM observation of hydroxyapatite whiskers obtained after the precursor was hydrothermally treated at 225 °C (a) for 3 hours and (b) for 96 hours.

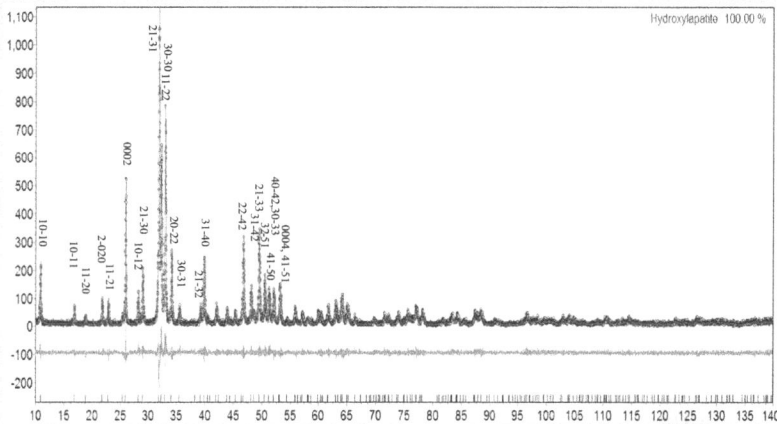

FIGURE 8.2   X-ray diffraction pattern of hydroxypaptite obtained after the precursor was hydrothermally treated at 225 °C for 12 hours. Some strong peaks are indexed.

X-ray diffraction analysis suggests that pure hydroxyapatite materials are obtained at 200 °C or above from the precursor prepared. The results were reported at the International Crystallography Meeting in Broome in 2003 (Tseng and Nalwa, 2009). For the hydroxyapatite obtained after hydrothermal treatment at 225 °C for 12 hours, the X-ray diffraction pattern is shown in Figure 8.2.

The corresponding crystal structure data are listed in Tables 8.1 and 8.2.

Based on the crystal structure data in Tables 8.1 and 8.2, the projected atom arrangements along different orientations are displayed in Figure 8.3. In this hydroxy-apatite structure, the P-O bond length is in the range of 1.542 to 1.567 Å. From the lattice constants and atom coordinates listed in Tables 8.1 and 8.2, the A(I)-$O_{1,2,3}$ and A(II)-$O_{1,2,3,4}$ bond lengths can also be obtained and the A(I)-O and A(II)-O polyhedra can be drawn using the method presented in Chapter 4.

TABLE 8.1

Crystal Structure Data and Refinement Parameters of Hydroxyapatite Obtained after the Precursor was Hydrothermally Treated at 225°C for 12 Hours

Phase Name	Hydroxylapatite
R-Bragg	4.517
R_{exp}	19.89
R_{wp}	21.88
GOF	1.10
Space group	$P 6_3/m$
Cell Mass	1004.620
Cell Volume (Å³)	530.06(18)
Crystal Density (g/cm³)	3.1472(11)
Twist Angle	24.0°
Lattice parameters	
a (Å)	9.4280(15)
c (Å)	6.8857(11)

TABLE 8.2

Atomic Coordinates in the Structure of Hydroxyapatite Obtained after the Precursor was Hydrothermally Treated at 225°C for 12 Hours

Atom	Site	x	y	z	Occupation	Beq
Ca1	4f (A^I)	1/3	2/3	0.00216(44)	1	0.31(17)
Ca2	6h (A^{II})	0.24644(25)	0.99173(28)	1/4	1	0.37(17)
P	6h	0.39867(34)	0.36868(32)	1/4	1	0.56(18)
O1	6h	0.32648(65)	0.48660(60)	1/4	1	0.87(22)
O2	6h	0.58752(70)	0.46271(70)	1/4	1	0.92(23)
O3	12i	0.34056(43)	0.25681(45)	0.06680(50)	1	0.85(19)
O4	4e	0	0	0.1952(21)	0.5	0.59(38)
H	4e	0	0	0.046(38)	0.5	0.2(59)

(a)

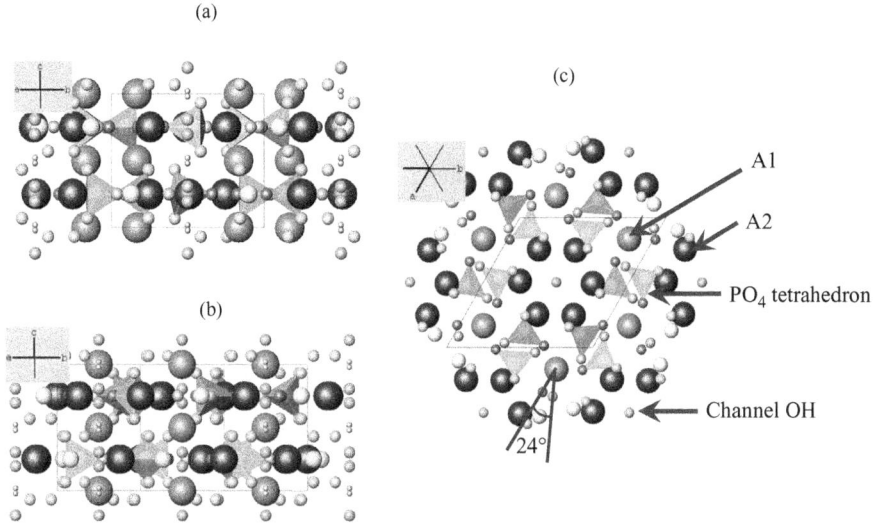

(c)

(b)

A1

A2

PO_4 tetrahedron

Channel OH

24°

FIGURE 8.3 Crystal structure projections of the hydroxyapatite obtained after the precursor was hydrothermally treated at 225 °C for 6 hours. (a) Along the [2-1-1 0] direction, (b) along the [1 0-1 0] direction and (c) along the [0 0 0 1] direction. For simplicity, we used a hexagonal cell, or half of the monoclinic unit cell, to represent the hydroxyapatite structure. The crystal structure data were obtained from the Rietveld refinement analysis based on the space group $P6_3/m$. We could not obtain the ordering information for the (OH) groups in the channel using X-ray diffraction analysis.

SUMMARY

In this chapter, we have discussed the intensity of a diffracted beam for a polycrystal sample. In the case where there are two or more phases present, the peak intensities from each phase can also be calculated. The Rietveld refinement is a whole-pattern-fitting least-squares technique that uses the entire pattern rather than a few selected reflections to extract the crystallographic information.

The Rietveld refinement for an X-ray powder diffraction pattern allow us to:

I. Solve an unknown crystal structure, especially when a similar structure file is imported;
II. Calculate the site occupancies when two or more elements are located at the same Wyckoff site;
III. Determine the weight fraction of each phase in a multi-phase sample. Recently developed software is able to analyze the amorphous phase as well;
IV. Determine the size of crystallites from the broadening of diffraction peaks.

REFERENCES

Cheary, R. W. and Coelho, A. (1992) 'A fundamental parameters approach to X-ray line-profile fitting', *Journal of Applied Crystallography*. International Union of Crystallography (IUCr), 25(2), pp. 109–121. doi:10.1107/S0021889891010804.

Elliott, J. C., Mackie, P. E. and Young, R. A. (1973) 'Monoclinic hydroxyapatite', *Science*, 180(4090), pp. 1055–1057. doi:10.1126/science.180.4090.1055.

Hitmi, N., LaCabanne, C. and Young, R. A. (1988) 'Oh– reorientability in hydroxyapatites: effect of F– and Cl–', *Journal of Physics and Chemistry of Solids*, 49(5), pp. 541–550. doi:10.1016/0022-3697(88)90065-0.

Liu, H. S. et al. (1997) 'Hydroxyapatite synthesized by a simplified hydrothermal method', *Ceramics International*, 23(1), pp. 19–25. doi:10.1016/0272-8842(95)00135-2.

Pecharsky, V. and Zavalij, P. (2003) *Fundamentals of Powder Diffraction and Structural Characterization of Materials [electronic resource]/by Vitalij Pecharsky, Peter Zavalij*. 1st ed. 20. New York, NY: Springer US. doi: 10.1007/b106242.

Rietveld, H. M. (1969) 'A profile refinement method for nuclear and magnetic structures', *Journal of Applied Crystallography*. International Union of Crystallography (IUCr), 2(2), pp. 65–71. doi:10.1107/S0021889869006558.

Toby, B. H. (2006) 'R factors in Rietveld analysis: how good is good enough?', *Powder Diffraction*, 21, pp. 67–70. doi:10.1154/1.2179804.

Tseng, T.-Y. and Nalwa, H. S. (eds) (2009) Handbook of nanoceramics and their based nanodevices, Stevenson Ranch, CA: American Scientific Pub. (Nanotechnology Book Series; 24).

Young, R. A. (Robert A.) (1995) *The Rietveld method*/edited by R.A. Young. Chester, England: International Union of Crystallograhy (International Union of Crystallography monographs on crystallography; 5).

Zevin, L. S., Kimmel, G. and Mureinik, I. (1995) *Quantitative X-ray diffractometry [electronic resource]/by Lev S. Zevin, Giora Kimmel*; edited by Inez Mureinik. New York, NY: Springer US. doi:10.1007/978-1-4613-9535-5.

Part III

Transmission Electron Microscopy of Materials

The purpose of Part III is to explain the theories of transmission electron microscopy (TEM) for the analysis of materials. This part mainly covers the theories of (i) electron diffraction, (ii) diffraction contrast and (iii) phase contrast; this will help students understand how to apply electron microscopy technology to materials research. Other theories such as mass-thickness contrast, Z-contrast in scanning transmission electron microscopy (STEM) and electron energy loss spectroscopy are not included.

When starting TEM lectures, students often ask why we use an electron beam as an illuminating source for a microscope. Therefore, I provide some brief explanation on the advantages of using an electron source before teaching the TEM theories.

The wavelengths of visible light are in the range of 3900 to 7600 Å. Based on the theory of diffraction, the resolution of a light optical microscope can only reach to about 2000 Å, or about $\lambda/2$. This resolution is not enough to study the tiny structures of biological samples, like chromosomes and even smaller biomolecules. In the early twentieth century during the study of tiny substances of different types of samples, the search for an illuminating source with a shorter wavelength to obtain a higher resolving power led to the construction of electron microscopes, in which a shorter wavelength electron illuminating beam is generated by a high accelerating voltage.

Transmission electron microscopes require high vacuum levels inside the column. High-resolution ones are operated at a very high vacuum or low pressure (of the order of 10^{-5} Pa or higher) to avoid the collision of electrons with air molecules inside the column.

As electron beams are charged particles that can only penetrate very thin specimens of about 1000 Å, specimens as thin as a few nanometers or even thinner should be prepared to better show the electron diffraction patterns or images. Thin TEM specimens can be prepared manually or by a focused ion beam (FIB) machine. In the semiconductor industry, using a FIB to prepare thin TEM specimens is very common.

For metal alloys and ceramics used for various applications, the materials are used in the form of powders, fibers, coatings or large bulk pieces. However, an electron beam can only penetrate a thin specimen loaded in the sample holder. Therefore, specimen preparation is needed unless the sample materials are fine powders.

For bulk metals, cutting, mechanical grinding, mechanical polishing and electrochemical polishing are employed for the preparation of thin samples of about 3 mm in diameter, which can properly fit into different sample holders. Less than 3 mm is fine if properly secured by the holder. Smaller sized pieces can also be bonded to specially designed grids about 3 mm in diameter, which can perfectly fit into the sample holder.

If ceramic powders are fine particles, they can be dispersed into ethanol or acetone first, followed by dropping a small amount of solution onto a holey carbon grid or carbon film. If the particle sizes are large, they can be mixed into resin. After curing, a thin slice can be mechanically prepared. Fiber materials can be prepared in the same way. To prepare a TEM specimen from bulk ceramics, we can use a similar procedure as for metals. However, ceramics are insulating materials, so at the final step, electrochemical polishing is not suitable. Dimpling followed by ion beam thinning can be employed to achieve electron transparency. Prior to ion milling, dimpling to a thickness of about 25 μm is possible. TEM specimens of metallic materials can also be prepared through a dimpling and ion milling approach if electrochemical polishing is not chosen.

Precaution needs to be taken during mechanical grinding and polishing for metallic samples. A deformation layer can alter the microstructure, and the final step of ion milling or electrochemical polishing should be carried out before the central part of the metal sample is mechanically deformed. In such a case, the thin slice obtained after the final polishing is from the non-deformed central part.

It is challenging to prepare cross-sectional TEM specimens from coating/substrate interface regions. Some engineering materials are coated with a functional layer, and some microelectronic devices have a few coating layers on silicon substrates. The cross-sectional view of the interface using HRTEM is an effective method to understand the interfacial microstructure. To prepare such types of TEM specimen, a face-to-face sandwich method is effective. The descriptions of this method can be found in Williams (1996) and many other scientific papers, as well as in some of my research projects (Nutting *et al.*, 1995; Li *et al.*, 2012).

In Chapter 8, there was a reminder to our materials science and engineering students to conduct composition analysis prior to qualitative and quantitative X-ray diffraction analysis. In TEM analysis, the composition analysis, phase structure and defect analysis can be carried out in the same region of the specimen during observation if the microscope is equipped with EDX and EELS. It is suggested that X-ray diffraction analysis be conducted to get the phase information before TEM structural analysis is carried out, because X-ray diffraction can provide overall structure information, which can be a useful guide when conducting a local structural analysis using TEM.

The period from the discovery of the electron to the construction of the first transmission electron microscope was only about 35 years. The electron was discovered by British physicist J. J. Thomson in 1897. When he was studying the properties of cathode rays at the Cavendish Laboratory, he showed that they were composed of previously unknown negatively charged particles, later called electrons. Thomson won the Nobel Prize in Physics in 1906 "for his theoretical and experimental investigations on the conduction of electricity by gases". In 1909, the American experimental physicist R. Millikan measured the charge of an electron using negatively charged oil droplets. Millikan received the Nobel Prize in Physics in 1923 "for his work on the elementary charge of electricity and on the photoelectric effect". In 1923–1924 French scientist Prince Louis de Broglie postulated wave-like properties for matter, with the wavelength λ related to momentum p in the same way as for light, or electromagnetic waves. He was awarded the Nobel Prize in Physics in 1929 "for his discovery of the wave nature of electrons". G. P. Thomson, son of J. J. Thomson, and also the American scientists Davisson and Germer, independently demonstrated these electron diffraction experiments in 1927. Thomason and Davison received the Nobel Prize in Physics in 1937 "for their experimental discovery of the diffraction of electrons by crystals".

During the period from 1926 to 1927, German scientist Hans Busch demonstrated theoretically that a coaxial magnetic or electric field could focus an electron beam, which provided theoretical support for producing a lens for electrons, a critical component for constructing electron microscopes. In about 1931, Knoll and Ernst Ruska completed building the first transmission electron microscope. In 1986, Ernst Ruska won the Nobel Prize in Physics "for his fundamental work in electron optics, and for the design of the first electron microscope". Modern transmission electron microscopes with improved resolutions are computerized and more advanced in acquiring electron diffraction patterns, amplitude contrast and phase contrast images. Analytical capability has been improved due to the development of a high quality STEM image collection system and electron energy loss spectrum analysis. Point resolution has reached the sub-angstrom level due to the incorporation of spherical aberration correctors. In fact, many scientists had been working on the design of Cs correctors, and based on a suggestion by Harald Rose, a German team figured out a solution for spherical aberration correction (Haider *et al.*, 1998), which propelled the development of TEM techniques into a new era with sub-angstrom point resolution.

Part III mainly explains the theories of electron diffraction, diffraction contrast and the phase contrast of TEM. Other theories related to STEM, EELS, holography and so on are not covered. It is known that the TEM operation mode employs a broad stationary beam to transmit the specimen, whereas STEM uses a focused electron beam scanning across the specimen, collecting various electron beam signals from below the specimen with detectors. STEM is also different from SEM, as SEM can study bulk sample and collect various signals above the sample. Figure III.1 schematically shows the beam and sample geometry of TEM, STEM and scanning electron microscopy (SEM). For a modern transmission electron microscope, it is easy to

TEM:	STEM:	SEM:
Broad parallel beam Stationary beam	Fine probe Scanning probe	Fine probe Scanning probe
Thin specimen	Thin specimen	Bulk specimen

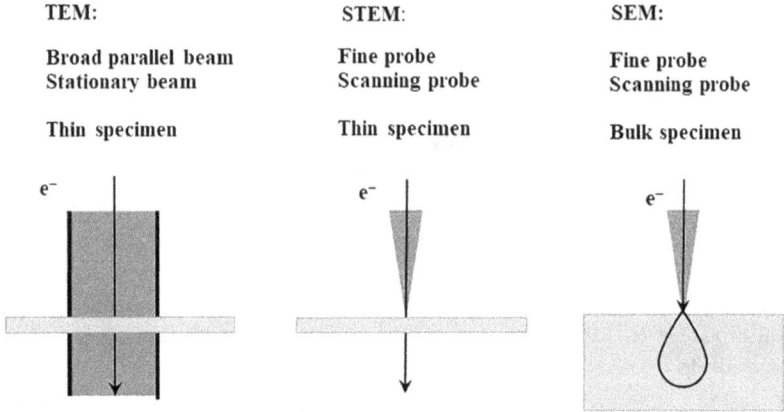

FIGURE III.1   Schematic illustration of e-beam and sample geometries for TEM, STEM and SEM operating modes.

switch between the TEM operating mode and the STEM operating mode during operations.

As described earlier, the electron beam source can have much shorter wavelengths compared with those of visible light spectra. The wavelengths for electron beams can be calculated from the special theory of relativity and the de Broglie relationship when the accelerating voltage for electron beams is known.

Based on the special theory of relativity, the electron total energy E, momentum p and the rest mass of electron m_0 can be expressed by:

$$E^2 = c^2 p^2 + m_0^2 c^4 \tag{III.1}$$

The electron total energy can also be expressed by kinetic energy E_k plus the rest energy $m_0 c^2$:

$$E = E_k + m_0 c^2 \tag{III.2}$$

Considering that $E_k = eV$ when electrons are accelerated by a high voltage V, we rewrite Equation (III.2) as

$$E = eV + m_0 c^2 \tag{III.3}$$

and

$$E^2 = \left(eV + m_0 c^2\right)^2 = \left(eV\right)^2 + 2eVm_0 c^2 + m_0^2 c^4 \tag{III.4}$$

Comparing Equations (III.1) and (III.4), we obtain

$$c^2 p^2 = (eV)^2 + 2eVm_0c^2$$

Hence, the momentum expression is obtained and expressed by:

$$p = \left[2m_0eV \left(1 + \frac{eV}{2m_0c^2} \right) \right]^{1/2} \tag{III.5}$$

Using the de Broglie relationship, the electron wavelength after being accelerated through a voltage V is

$$\lambda = \frac{h}{p} = \frac{h}{\left[2m_0eV \left(1 + \frac{eV}{2m_0c^2} \right) \right]^{1/2}} \tag{III.6}$$

The non-relativistic wavelength is

$$\lambda = \frac{h}{p} = \frac{h}{\left[2m_0eV \right]^{1/2}} \tag{III.7}$$

Another approach for calculating the relativistic wavelength is presented below. Based on the special theory of relativity, we write the kinetic energy as

$$E_k = eV = mc^2 - m_0c^2$$

and the relativistic mass as

$$m = \frac{m_0}{\sqrt{1 - \left(\dfrac{v}{c} \right)^2}}$$

Relativistic mass m can be obtained from the first equation, after which the relativistic velocity v can be obtained from the second equation, which are:

$$m = m_0 + \frac{eV}{c^2}$$

$$v = c\sqrt{1 - \left(\frac{m_0c^2}{eV + m_0c^2} \right)^2}$$

Using the de Broglie relation, the relativistic wavelength can be obtained, which is the same as (III.6):

$$\lambda = \frac{h}{p} = \frac{h}{mv} = \frac{h}{\left[2m_0eV\left(1+eV/2m_0c^2\right)\right]^{1/2}}$$

Table III.1 is a list of the calculated electron beam wavelengths with and without relativistic corrections. In the calculations, the following fundamental physical constants are employed (Young *et al.*, 2012): rest mass of electron $m_0 = 9.10938 \times 10^{-31}$kg, magnitude of change of electron $e = 1.60217 \times 10^{-19}$C, speed of light in a vacuum $c = 2.99792 \times 10^8$m/s.

In order to explain the working principles and theories of transmission electron microscopes, we need to introduce the Abbe imaging theorem in diffraction optics to students with a materials science and engineering background.

The image formation by the objective lens in a TEM is schematically shown in Figure III.2. For the explanation of electron diffraction and image formation, the Fourier transform is an effective mathematical tool.

According to the Abbe imaging theorem, the image formation by a lens can be explained as below. If the electron wave leaving the lower surface of the specimen is represented by $\psi_e(x,y)$, rays leaving the specimen at the same angle are brought together to a focus at one point in the back-focal plane of the lens. This is equivalent to interference at a point at infinity, which is Fraunhofer diffraction (Cowley, 1995). In mathematics, the process is presented by a Fourier transform:

$$\Psi_d(u,v) = FT\left[\psi_e(x,y)\right] = \iint \psi_e(x,y)\exp\left[-i2\pi(ux+vy)\right]dxdy \qquad \text{(III.8)}$$

TABLE III.1
Wavelengths of Electron Beam as a Function of Accelerating Voltage

Voltage	Mass (m/m_0)	Velocity (v/c)	Relativistic Wavelength (Å)	Non-relativistic Wavelength (Å)
80 kV (also suitable to study polymers)	1.16	0.502	0.0417	0.0433
100 kV	1.20	0.548	0.0370	0.0388
200 kV	1.39	0.695	0.0251	0.0274
300 kV	1.59	0.777	0.0197	0.0224
1000 kV (ultrahigh voltage TEM)	2.96	0.941	0.0087	0.0123
3000 kV (ultrahigh voltage TEM)	6.87	0.989	0.0036	0.0071

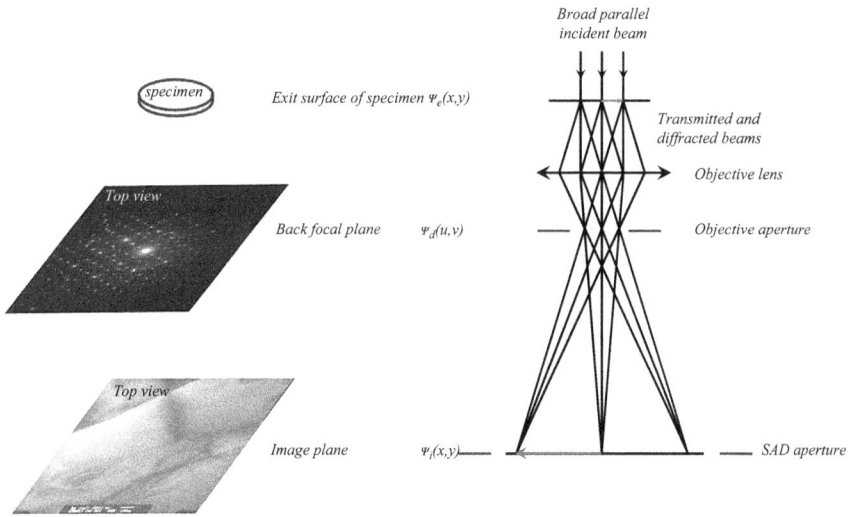

FIGURE III.2 Electron diffraction pattern and image formed at the back focal plane and image plane of the objective lens. The diffraction pattern formed at the back focal plane can be magnified by the intermediate and projector lenses and viewed on the screen. Similarly, the image formed at the image plane can be magnified by the intermediate and projector lenses and viewed on the screen. The intermediate lens current and focal length are different in the above two modes, one of which is to enlarge the diffraction pattern, and the other is to enlarge the image.

where (u,v) are the coordinates in the back focal plane, or the coordinates in reciprocal space.

The radiation from the back focal plane again forms an interference pattern on the image plane of the lens. In the case where the distance between the back focal plane and image plane is large enough, as an approximation, the process is regarded as Fraunhofer diffraction, and again it is expressed by a Fourier transform:

$$\psi_i(x,y) = FT\left[\Psi_d(u,v)\right] = \iint \Psi_d(u,v)\exp\left[-i2\pi(ux+vy)\right]dudv$$
$$= \iint \Psi_d(u,v)\exp\left\{i2\pi\left[u(-x)+v(-y)\right]\right\}dudv = \psi_e(-x,-y) \quad \text{(III.9)}$$

From Figure III.2 and Equation (III.9), we see that rays from a common point in the object converge on a common point in the image plane. The negative sign in the wave function indicates that the image is inverted. The magnification, M, of the lens is missing in the Fourier transforms. As the scattering angles considered are relatively small in TEM, and the distance from the back focal plane to the image plane is relatively large, it is good to consider the process through Fraunhofer diffraction, which can be mathematically treated through the Fourier transform.

Some textbooks just use the inverse Fourier transform to represent the second step, in which case there is no inversion of the image, and in general it does not cause problems with image formation discussions.

To analyze the structure of a material by electron diffraction, we need to collect the electron diffraction patterns from the back focal plane of the objective lens. In a transmission electron microscope, this can be realized by adjusting the intermediate lens strength and enlarge the electron diffraction patterns formed on the back focal plane through the intermediate lens and projector lenses.

To get images and display diffraction contrast or phase contrast, the image formed on the imaging plane can be enlarged by adjusting the intermediate lens current, so that the image can be magnified by the intermediate lens and the projector lenses.

Before studying electron diffraction theories, we need to discuss the atomic scattering factor for an electron beam. We use some theories from physics textbooks, and during some discussions we neglect the influence of higher order terms and only keep first order terms as an approximation.

Students are strongly recommended to read the textbook *Electron Microscopy of Thin Crystals* (Hirsch *et al.*, 1977) and the textbook *Transmission Electron Microscopy: A Textbook for Materials Science* (Williams, 1996). Our discussions on the two-beam dynamical theory in the diffraction contrast are based on the approach presented in the textbook Electron Microscopy of Thin Crystals.

REFERENCES

Cowley, J. M. (John M.) (1995) *Diffraction physics [electronic resource]/John M. Cowley*. 3rd rev. e. Amsterdam: Elsevier Science B.V. (North-Holland personal library).

Haider, M. et al. (1998) 'Electron microscopy image enhanced [7]', *Nature*, 392(6678), pp. 768–769. doi:10.1038/33823.

Hirsch, P. B., Howrie, A., Nicholason, R. B., Pashley, D. W. and Whelan, M. J. (1977) *Electron microscopy of thin crystals/*. Malaber, FL: Krieger.

Li, Z. et al. (2012) 'Interface and surface cation stoichiometry modified by oxygen vacancies in epitaxial manganite films', *Advanced Functional Materials*, 22(20), pp. 4312–4321. doi:10.1002/adfm.201200143.

Nutting, J., Guilemany, J. M. and Dong, Z. (1995) 'Substrate/coating interface structure of we-co high velocity oxygen fuel sprayed coating on low alloy steel', *Materials Science and Technology*. Taylor & Francis, 11(9), pp. 961–966. doi:10.1179/mst.1995.11.9.961.

Williams, D. B. (1996) *Transmission Electron Microscopy [electronic resource]: A Textbook for Materials Science/by David B. Williams, C. Barry Carter*. Edited by C. B. Carter. Boston, MA: Springer US. doi:10.1007/978-1-4757-2519-3.

Young, H. D. (Hugh D.) et al. (2012) *University physics: with modern physics/Hugh D. Young, Roger A. Freedman; contributing author, A. Lewis Ford*. 13th ed., *Sears and Zemansky's University physics*. 13th ed. San Francisco: Addison-Wesley.

9 Atomic Scattering Factors for Electrons and X-rays

In the following section, we will discuss the atomic scattering factor for electrons, and its correlation with the atomic scattering factor for X-rays.

9.1 ATOMIC SCATTERING FACTOR FOR AN ELECTRON

The scattering of the electron wave by an atom can be expressed as

$$\psi_{tot} = \psi + i\psi_{sc} = \psi_0 \left[e^{2\pi i \vec{k}\cdot\vec{r}} + if(\theta)\frac{e^{2\pi i \vec{k}\cdot\vec{r}}}{r} \right]$$

where $f(\theta)$ is defined as the atomic scattering factor for electrons and i is due to the $\frac{\pi}{2}$ difference for the scattered wave (Williams, 1996). The atomic scattering factor $f(\theta)$ depends on the atomic number of the atom and the scattering angle θ_S. When the Bragg condition is satisfied, $\theta_S = 2\theta_B$. In other cases, we use $\theta_S = 2\theta$.

Electrons that pass through an area element $d\sigma$ of the parallel incident beam will be scattered into a cone of solid angle element $d\Omega$ as shown in Figure 9.1. The ratio $d\sigma/d\Omega$ is known as the differential scattering cross-section and is a function of the scattering angle θ_S. j_0 and j_{SC} are the electron flux densities. j_0 is the flux per unit area for the incident beam and j_{SC} is the flux per unit area for the scattered beam. As the particles are electron beams, $-ej_0$ represents the current density.

Since the electrons scattered into the solid angle $d\Omega$ are observed when these electrons hit the fraction $d\sigma$, the scattered flux that passes through the area element dS or $r^2 d\Omega$ is

$$dn = j_{sc} r^2 d\Omega = j_0 d\sigma$$

which implies that

$$j_{sc} = \frac{j_0}{r^2} \left(\frac{d\sigma}{d\Omega} \right) \tag{9.1}$$

DOI: 10.1201/9780429351662-13

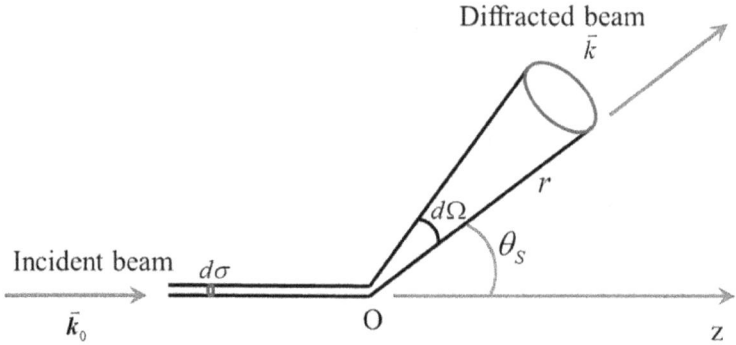

FIGURE 9.1 Scattering of incident electron beam into an element of solid angle $d\Omega$ by an atom.

Therefore, the scattered flux passing through the area element dS or $r^2 d\Omega$ at distance r is

$$dn = j_{sc} r^2 d\Omega = \frac{j_0}{r^2}\left(\frac{d\sigma}{d\Omega}\right) r^2 d\Omega = j_0\left(\frac{d\sigma}{d\Omega}\right) d\Omega$$

where $\dfrac{d\sigma}{d\Omega}$ is the differential scattering cross-section. For a parallel electron beam with N electrons per unit volume and with a velocity v, its flux density can be represented as $j_0 = Nv$, which is the number of electrons traveling through a unit area per unit time.

When substituting the incident plane wave $\psi_0 e^{2\pi i \bar{k}\cdot\bar{r}}$ into the quantum mechanical formula for a flux of particles

$$j = \frac{i\hbar}{2m}\left(\psi\nabla\psi * -\psi * \nabla\psi\right)$$

We have

$$j_0 = \frac{\hbar}{m} 2\pi k |\psi_0|^2 = v|\psi_0|^2 \tag{9.2}$$

$$\text{since } \frac{\hbar}{m} 2\pi k = \frac{h}{2\pi m} 2\pi k = \frac{p}{m} = v$$

Substituting the scattered wave amplitude $\psi_{sc} = \psi_0 f(\theta)\dfrac{e^{2\pi i \bar{k}\cdot\bar{r}}}{r}$ into the quantum mechanical expression for the flux density, we have

$$j_{sc} = v|\psi_0|^2 \frac{|f(\theta)|^2}{r^2} \tag{9.3}$$

Comparing Equations (9.2) and (9.3), we get

$$j_{SC} = j_0 \frac{\left| f(\theta) \right|^2}{r^2} \tag{9.4}$$

From Equation (9.1), we also know that

$$j_{sc} = \frac{j_0}{r^2} \left(\frac{d\sigma}{d\Omega} \right)$$

Therefore, we have

$$\frac{d\sigma}{d\Omega} = \left| f(\theta) \right|^2 \tag{9.5}$$

This is the relationship between the differential cross-section and the atomic scattering factor for electrons. Therefore, we can obtain the atomic scattering factor for electrons via a differential scattering cross-section calculation, which is presented below.

In quantum mechanics, Fermi's golden rule in the time-dependent perturbation theory shows the transition probability rate, or transition probability per unit time from an initial state $h\vec{k}_0$ to a final state $h\vec{k}$. Instead of considering the transition to a particular state, it is often necessary to deal with transition to a group of final states, where the final states lie on a continuum. The transition probability rate, or the probability rate of transition $w_{\vec{k}_0 \vec{k}}$, tells us the rate at which the initially empty levels become populated (Liboff, 2003):

$$w_{\vec{k}_0 \vec{k}} = \frac{dP_{\vec{k}_0 \vec{k}}}{dt} = \frac{2\pi}{\hbar} \left| H'_{\vec{k} \vec{k}_0} \right|^2 g(E_k) \tag{9.6}$$

where $H'_{\vec{k} \vec{k}_0}$ is the matrix element of the perturbation between the initial and final states, $P_{\vec{k}_0 \vec{k}}$ is the probability that a transition occurs and $g(E_k)$ is the density of the final states.

Assuming that an electron is inside a cubic box with edge length L, then we know that the normalized wave function is

$$\phi_k(\vec{r}) = L^{-\frac{3}{2}} \exp\left(\frac{i}{\hbar} \vec{p} \cdot \vec{r} \right)$$

By imposing the periodic boundary conditions, we find that the components of momentum along the three axes are given by $p_x = \dfrac{2\pi \hbar n_x}{L}$, $p_y = \dfrac{2\pi \hbar n_y}{L}$ and $p_z = \dfrac{2\pi \hbar n_z}{L}$, where n_x, n_y and n_z are integers.

It is noticed that one state occupies a volume of $\left(\dfrac{2\pi\hbar}{L}\right)^3$ in momentum space, and the number of the states per unit volume in that space is

$$\frac{L^3}{(2\pi\hbar)^3}$$

In the above case, we only consider one particle in a cubic box L^3.

Hence, the density of the state expressed by the spherical coordinates in momentum space is given by

$$\left(\frac{L}{2\pi\hbar}\right)^3 p^2 d\Omega dp = \left(\frac{L}{2\pi\hbar}\right)^3 p^2 dp \sin\theta d\theta d\phi$$

since we know that a volume element in spherical coordinates in momentum space is $(dp)(pd\theta)(p\sin\theta d\phi)$. Let

$$\left(\frac{L}{2\pi\hbar}\right)^3 p^2 d\Omega dp = \left(\frac{L}{2\pi\hbar}\right)^3 p^2 dp \sin\theta d\theta d\phi = g(E_k)dE_k$$

where $g(E_k)$ is the density of the state about E_k.

As $E_k = \dfrac{p^2}{2m_e}$, we have $dE_k = \dfrac{p}{m_e}dp$, or $dp = \dfrac{m_e}{p}dE_k$. The number of energy states in the interval E_k to $E_k + dE_k$ can be expressed as (Zhou, 1979):

$$g(E_k)dE_k = \left(\frac{L}{2\pi\hbar}\right)^3 m_e p \sin\theta d\theta d\phi dE_k$$

or

$$g(E_k) = \left(\frac{L}{2\pi\hbar}\right)^3 m_e p \sin\theta d\theta d\phi = \left(\frac{L}{2\pi\hbar}\right)^3 m_e p d\Omega$$

Generally, materials scientists use transmission electron microscopes with high acceleration voltages V, such as 200 and 300 kV. In such cases, the electrostatic potential $V(r)$ near an atom can be considered as a perturbance.

Based on the first Born approximation in quantum mechanics, the Hamiltonian is

$$H = H_0 + H' = H_0 + U(r)$$

where $U(r)$ is the perturbance, which is the interaction potential energy $-eV(r)$ of an electron.

The interaction potential energy $U(r) = -eV(r)$ is viewed as being "turned on" during the time that the incident electron beam is within the range of the potential field of the associated atom (Liboff, 2003). With the perturbation, the electrons transit from $h\bar{k}_0$ states to $h\bar{k}$ states. Or in other words, the incident particle enters the range of interaction with momentum $h\bar{k}_0$ and leaves the range of interaction with momentum $h\bar{k}$.

Students may find that many physics textbooks use $\bar{p} = \hbar\bar{k}$, but $h\bar{k}$ to represent the momentum. In that case, $\hbar = \dfrac{h}{2\pi}$ and $|k| = \dfrac{2\pi}{\lambda}$, but $\dfrac{1}{\lambda}$. In the discussion below, we use $|k| = |k_0| = \dfrac{1}{\lambda}$.

When we only consider one electron in the cube sample with a volume of L^3, the density of the energy state is expressed as $g(E_k) = \left(\dfrac{L}{2\pi\hbar}\right)^3 m_e(hk)d\Omega$. Hence, the probability for an electron from $h\bar{k}_0$ state to transit to $h\bar{k}$ state per unit time is given by

$$w_{\bar{k}_0\bar{k}} = dn = \frac{2\pi}{\hbar}\left|H'_{\bar{k}\bar{k}_0}\right|^2 g(E_k)$$

$$= \frac{2\pi}{\hbar}\left|-L^{-3}\int U(\bar{r})e^{-i2\pi(\bar{k}-\bar{k}_0)\cdot\bar{r}}d\bar{r}\right|^2 \frac{L^3 m_e hk}{8\pi^3\hbar^3}d\Omega \qquad (9.7)$$

$$= \left(vL^{-3}\right)\frac{m_e k}{2\pi\hbar^3 v}\left|-\int U(\bar{r})e^{-i2\pi(\bar{k}-\bar{k}_0)\cdot\bar{r}}d\bar{r}\right|^2 d\Omega$$

Let us have a look again at the scattering of an electron beam by an atom. In this case we only consider one electron inside the cube sample with a volume L^3, which means we employ the same cube sample assumption as in the above discussion for the time-dependent perturbation; then the density of the particle is $\dfrac{1}{L^3}$, and the flux of the incident beam is $j_0 = vL^{-3}$. Therefore, the number of particles scattered through $d\sigma$ to the solid angle element $d\Omega$ per unit time is

$$dn = j_{scr}r^2 d\Omega = \frac{j_0}{r^2}\left(\frac{d\sigma}{d\Omega}\right)r^2 d\Omega,$$

or

$$dn = \left(vL^{-3}\right)\left(\frac{d\sigma}{d\Omega}\right)d\Omega \qquad (9.8)$$

If we compare Equations (9.7) and (9.8), we notice that

$$\frac{d\sigma}{d\Omega} = \frac{m_e k}{2\pi\hbar^3 v}\left|-\int U(\bar{r})e^{-i2\pi(\bar{k}-\bar{k}_0)\cdot\bar{r}}d\bar{r}\right|^2 \qquad (9.9)$$

When we use the relationship $v = \dfrac{hk}{m_e}$, we have

$$\frac{d\sigma}{d\Omega} = \frac{m_e^2}{4\pi^2\hbar^4}\left| -\int U(\vec{r})e^{-i2\pi\left(\vec{k}-\vec{k}_0\right)\cdot\vec{r}}\,d\vec{r}\right|^2,$$

or

$$f(\theta)^2 = \frac{m_e^2}{4\pi^2\hbar^4}\left| -\int U(\vec{r})e^{-i2\pi\left(\vec{k}-\vec{k}_0\right)\cdot\vec{r}}\,d\vec{r}\right|^2$$

Therefore,

$$f(\theta) = \frac{m_e}{2\pi\hbar^2}\left(-\int U(\vec{r})e^{-2\pi i\left(\vec{k}-\vec{k}_0\right)\cdot\vec{r}}\,d\vec{r}\right) \tag{9.10}$$

since

$$-eV(r) = U(r),$$

where $V(r)$ is the electric potential of an atom, and $U(r)$ is the potential energy of the electron associated with the atom.

We can rewrite Equation (9.10) as

$$f(\theta) = \frac{m_e e}{2\pi\hbar^2}\int V(\vec{r})e^{-2\pi i\left(\vec{k}-\vec{k}_0\right)\cdot\vec{r}}\,d\vec{r} \tag{9.11}$$

If assuming the electrostatic potential of an atom $V(r)$ or the electric potential energy of an electron $U(r)$ has spherical symmetry, then

$$
\begin{aligned}
\int U(\vec{r})e^{-2\pi i\vec{q}\cdot\vec{r}}\,d\vec{r} &= \int_{r=0}^{\infty}\int_{\varphi=0}^{2\pi}\int_{\alpha=0}^{\pi} e^{-i(2\pi q)r\cos\alpha}U(r)r^2\sin\alpha\cdot d\alpha\cdot d\varphi\cdot dr\\
&= \int_{r=0}^{\infty}\int_{\varphi=0}^{2\pi}U(r)r^2\left[\frac{1}{i(2\pi q)r}\int_{\alpha=0}^{\pi}e^{-i2\pi qr\cos\alpha}d(-i2\pi qr\cos\alpha)\right]\cdot d\varphi\cdot dr\\
&= \int_{r=0}^{\infty}\int_{\varphi=0}^{2\pi}U(r)r^2\left[\frac{1}{i(2\pi q)r}\left(e^{i2\pi qr}-e^{-i2\pi qr}\right)\right]\cdot d\varphi\cdot dr\\
&= \int_{r=0}^{\infty}\int_{\varphi=0}^{2\pi}U(r)r^2\left[\frac{i2\sin(2\pi qr)}{i2\pi qr}\right]\cdot d\varphi\cdot dr\\
&= \int_{r=0}^{\infty}4\pi U(r)r^2\,\frac{\sin(2\pi qr)}{2\pi qr}\cdot dr
\end{aligned}
$$

Therefore,

$$
\begin{aligned}
f(\theta) &= \frac{m_e}{2\pi\hbar^2}\left(-\int U(\vec{r})e^{-2\pi i(\vec{k}-\vec{k}_0)\cdot\vec{r}}\,d\vec{r}\right)\\
&= \frac{2m_e}{\hbar^2}\int_0^\infty -r^2 U(r)\frac{\sin(2\pi qr)}{2\pi qr}\,dr \qquad (9.12)\\
&= \frac{2m_e e}{\hbar^2}\int_0^\infty r^2 V(r)\frac{\sin(2\pi qr)}{2\pi qr}\,dr
\end{aligned}
$$

or

$$
f(\theta) = \frac{2m_e e}{\hbar^2}\int_0^\infty r^2 V(r)\frac{\sin(4\pi sr)}{4\pi sr}\,dr \qquad (9.13)
$$

where $q = 2s$.

There are different ways to derive the atomic scattering factor for electrons, and in the above discussion we employed the scattering theory and the time-dependent perturbation theory (Liboff, 2003). In both cases, (i) it is assumed that there is only one electron inside a cubic box with edge length L, or volume L^3; (ii) the transition is from the initial state $\hbar\vec{k}_0$ to the final state $\hbar\vec{k}$; (iii) electrons are scattered into the same differential solid angle $d\Omega$. By comparing the results from the two approaches, the expressions for the differential cross-section and atomic scattering factors are obtained.

The X-ray and electron scattering factors can be found in different references, for example in Doyle and Turner (1968).

9.2 RELATIONSHIPS BETWEEN THE ATOMIC SCATTERING FACTOR FOR X-RAYS AND ELECTRONS

For an atom with an atomic number Z, the atomic electrostatic potential at position $\vec{r}$ is given by

$$
V(\vec{r}) = \frac{1}{4\pi\varepsilon_0}\left\{\frac{Ze}{r} - \int\frac{e\rho(\vec{r}')}{|\vec{r}-\vec{r}'|}\,d\vec{r}'\right\}
$$

where $\rho(\vec{r}')$ is the electron density at the point $\vec{r}'$.

From the atomic scattering factor expression (9.11), we have

$$f(\theta) = \frac{m_e e}{2\pi\hbar^2} \int V(\vec{r}) e^{-2\pi i(\vec{k}-\vec{k}_0)\cdot\vec{r}} d\vec{r}$$

$$= \frac{2\pi m_e e}{h^2} \int V(\vec{r}) e^{-2\pi i\vec{q}\cdot\vec{r}} d\vec{r}$$

$$= \frac{m_e e^2}{2h^2\varepsilon_0} \left\{ Z \int \frac{e^{-2\pi i\vec{q}\cdot\vec{r}}}{r} dr - \int \left(\int \frac{\rho(\vec{r}')}{|\vec{r}-\vec{r}'|} d\vec{r}' \right) e^{-2\pi i\vec{q}\cdot\vec{r}} dr \right\}$$

$$= \frac{m_e e^2}{2h^2\varepsilon_0} \left\{ Z \int \frac{e^{-2\pi i\vec{q}\cdot\vec{r}}}{r} dr - \int \rho(\vec{r}') e^{-2\pi i\vec{q}\cdot\vec{r}'} d\vec{r}' \int \frac{e^{-2\pi i\vec{q}\cdot(\vec{r}-\vec{r}')}}{|\vec{r}-\vec{r}'|} d(\vec{r}-\vec{r}') \right\}$$

$$= \frac{m_e e^2}{8\pi h^2\varepsilon_0} \frac{[Z-f_x]}{s^2}$$

where $\int d\vec{r} \exp(-2\pi i\vec{q}\cdot\vec{r})/r = \dfrac{4\pi}{(2\pi q)^2}$ (Peng, 1999) and $q = 2s$.

Therefore, the atomic scattering factor for electron $f(\theta)$ is related to the atomic scattering factor for X-ray f_X through the relationship

$$f(\theta) = \frac{m_e e^2}{8\pi h^2\varepsilon_0} \frac{(Z-f_x)}{s^2} \tag{9.14}$$

For the explanation of the relationships between the atomic scattering factor for electrons and the atomic scattering factor for X-rays, the Mott formula is often employed, which can be derived in the following method. In the discussion below, we need to use the same convention for f_e and f_X, either as a crystallographic or a quantum mechanical one.

To simply the calculation, we just use the non-vector one-dimensional expressions for the two integrals:

$$f_e(u) = \int V(r) \exp\{2\pi i u \cdot r\} dr$$

$$f_X(u) = \int \rho_e \exp\{2\pi i u \cdot r\} dr$$

More detailed discussion is provided in Cowley (1995).

Based on the relationship between the Fourier transform and the inverse Fourier transform, we can obtain the following expressions:

$$V(r) = \int f_e(u) \exp\{-2\pi i u \cdot r\} du$$
$$\rho_e = \int f_x(u) \exp\{-2\pi i u \cdot r\} du \tag{9.15}$$

Based on Poisson's equation, the relationship between electric potential and charge density is:

$$\nabla^2 V(r) = \frac{-e}{\varepsilon_0} \{ \rho_n(r) - \rho_e(r) \}$$ (9.16)

where $e\rho_n$ is the charge density due to the atomic nuclei, which can be expressed by a delta function of weight Z, $-e\rho_e$ is the charge density due to the electrons, ρ_e is the electron density, and $V(r)$ is the electrostatic potential distribution.

Substituting Equation (9.15) into (9.16), we have

$$\nabla^2 \left[\int f_e(u) \exp\{-2\pi i u \cdot r\} du \right] = \frac{e}{\varepsilon_0} \int f_x(u) \exp\{-2\pi i u \cdot r\} du - \frac{e}{\varepsilon_0} \int Z \exp\{-2\pi i u \cdot r\} du$$

where the final integral is a delta function of weight Z, the atomic number, due to the positive charge on the nucleus (Cowley, 1995).

The left-hand side can be expressed as

$$\int \left(-2\pi i |u| \right)^2 f_e(u) \exp\{-2\pi i u \cdot r\} du$$

Therefore, the following relationship is obtained, which is known as the Mott formula:

$$f_e(u) = \frac{e}{4\pi^2 \varepsilon_0 u^2} (Z - f_X) = \frac{e}{16\pi^2 \varepsilon_0 s^2} (Z - f_X)$$ (9.17)

If we compare Equation (9.14), the result from the first Born approximation, with Equation (9.17), the Mott formula, we find that the difference between them is the constant $2\pi m_e e/h^2$, or

$$f_{FB}(u) = \left(2\pi m_e e / h^2 \right) f_{Mott}(u)$$ (9.18)

A table listing the atomic scattering amplitudes (Å) for the electrons of neutral atoms can be found in the *International Tables for Crystallography, Volume C*, pp. 259–429.

SUMMARY

In Section 6.2, the atomic scattering factor for X-rays was expressed using an integral of the electron density associated with an atom. In this chapter, the linkage between the atomic scattering factors for X-rays and electron beams was derived.

As this textbook is for materials scientists and engineers, we have only employed some basic concepts from physics to derive the atomic scattering factor for high energy electron beams.

The atomic scattering factor of electrons is related to the electrostatic potential, whereas the atomic scattering factor for X-rays depends on the electron density. Since the electrostatic potential of an atom can be expressed by the charges carried by the nucleus and electrons of the atom, the atomic scattering factors for X-rays and electron beams can be correlated.

The atomic scattering factor expression obtained from the first Born approximation and that from the Mott formula are different, and they can be correlated through the formula.

$$f_{FB}(u) = \left(2\pi m_e e / h^2\right) f_{Mott}(u)$$

REFERENCES

Cowley, J. M. (John M.) (1995) *Diffraction physics [electronic resource]/John M. Cowley*. 3rd rev. e. Amsterdam: Elsevier Science B.V. (North-Holland personal library).

Doyle, P. A. and Turner, P. S. (1968) 'Relativistic Hartree–Fock X-ray and electron scattering factors', *Acta Crystallographica Section A*. International Union of Crystallography (IUCr), 24(3), pp. 390–397. doi:10.1107/S0567739468000756.

Liboff, R. L. (2003) *Introductory quantum mechanics/Richard L. Liboff*. 4th ed., *Quantum mechanics*. 4th ed. San Francisco: Addison-Wesley.

Peng, L. (1999) 'Electron atomic scattering factors and scattering potentials of crystals', *Micron* 30, pp. 625–648.

Williams, D. B. (1996) *Transmission electron microscopy [electronic resource]: A textbook for materials science/by David B. Williams, C. Barry Carter*. Edited by C. B. Carter. Boston, MA: Springer US. doi:10.1007/978-1-4757-2519-3.

Zhou, S. X. (1979) *Quantum mechanics/Zhou, S. X*. Beijing: Chinese Higher Education Publisher.

10 Electron Diffraction in Transmission Electron Microscopes

In the previous chapter, we studied the relationships between the atomic scattering factor for X-rays f_X and the atomic scattering factor for electron beams f_e. The scattering factor for electron beams is much higher than that for X-rays: f_e is of the order of 10^4 times larger (Hirsch, 1977).

We should also know the difference in geometry between electron scattering and X-ray scattering. The wavelengths for electron beams are normally at the 10^{-2} angstrom level; for instance, the wavelength of the electron beams accelerated by 200 kV is 0.0251 Å, whereas the wavelengths for X-rays are at the angstrom level, for example, $\lambda_{CuK\alpha} = 1.5418$ Å. When we draw the Ewald sphere in X-ray diffraction, its radius is comparable with the lattice constants for most of the crystalline materials. However, for electron diffraction, the radius of the Ewald sphere is much larger in comparison with the reciprocal lattice unit cell edges, and the portion of the Ewald sphere surface cutting the relrods near the reciprocal lattice origin O* looks very flat. In general, a diffraction pattern reflects the geometry of a weighted reciprocal lattice plane. In transmission electron microscope (TEM) analysis, the electron diffraction pattern should be indexed, and the zone axis should be calculated.

10.1 GEOMETRY OF ELECTRON DIFFRACTION IN TEMs

When an electron beam travels antiparallel to the zone axis direction of a thin crystalline specimen, or close to the antiparallel direction, a spotty pattern is formed on the back focal plane of the objective lens in a TEM. Each diffracted beam travels along a direction defined by a Bragg angle and contributes to the formation of a diffraction spot in the two-dimensional diffraction pattern at the back focal plane of the objective lens. In fact, each reciprocal lattice point of the crystal is associated with a domain, called a relrod, due to the crystal size effect; the Ewald sphere may cut many relrods with non-zero excitation errors, resulting in slight diffraction angle deviations from the exact Bragg conditions. A typical two-dimensional electron diffraction pattern is shown in Figure 10.1. The geometry of the diffraction pattern on the back focal plane and the spot intensities carry the structural information of the specimen.

For a TEM, the focal length of the intermediate lens can be modified through changing the intermediate lens current, which enables either the diffraction pattern in the back focal plane of the objective lens, or the image in the image plane of the objective lens, to be magnified by the intermediate lens and projector lenses.

DOI: 10.1201/9780429351662-14

Therefore, through tuning the intermediate lens current, either the magnified diffraction pattern or the magnified image can be viewed on the screen or recorded by the CCD camera.

To study an image and the electron diffraction pattern from the same region, selected electron diffraction (SAED or SAD) is a suitable method. In order to study the electron diffraction pattern from an area of interest, a selected area aperture can be inserted before switching to the electron diffraction mode. The position of the selected area aperture lies in the image plane of the objective lens (Figure III.2). During image observation, firstly, the intermediate lens current should be adjusted, so that the edge of the selected aperture can be clearly observed on the screen. Then the objective lens focus can be tuned to get a clear image of the features in the specimen. The above two steps enable the image formed by the objective lens to fall on the plane with the selected area aperture, and to be enlarged by the intermediate lens. When switching to electron diffraction mode, the objective lens current must not be changed, or the focus of the objective lens must be kept unchanged, which means the beam path above the selected area aperture is unchanged. A clear diffraction pattern can be obtained by adjusting the intermediate lens current, but not by changing the beam path above the selected area aperture. Such operations ensure that the electron diffraction pattern comes from the area of the image selected. A small error can occur due to the existence of spherical aberration and defocus.

In the conventional selected area of electron diffraction, the incident beam which strikes the specimen is a broad parallel beam, and the diffraction pattern obtained at the back focal plane is a spotty pattern. When the selected region is from a single grain aligned along a zone axis, the selected area electron diffraction pattern is associated with a weighted reciprocal lattice of that grain, such as shown in Figure 10.1. However, if the incident beam is converged onto a small region of the crystal, each spot of the diffraction pattern becomes a disk on the back focal plane, and such diffraction is called a convergent beam electron diffraction (CBED).

The electron beam employed in a TEM has short wavelengths, if compared with those of X-rays. Hence the diffraction angles are very small, due to which Bragg's equation, $2d\sin\theta = \lambda$, can be expressed in another way, or $Rd = L\lambda$, as explained below.

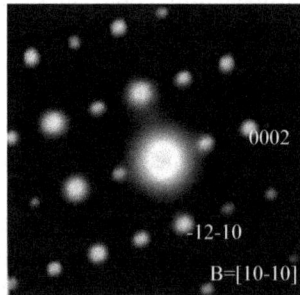

FIGURE 10.1 Electron diffraction pattern from the hydroxypaptite obtained after the precursor was hydrothermally treated at 225 °C for 6 hours.

It is known that:

- The interplanar spacing d: at the level of Å;
- The amplitude of the reciprocal lattice vectors $g = 1/d$: at the level of Å^{-1};
- The electron wavelength λ: at the level of 10^{-2} Å;
- The radius of the Ewald sphere $k_0 = 1/\lambda$: at the level of 10^2 Å^{-1}.

Therefore, typically θ_B is around the 10^{-2} radian level. When θ is small,

$$\sin\theta \approx \theta, \tan 2\theta \approx 2\theta$$

Let us look at the geometry electron diffraction in the TEM when θ is small, as shown in Figure 10.2.

The distance between the transmitted beam and the diffracted beam is shown in the diffraction pattern

$$R = L\,tg(2\theta_B) \approx L\,2\theta_B \approx L\,2\sin\theta_B = L(\lambda/d)$$

Therefore, $R = \dfrac{1}{d}(L\lambda)$, or

$$Rd = L\lambda \qquad (10.1)$$

where R stands for the distance from the origin to a diffracted spot on the diffraction pattern; L refers to the effective camera length; and $L\lambda$ represents the magnification factor for the electron microscope, also called the camera constant (see Figure 10.2).

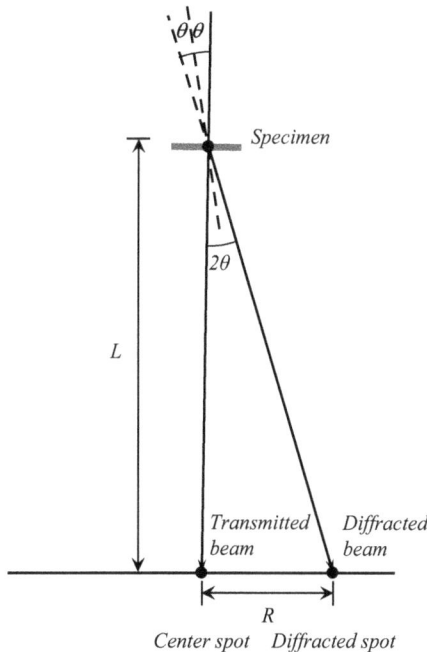

FIGURE 10.2 Geometry of electron diffraction in a TEM.

R also represents the magnified reciprocal lattice vectors, if the equation is rewritten as:

$$R = (L\lambda)\frac{1}{d} = (L\lambda)g \qquad (10.2)$$

This formula can be used to calculate the d spacing, if R is measured, although the value is not that accurate without calibration.

When indexing a two-dimensional electron diffraction viewing from the incident beam direction, for example, the pattern shown in Figure 10.1, the d spacings can be calculated based on Equation (10.2), and the angle between the two diffracting planes can be calculated based on a relevant equation of the interplanar angles, which can be found in Table 3.2. The example below will show the steps in indexing a two-dimensional electron diffraction pattern.

Example 10.1

$BaTiO_3$, having a primitive cubic lattice with $a = 4.02$ Å was examined using a TEM operating at 200 kV. Figure 10.3 is a selected area electron diffraction pattern of a zero order Laue zone (ZOLZ) from a $BaTiO_3$ crystal. Given the camera constant $L\lambda = 25.1$ mm Å, index diffraction spots A, B, and C, and determine the zone axis of the diffraction pattern.

OA = 14.0 mm, OB = 8.8 mm and OC = 14.0 mm. The angle between OA and OB is 71.6°, and the angle between OB and OC is also 71.6°.

Solution: Using the d spacing formula $d = \dfrac{a}{\sqrt{h^2 + k^2 + l^2}}$, the interplanar spacings for different crystallographic planes can be calculated. The planes and corresponding d spacings are as presented below.

hkl	d (Å)
001	4.02
011	2.84
111	2.32
002	2.01
012	1.79
112	1.64

From the electron diffraction pattern, we know that

$$d_A = (L\lambda)/R_A = \left(25.1\text{mm}\times\text{Å}\right)/14\text{ mm} = 1.79\,\text{Å},$$

suggesting that spot A is due to the diffraction from a plane in the {012} family. Similarly we have

$$d_B = (L\lambda)/R_B = \left(25.1\text{ mm}\times\text{Å}\right)/8.8\text{mm} = 2.85\,\text{Å}$$

and

$$d_C = (L\lambda)/R_C = \left(25.1\text{mm}\times\text{Å}\right)/10.1\text{mm} = 1.79\,\text{Å},$$

meaning that a plane in the {110} family gives rise to spot B, and a plane from the {012} family gives rise to spot C.

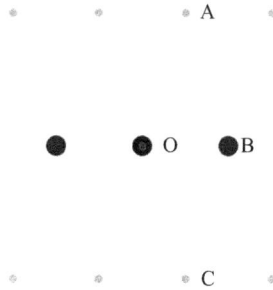

FIGURE 10.3 Selected area electron diffraction pattern from a BaTiO$_3$ crystal.

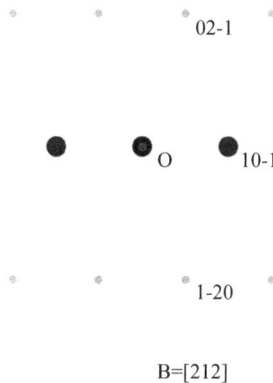

02-1

O 10-1

1-20

B=[212]

FIGURE 10.4 Selected area electron diffraction pattern of a BaTiO$_3$ crystal from zone axis [212].

In fact, there is no need to calculate d_C as the associated diffraction plane can be derived by vector operations.

Using $\cos\phi = \dfrac{\vec{g}_1 \cdot \vec{g}_2}{|\vec{g}_1||\vec{g}_2|} = \dfrac{h_1 h_2 + k_1 k_2 + l_1 l_2}{\sqrt{h_1^2 + k_1^2 + l_1^2}\ \sqrt{h_2^2 + k_2^2 + l_2^2}}$, the interplanar angles can be calculated.

The angle between $\left(10\bar{1}\right)$ and $\left(02\bar{1}\right)$ is 71.6°. This pair of planes is only one of the correct answers as a few other pairs of planes can also satisfy the condition. In TEM analysis, it is recommended to collect more diffraction patterns through tilting the crystal to different zone axes, in which case the unique correct answer can be obtained.

The zone axis calculated from $\left(10\bar{1}\right)$ and $\left(02\bar{1}\right)$ is [212], and the indexed pattern is given in Figure 10.4.

For electron diffraction, the zone axis can be calculated using the Weiss zone law

$$hu + kv + lw = 0 \tag{10.3}$$

The zone axis can be expressed as a direction vector in the real lattice, or

$$\vec{B} = u\vec{a} + v\vec{b} + w\vec{c}$$

whereas a plane giving rise to Bragg diffraction can be expressed as a vector in the reciprocal lattice, or

$$\vec{g}_{hkl} = h\vec{a}^* + k\vec{b}^* + l\vec{c}^*$$

We know that the zone axis is parallel to plane (hkl), or $\vec{B} \perp \vec{g}_{hkl}$. Therefore, $\vec{B} \cdot \vec{g}_{hkl} = 0$, or

$$hu + kv + lw = 0$$

Once the indices of two diffraction spots $(h_1 k_1 l_1)$ and $(h_2 k_2 l_2)$ in a diffraction pattern are known, the relationships

$$h_1 u + k_1 v + l_1 w = 0 \text{ and } h_2 u + k_2 v + l_2 w = 0$$

can be used for the calculation of the zone axis, which is expressed by

$$u : v : w = \left(k_1 l_2 - k_2 l_1 \right) : \left(l_1 h_2 - l_2 h_1 \right) : \left(h_1 k_2 - h_2 k_1 \right) \tag{10.4}$$

In the case where there is a common factor for u, v and w, the common factor should be removed, and the zone axis is written as $[uvw]$, or $B = [uvw]$.

The zone axis is antiparallel to the electron beam direction. If the zone axis direction is just slightly deviated from the antiparallel direction of the electron beam due to the orientation variation of the crystal under study, a diffraction pattern can still be observed. If the electron beam in the microscope travels downward, the zone axis points upward. Students should not forget that the right-hand rule is employed when choosing $\vec{g}_1$ and $\vec{g}_2$ from an electron diffraction pattern, so that $\vec{g}_1 \times \vec{g}_2$ is pointing upwards.

In the textbook *Essentials of Crystallography* (Wahab, 2009), the reciprocal lattice expression $\dfrac{\left(h_1 \vec{a}^* + k_1 \vec{b}^* + l_1 \vec{c}^* \right) \times \left(h_2 \vec{a}^* + k_2 \vec{b}^* + l_2 \vec{c}^* \right)}{V^*}$ is used to calculate the direct lattice expression for the zone axis $[uvw]$. Students are encouraged to derive the expression

$$\vec{B} = u\vec{a} + v\vec{b} + w\vec{c} = \left(k_1 l_2 - k_2 l_1 \right)\vec{a} + \left(l_1 h_2 - l_2 h_1 \right)\vec{b} + \left(h_1 k_2 - h_2 k_1 \right)\vec{c}$$

from $\vec{B} = \dfrac{\left(h_1 \vec{a}^* + k_1 \vec{b}^* + l_1 \vec{c}^* \right) \times \left(h_2 \vec{a}^* + k_2 \vec{b}^* + l_2 \vec{c}^* \right)}{V^*}$ independently. During derivation, you need to use the relationships $\dfrac{\vec{b}^* \times \vec{c}^*}{V^*} = \vec{a}$, $\dfrac{\vec{c}^* \times \vec{a}^*}{V^*} = \vec{b}$ and $\dfrac{\vec{a}^* \times \vec{b}^*}{V^*} = \vec{c}$.

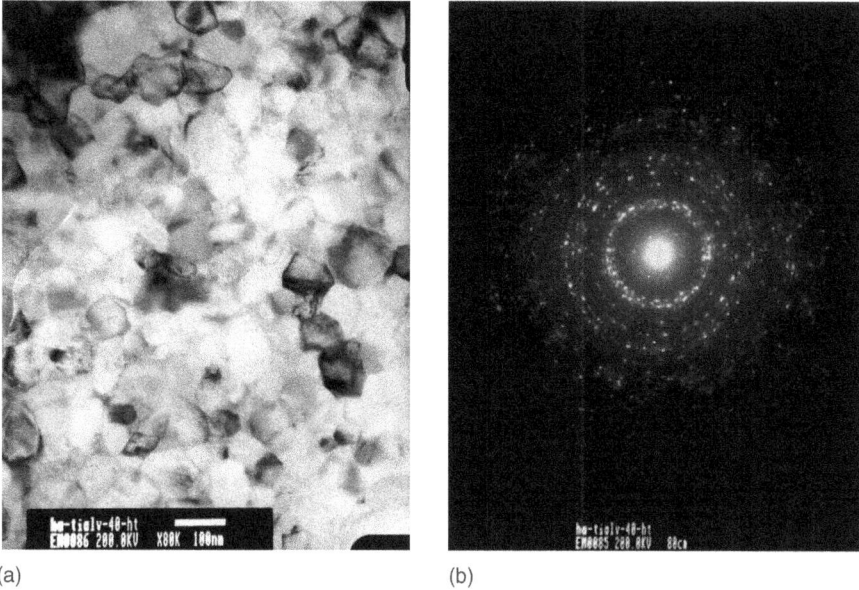

(a)   (b)

FIGURE 10.5   (a) Polycrystalline Ti6Al4V region in an HA/Ti6Al4V composite. (b) Spotty ring electron diffraction pattern from region of (a), showing the diffractions from the (1010), (0002), (1011), (1012), ... planes of the hexagonal phase in Ti6Al4V. Some weak diffraction spots are from HA nearby in the HA/Ti6Al4V composite.

(Source: Reproduced from Khor et al. (2000), with the permission of Elsevier.)

In the case where the selected area electron diffraction pattern is collected from a polycrystalline region, many single crystalline patterns from those grains that satisfy the Bragg conditions overlap, and a spotty ring pattern is formed. A spotty ring diffraction pattern from the hexagonal phase of a polycrystalline Ti6Al4V region in an HA/Ti6Al4V composite is displayed in Figure 10.5.

The diffraction pattern shown in Example 10.1 is an SAED pattern obtained using a parallel incident electron beam. For convergent beam electron diffraction, each spot becomes a disk within which variations in intensity can be seen. The fringes formed inside the disks from the transmitted beam and the diffracted beams contain very rich information about crystal symmetry.

It has been noticed that there are different ways to present the geometry of convergent beams with respect to a reciprocal lattice that is coupled with a direct lattice. We can fix the reciprocal lattice and change the Ewald sphere center (Williams, 1996; Wei, 1990) to explain the geometry of diffraction. In this case, the position of the Ewald sphere center changes while the orientation of the incident beam changes within the convergence angle 2α. As illustrated in Figure 10.6, (i) the reciprocal lattice origin O* is fixed, and (ii) the reciprocal lattice orientation is also fixed.

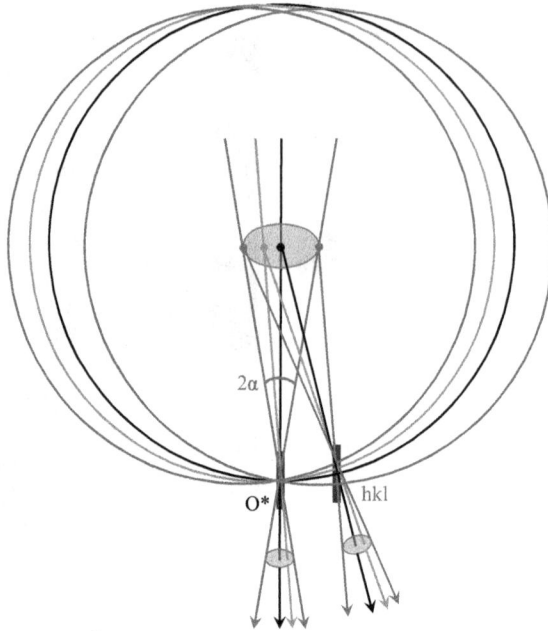

FIGURE 10.6 Schematic illustration of the beam geometry of convergent beam electron diffraction. The cone of the incident beams has the convergent angle 2α. The Ewald spheres and incident beams are coupled, and they vary together. The reciprocal lattice origin O* and the orientation of the reciprocal lattice are fixed. The central spot and *hkl* spot become disks when the incident beam is convergent. The diffracted beams pass through different parts of the *hkl* relrod with different *s* values. As the excitation error *s* is very small in comparison with the Ewald sphere radius *k*, in approximation, the diffracted beams are considered to travel through a cone.

Based on the two-beam dynamical theory to be discussed in Chapter 11, the intensity of the diffracted beam is a function of excitation error *s* when the specimen has a uniform thickness *T*:

$$I_g(T,s) = I_g(s) = \frac{\pi^2}{\xi_g^2} \frac{\sin^2(\pi s_{eff} T)}{(\pi s_{eff})^2}$$

$$I_o(s) = 1 - I_g(s)$$

Where $s_{eff} = \sqrt{s^2 + \xi_g^{-2}}$.

The above equation $I_g(s)$ can be used to explain the fringes formed inside the diffraction disk. For a known excitation error *s*, s_{eff} can be obtained through $s_{eff} = \sqrt{s^2 + \xi_g^{-2}}$. Therefore, the intensity of the diffracted beam $I_g(s)$ can be calculated. When $s_{eff}T = T\sqrt{s^2 + \xi_g^{-2}}$ is an integer, $I_g(s) = 0$, hence a dark fringe is formed in the corresponding position in the diffraction disk *hkl*.

Normally, we can measure the distance between the middle of the central bright fringe and each of the dark fringes x_i, and subsequently use $s_i = \dfrac{x_i \lambda}{R d^2}$ to calculate the excitation error. Each dark fringe in the diffraction disk hkl is a straight line perpendicular to the scattering vector $\vec{q}\left(= \vec{g}_{hkl} + \vec{s}\right)$.

To better understand the linkage between values of excitation error s and the positions x of the fringes in the diffracted disk hkl, we further use Figure 10.7 to assist the explanation.

The scattering vector $\vec{q}$ intercepts with the relrod, and $\vec{s} = \vec{q} - \vec{g}_{hkl}$. When $s_{eff} T = T\sqrt{s^2 + \xi_g^{-2}} = n_k$ (an integer number), the diffracted beams $I_g(s) = 0$. If in this condition, we fix the scattering vector $\vec{q}$ and $\vec{s}$, and only change the incident beam and diffracted beam directions, the incident beams coming from the directions of line AB as shown in the disk travel through a plane in the cone of convergent incident beams and converge to the reciprocal lattice origin O* (Figure 10.7(b)). The beam geometry shows that a dark fringe forms inside the diffracted disk hkl as $I_g(s) = 0$ at

(a)

(b)

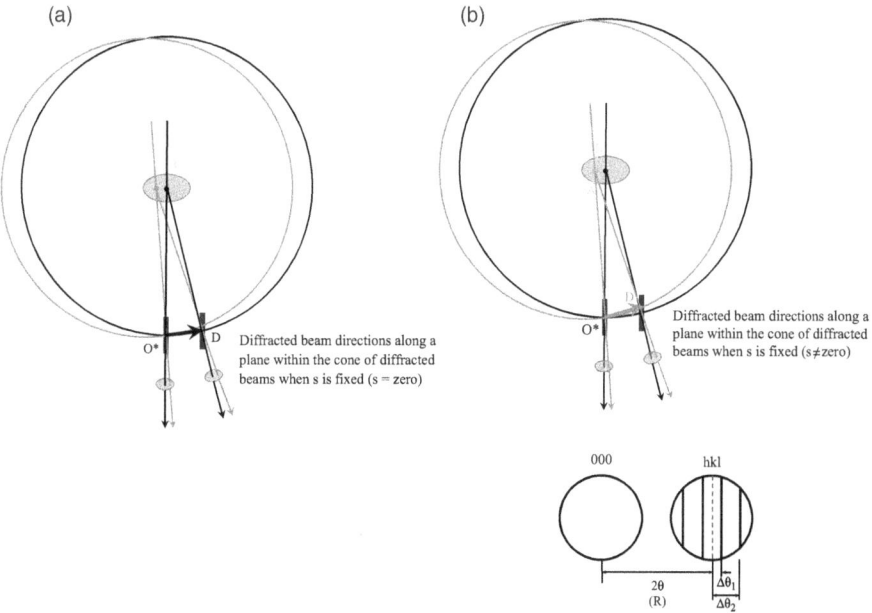

Diffracted beam directions along a plane within the cone of diffracted beams when s is fixed (s = zero)

Diffracted beam directions along a plane within the cone of diffracted beams when s is fixed (s≠zero)

000   hkl

2θ
(R) $\Delta\theta_1$
              $\Delta\theta_2$

FIGURE 10.7 Schematic diagram showing the variation of intensities with s values in the hkl disk of the K-S pattern through the relationship $I_g\left(T,s\right) = \dfrac{\pi^2}{\xi_g^2} \dfrac{\sin^2\left(\pi s_{eff} T\right)}{\left(\pi s_{eff}\right)^2}$, where $s_{eff} = \sqrt{s^2 + \xi_g^{-2}}$. (a) Beams passing through line AB satisfy the Bragg condition with $s = 0$ and $\vec{q} = \vec{g}_{hkl}$. (b) Beams passing through line AB satisfy the condition $\vec{k} - \vec{k}_0 = \vec{q} = \vec{g}_{hkl} + \vec{s}$, and s is fixed. As with any position $\Delta\theta$ with respect to the middle of the central bright fringe in the hkl disk, the s value can be calculated through the relationship $s = \dfrac{\Delta\theta \lambda}{\left(2\theta\right)d^2} = \dfrac{x\lambda}{Rd^2}$. Therefore, from the dark fringes at $\Delta\theta_1$ and $\Delta\theta_2$ with respect to the middle of the central bright fringe in the hkl disk, the s_1 and s_2 values can be calculated through the relationship $s_i = \dfrac{\Delta\theta_i \lambda}{\left(2\theta\right)d^2} = \dfrac{x_i \lambda}{Rd^2}$.

such $\bar{s}$ values associated with position D on the *hkl* relrod. In the cone of the diffracted beams, line AB converges to position D on the *hkl* relrod, and links to a dark fringe formed inside the diffracted disk. Line AB, position D and the dark fringe are within the same plane.

When the s values get large, the contrast becomes lower. The key to understanding the geometry is to fix the scattering vector $\bar{q}$, which ends at a fixed position on the relrod with a fixed deviation parameter $\bar{s}$. When the s values satisfy $T\sqrt{s^2 + \xi_g^{-2}} = n_k$ or is close to the condition, the diffracted beams with minimal intensity result in a straight dark fringe in the diffracted disk.

In geometry, this means (i) fixing the chord of the great circle of the Ewald sphere, and (ii) changing the center of the sphere. The center of the Ewald sphere moves along a line perpendicular to the chord within the range of incident beam convergence 2α. From the line, the diffracted beam directions with the fixed s value can be determined.

Using the same concept, when the Ewald sphere cuts the relrod at a HOLZ, we can also draw the scattering vector $\bar{q} = \bar{g}$. By fixing the scattering vector $\bar{g}$ on the relrod center in a HOLZ and find the diffracted beam directions within the cone of the convergent beam, the bright line position can be found in the HOLZ disk and the corresponding dark line can be found in the transmitted disk.

From CBED patterns, more information about the crystal structure and symmetries can be obtained, and detailed discussions can be found in the textbook recommended to students: Williams (1996).

10.2 INTENSITY OF DIFFRACTED BEAM

Concerning electron diffraction, we employ a similar methodology in deriving the diffracted beam intensity as in X-ray diffraction. The first step is to explain the atomic scattering factor for an electron beam; the second step is to analyze the structure factor, which is the scattering by a unit cell under the Bragg condition; and the third step is to find out the size effect of a crystallite, which is composed of many unit cells. The atomic scattering factor has been explained in detail in Chapter 9.

Using the SI system, the atomic scattering factor for an electron beam can be expressed as

$$f(\theta) = \frac{m_e e^2}{8\pi h^2 \varepsilon_0} \frac{(Z - f_X)}{s^2}$$

where $s = \dfrac{\sin\theta}{\lambda}$. In fact, electrons are negatively charged particles and they interact through Coulomb forces with the positively charged nuclei, as well as with the negatively charged electron clouds. Therefore, the atomic scattering factor for electrons contains two contributions. The first term is due to the interaction with the nucleus with an atomic number Z, whereas the second term is due to the contribution from the electron cloud associated with the atom. In Hirsch (1977), the discussions on the atomic scattering factor are based on CGS units. In the calculation, the amplitude of the electron wave scattered by an atom is about 10^4 times that of X-rays.

As discussed previously, although we can use either the crystallographic convention or the quantum mechanical convention to calculate structure factors in X-ray diffraction and electron diffraction, it is observed that the crystallographic convention is normally used in X-ray diffraction, and the quantum mechanical convention is employed in electron diffraction. The difference in the structure factor is that there is a "–" sign in the exponential term for the quantum mechanical convention.

When the electron wave is scattered by atoms at different locations in a unit cell, the phase difference can be expressed by:

$$\phi = -2\pi i \left(\vec{k} - \vec{k_0} \right) \cdot \vec{r} = -2\pi i \vec{q} \cdot \vec{r}$$

where $\vec{q} = \vec{k} - \vec{k_0} = \vec{g} + \vec{s}$ and the meaning of each parameter is indicated in Figure 10.8.

During the explanation of the structure factor in electron diffraction, deviation parameter $\vec{s}$ is introduced, which indicates how far away the electron diffraction has deviated from the Bragg condition. $\vec{s}$ is also called the excitation error, and its value range is small when the size of the crystallite is large. The rules to determine the signs of s are:

(I) s is positive when it is parallel to the beam direction, $\vec{k_0}$. This is the definition for the sign of s.

(II) s is positive when $\theta > \theta_B$.

(III) s is positive when the reciprocal lattice point is within the Ewald sphere.

(IV) s is positive when the excess hkl Kikuchi line lies just outside the hkl spot.

When $\theta = \theta_B$, $\vec{s} = 0$. In this case, $\vec{q} = \vec{g} = h\vec{a}^* + k\vec{b}^* + l\vec{c}^*$ and $\phi = -2\pi i \left(\vec{k} - \vec{k_0} \right) \cdot \vec{r} = -2\pi i \vec{g} \cdot \vec{r}$, and in geometry, it is at the exact Bragg condition.

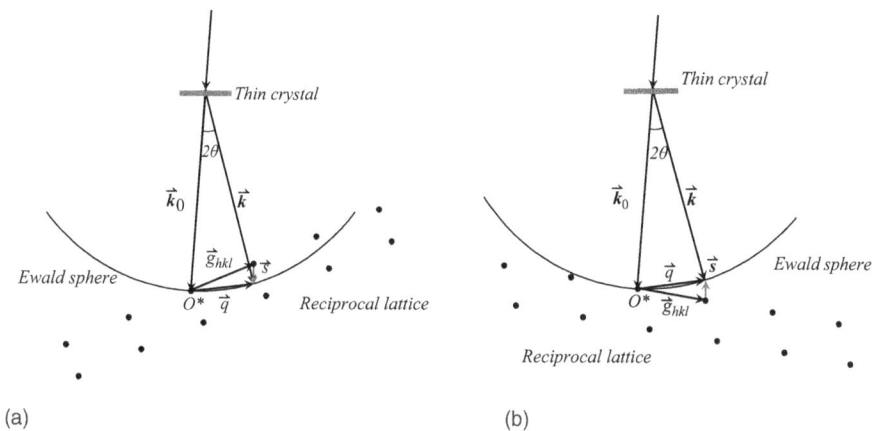

FIGURE 10.8   The Ewald sphere construction, defining incident wave vector $\vec{k_0}$, diffracted wave vector $\vec{k}$, diffraction vector $\vec{q}$ and deviation vector $\vec{s}$ when the beam is slightly away from the exact Bragg condition. (a) s is positive and (b) s is negative.

Under the Bragg condition, the amplitude of the electron wave scattered by a unit cell is defined as the structure factor. And the structure factor is expressed as

$$F_{\bar{g}} = \sum_{\substack{all \ atoms \\ per \ cell}} f_j(\theta_B)\left[\exp\left(-2\pi i \bar{g} \cdot \bar{r}_j\right)\right] \text{ or}$$

$$F_{hkl} = \sum_{\substack{all \ atoms \\ per \ cell}} f_j(\theta_B)\left\{\exp\left[-2\pi i\left(hx_j + ky_j + lz_j\right)\right]\right\} \quad (10.5)$$

where f_j is the atomic scattering factor for atom j, and $\bar{r}_j$ is the position of atom j in a unit cell.

When $\theta \neq \theta_B$, $\bar{s} \neq 0$. Then the amplitude of the electron wave scattered by a unit cell becomes

$$F_{\bar{q}} = \sum_{\substack{all \ atoms \\ per \ cell}} f_j(\theta)\exp\left(-2\pi i \bar{q} \cdot \bar{r}_j\right)$$

$$= \sum_{\substack{all \ atoms \\ per \ cell}} f_j(\theta)\exp\left[-2\pi i\left(\bar{g} + \bar{s}\right)\cdot \bar{r}_j\right] \quad (10.6)$$

When $|\bar{s}|$ is small, $\left|F_{\bar{q}}\right| \approx \left|F_{\bar{g}}\right|$.

The structure factor expressions in X-ray diffraction and electron diffraction look very similar except that there is a negative sign inside the exponential term for the structure factor in electron diffraction. However, if we compare the atomic scattering factors $f_j(\theta)$ for X-rays and electron beams, it is noticed that in X-ray scattering, the atomic scattering factor is the ratio of the amplitude of the wave scattered by one atom to the amplitude of the wave scattered by one electron. It is dimensionless for the atomic scattering factor for X-rays. For electron diffraction, the unit for the atomic scattering factor is length.

For both X-ray diffraction and electron diffraction, the allowed reflections and forbidden reflections can be found by calculating the structure factors for the associated reflections (*hkl*). Since the magnitude of the atomic scattering factor for electron beams is much higher than that for X-rays, double diffraction can occur in an electron diffraction pattern. Double diffraction means that the diffracted beam from the upper layer of the specimen can be further diffracted in the lower layer. Or we can say that the lower layer not only diffracts the direct beam that has penetrated the upper layer, but also rediffracts the diffracted beam from the upper layer. In this case, the diffracted beam in the upper layer acts as another incident beam for the lower layer.

Double diffraction can occur within one grain and between grains. In the case where it happens to the same grain, the diffraction may be observed from the forbidden reflection plane. Figure 10.9 shows an electron diffraction pattern from a ZnO nanobelt with double diffraction spots.

If the upper and lower layers are two different grains, and both of them are along their zone axes, the double diffraction pattern can occur as well. Figure 10.10 shows

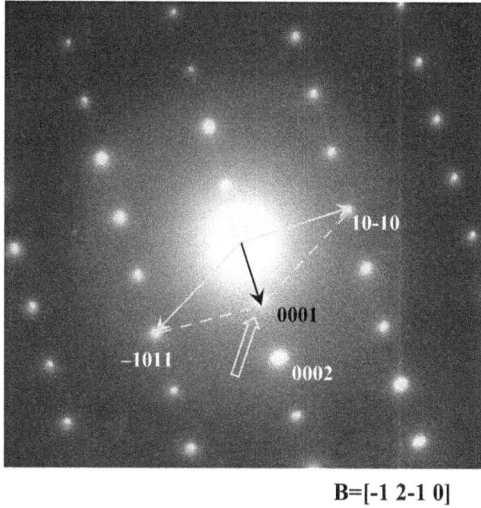

B=[-1 2-1 0]

FIGURE 10.9 Electron diffraction pattern from a [-1 2-1 0] zone axis of a zinc oxide nano-belt showing double diffraction. A bright spot appears at forbidden reflection 0001 which is due to double diffraction. 10-10 and -1011 are allowed diffractions.

(Source: Xu et al. (2005), with the permission of AIP Publishing.)

(a)   (b)   (c)

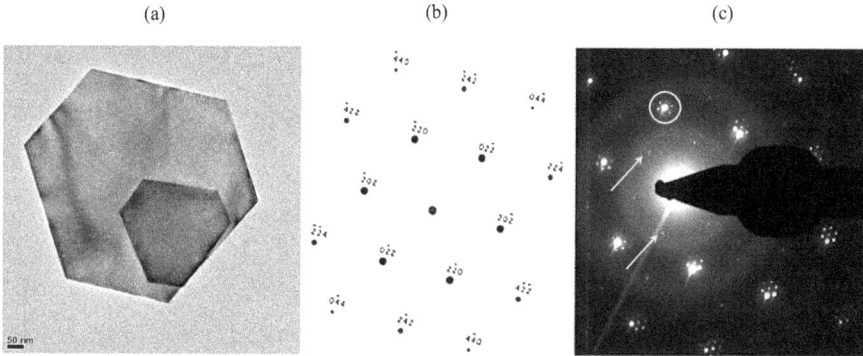

FIGURE 10.10 (a) Two gold thin plates from the [111] zone axis. (b) Indexed electron diffraction pattern of a gold single crystal. The zone axis is [111]. (c) Small hexagons (circle) formed are due to double diffraction. Spots from the higher order Laue zone (arrow) are due to the fact that plates are thin.

a double diffraction pattern from two gold nano-plates, and the synthesis method of gold plates is described in Xia et al. (2018).

As discussed previously, double diffraction is a common phenomenon in electron diffraction experiments. Therefore, caution needs to be taken during diffraction pattern indexing. When spots are observed in the forbidden planes, it is better to check whether double diffraction has occurred.

For a small crystallite, the amplitude of the scattered wave can be derived as well. When $\theta = \theta_B$, or $\vec{q} = \vec{g}$, we have $s = 0$. If the diffraction is slightly deviated from the Bragg condition, then $\vec{q} = \vec{g} + \vec{s} \neq \vec{g}$, as $s \neq 0$.

Assuming that the incident wave has a unit amplitude, or $\phi_0 = 1$, we have

$$\phi_g = \sum_{\substack{all \\ atoms}} f \exp\left(-2\pi i \vec{q} \cdot \vec{r}\right)$$

$$= \sum_{\substack{all \\ atoms}} f_j \exp\left[-2\pi i \left(\vec{g} + \vec{s}\right) \cdot \left(\vec{r}_j + \vec{r}_n\right)\right]$$

$$= \sum_{\substack{all \\ unit \\ cell}} \left\{ \sum_{\substack{all \ atoms \\ per \ cell}} f_j \exp\left[-2\pi i \left(\vec{g} + \vec{s}\right) \cdot \vec{r}_j\right] \cdot \exp\left[-2\pi i \left(\vec{g} + \vec{s}\right) \cdot \vec{r}_n\right] \right\} \qquad (10.7)$$

$$= \left\{ \sum_{\substack{all \ atoms \\ per \ cell}} f_j \exp\left[-2\pi i \left(\vec{g} + \vec{s}\right) \cdot \vec{r}_j\right] \right\} \left\{ \sum_{\substack{all \\ unit \\ cell}} \exp\left[-2\pi i \left(\vec{g} + \vec{s}\right) \cdot \vec{r}_n\right] \right\}$$

where $\vec{r}_n$ is the position of the nth unit cell with respect to the origin of the crystal.
    We know that

$$F_{\vec{g}} = \sum_{\substack{all \ atoms \\ per \ cell}} f_j \exp\left[-2\pi i \vec{g} \cdot \vec{r}_j\right]$$

and

$$F_{\vec{q}} = \sum_{\substack{all \ atoms \\ per \ cell}} f_j \exp\left(-2\pi i \vec{q} \cdot \vec{r}_j\right).$$

    When $|\vec{s}|$ is small, $|F_{\vec{q}}| \approx |F_{\vec{g}}|$. A detailed discussion on the structure factor calculation can be found in Section 6.3 in the X-ray diffraction part.
    For very small deviation parameter s, we can use $\sum_{\substack{all \ atoms \\ per \ cell}} f_j \exp\left[-2\pi i \vec{g} \cdot \vec{r}_j\right]$ to replace $\sum_{\substack{all \ atoms \\ per \ cell}} f_j \exp\left[-2\pi i \left(\vec{g} + \vec{s}\right) \cdot \vec{r}_j\right]$ as an approximation; then the error is negligible. Therefore, the expression of the amplitude of an electron beam scattered by a crystal can be rewritten as

$$\varphi_g \approx F_g \sum_{\substack{all \\ unit \\ cell}} \exp\left[-2\pi i \left(\vec{g} + \vec{s}\right) \cdot \vec{r}_n\right] \} \qquad (10.8)$$

The term, $\sum\limits_{\substack{all \\ unit \\ cell}} \exp\left[-2\pi i\left(\vec{g}+\vec{s}\right)\cdot\vec{r}_n\right]$, reflects the influence of the size and the

external shape of a crystal on the amplitude of the diffracted beam. In X-ray diffraction, the crystal shape and size effects are calculated via a summation approach. In electron diffraction, an integration approach is preferred in the analysis of the crystal shape and size effects.

To derive the expression of the crystal size effect in electron diffraction, again we can employ an idealized crystallite modeled as a parallelepiped having interaxial angles α, β and γ made up of triclinic parallelepiped unit cells with the same interaxial angles α, β and γ.

Assume that L_1, L_2 and L_3 are the edge lengths along the three cell-axes for a crystallite, and a, b and c are the edge lengths of the unit cell. Using the expression $G = \sum\limits_{\substack{all \\ unit \\ cells}} \exp\left[-2\pi i\left(\vec{g}+\vec{s}\right)\cdot\vec{r}_n\right]$, and considering $\vec{g}\cdot\vec{r}_n$ to be an integer, we have

$$G = \sum_{\substack{all \\ unit \\ cells}} \exp\left[-2\pi i\left(\vec{g}+\vec{s}\right)\cdot\vec{r}_n\right]$$

$$= \sum_{\substack{all \\ unit \\ cells}} \exp\left[-2\pi i\vec{s}\cdot\vec{r}_n\right]$$

No matter for which crystal system, $\dfrac{dxdydz}{abc}$ means the number of unit cells in the element $dxdydz$. Therefore, we can employ an integration approach to calculate G, and

$$G = \sum_{\substack{all \\ unit \\ cells}} \exp\left[-2\pi i\left(\vec{g}+\vec{s}\right)\cdot\vec{r}_n\right]$$

$$= \sum_{\substack{all \\ unit \\ cells}} \exp\left[-2\pi i\vec{s}\cdot\vec{r}_n\right]$$

$$= \int_{-L_1/2}^{L_1/2}\int_{-L_2/2}^{L_2/2}\int_{-L_3/2}^{L_3/2} \exp\left[-2\pi i\left(s_x x+s_y y+s_z z\right)\right]\frac{dxdydz}{abc}$$

$$= \frac{1}{abc}\int_{-L_1/2}^{L_1/2}\int_{-L_2/2}^{L_2/2}\int_{-L_3/2}^{L_3/2} \exp\left[-2\pi i\left(s_x x+s_y y+s_z z\right)\right]dxdydz$$

The calculation of the integral yields

$$G = \frac{1}{abc} \frac{\sin(\pi s_x L_1)}{\pi s_x} \frac{\sin(\pi s_y L_2)}{\pi s_y} \frac{\sin(\pi s_z L_3)}{\pi s_z} \qquad (10.9)$$

We can also use the number of unit cells along the cell axis directions to express G as:

$$G = G_x G_y G_z = \frac{\sin(\pi s_x L_1)}{\pi s_x a} \frac{\sin(\pi s_y L_2)}{\pi s_y b} \frac{\sin(\pi s_z L_3)}{\pi s_z c}$$

$$= \frac{\sin(\pi s_x N_1 a)}{\pi s_x a} \cdot \frac{\sin(\pi s_y N_2 b)}{\pi s_y b} \cdot \frac{\sin(\pi s_z N_3 c)}{\pi s_z c} \qquad (10.10)$$

where N_1, N_2 and N_3 are the number of unit cells along $\bar{a}$, $\bar{b}$ and $\bar{c}$ directions respectively, and

$$G_x = \frac{1}{a} \int_{-L_1/2}^{L_1/2} \exp(-2\pi i s_x x)\,dx = \frac{\sin(\pi s_x N_1 a)}{\pi s_x a}$$

$$G_y = \frac{1}{b} \int_{-L_2/2}^{L_2/2} \exp(-2\pi i s_y y)\,dy = \frac{\sin(\pi s_y N_2 b)}{\pi s_y b} \qquad (10.11)$$

$$G_z = \frac{1}{c} \int_{-L_3/2}^{L_3/2} \exp(-2\pi i s_z z)\,dz = \frac{\sin(\pi s_z N_3 c)}{\pi s_z c}$$

For orthorhombic, tetragonal and cubic systems, $\alpha = \beta = \gamma = 90°$, and abc is the unit cell volume V_C. Hence, G can also be written as

$$G = \frac{1}{V_c} \frac{\sin(\pi s_x L_1)}{\pi s_x} \frac{\sin(\pi s_y L_2)}{\pi s_y} \frac{\sin(\pi s_z L_3)}{\pi s_z} \qquad (10.12)$$

Equation (10.12) cannot be used for other crystal systems. In other crystal systems, for instance for the triclinic system,

$$abc = \frac{V_C}{\sqrt{1 - \cos^2\alpha - \cos^2\beta - \cos^2\gamma + 2\cos\alpha \cos\beta \cos\gamma}} = V_C / K_{cell}$$

Based on the parallelepiped crystal shape assumption, the intensity of the diffracted beam is:

$$I = |F|^2 |G|^2 = \frac{|F|^2}{(abc)^2} \frac{\sin^2(\pi s_x L_1)}{(\pi s_x)^2} \frac{\sin^2(\pi s_y L_2)}{(\pi s_y)^2} \frac{\sin^2(\pi s_z L_3)}{(\pi s_z)^2}$$

$$= |F|^2 \frac{\sin^2(\pi s_x N_1 a)}{(\pi s_x a)^2} \frac{\sin^2(\pi s_y N_2 b)}{(\pi s_y b)^2} \frac{\sin^2(\pi s_z N_3 c)}{(\pi s_z c)^2} \qquad (10.13)$$

Unlike in X-ray diffraction, the calculation of factor G in electron diffraction is via an integration approach instead of a summation approach.

Assuming that the incident electron wave amplitude $\varphi_0 = 1$, then the scattered wave from a crystallite is expressed as

$$\psi_g(r) = F_g G \frac{\exp(2\pi i k r)}{r} \tag{10.14}$$

F depends on the positions and type of atoms inside the unit cell. G depends on the size and the external shape of the crystal. The domain around each reciprocal lattice point is expressed by the factor, G^2. In the X-ray diffraction part, a spherical crystallite is used to express the effects of crystal size and shape (see Figure 6.12). In the case where the diffraction is from a thin crystalline disk, each reciprocal lattice point becomes a reciprocal lattice rod as shown in Figure 10.11.

According to the geometry of diffraction and the intensity of diffracted beams discussed above, it can be seen that the structure factor calculation indicates the forbidden and allowed diffraction planes for electron diffraction, similar to the case for X-ray diffraction. The discussion in this chapter is very helpful for understanding the different types of diffraction patterns, such as: (i) diffraction patterns from single crystals, including HOLZ patterns, super lattice patterns, double diffraction patterns, Kikuchi patterns and CBED patterns; (ii) diffraction patterns from twins; and (iii) diffraction patterns from polycrystalline materials, including materials with preferred orientations. The diffraction from amorphous materials shows diffused ring patterns, and the interpretation for such ring patterns is different from that for crystalline materials.

The electron diffraction pattern from a polycrystalline area is just the overlap of many single crystal diffraction patterns, resulting in the formation of the spotty ring pattern.

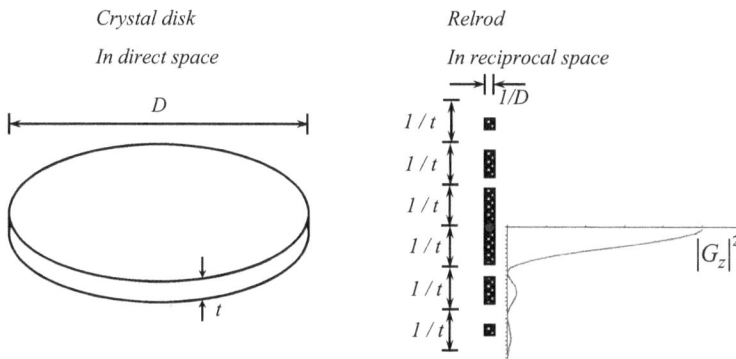

FIGURE 10.11 Schematic representation of the relationships between a crystalline disk in direct space and the relrod associated with the reciprocal lattice points. When the Ewald sphere cuts the extended reciprocal lattice points, diffraction occurs.

Example 10.2

A final year undergraduate student wants to identify whether the powder sample he received is $BaTiO_3$ or $CaTiO_3$. $BaTiO_3$ and $CaTiO_3$ have a primitive cubic lattice with lattice parameters a = 4.02 Å and 3.84 Å respectively. In a selected area electron diffraction study, a ring pattern was obtained at a camera length of L = 1000 mm, as shown in Figure 10.12. Determine whether the electron diffraction pattern is from $BaTiO_3$ or from $CaTiO_3$, and index the pattern. The accelerating voltage of the TEM is 200 kV (λ = 0.0251 Å).

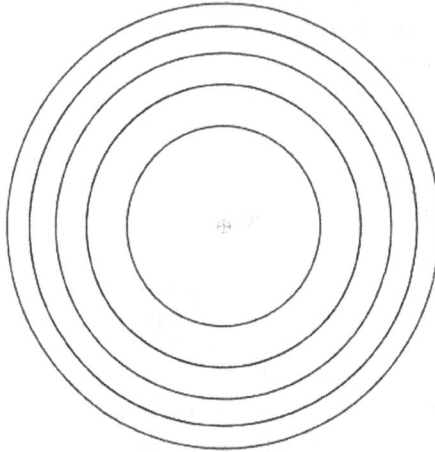

FIGURE 10.12 Selected area electron diffraction from a powder sample and the measured radius values for the first four rings (from inner to outer ring): $R1$ = 6.5 mm, $R2$ = 9.2 mm, $R3$ = 11.3 mm and $R4$ = 13.1 mm.

Solution: Based on the ring pattern obtained from the electron diffraction experiment, the d spacings can be calculated using $Rd = L\lambda$ when the R values are measured from the ring pattern. The results are:

$$d_1 = (L\lambda)/R_1 = 25.1/6.5\,\text{Å} = 3.86\,\text{Å}$$

$$d_2 = (L\lambda)/R_2 = 25.1/9.2\,\text{Å} = 2.73\,\text{Å}$$

$$d_3 = (L\lambda)/R_3 = 25.1/11.3\,\text{Å} = 2.22\,\text{Å}$$

$$d_4 = (L\lambda)/R_4 = 25.1/13.1\,\text{Å} = 1.92\,\text{Å}$$

For a cubic lattice, we can use $d = \dfrac{a}{\sqrt{h^2+k^2+l^2}}$ to obtain the theoretical d values, and the results are:

{*hkl*}	*d* ($BaTiO_3$)/Å	*d* ($CaTiO_3$)/Å
100	4.02	3.84
110	2.84	2.71
111	2.32	2.21
200	2.01	1.92

To compare the experimental d spacings with the theoretical values, it can be seen that the ring pattern is from $CaTiO_3$, not from $BaTiO_3$.

Figure 10.13 shows the indexed spotty ring pattern.

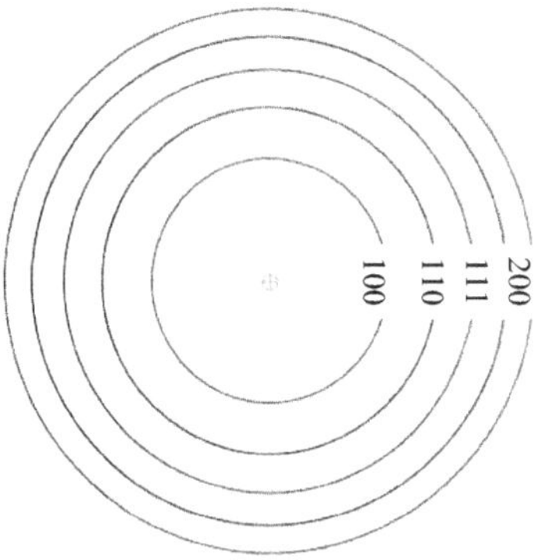

FIGURE 10.13 Indexed spotty ring pattern from $CaTiO_3$ powder sample.

Example 10.3

Hydroxyapatite (HA) is an important inorganic biomaterial. During materials processing and materials characterization, an impurity calcium oxide (CaO) phase is formed. In the TEM study, selected area electron diffraction was conducted. In one of the regions studied, a ring diffraction pattern was observed as shown in Figure 10.14(a). In its nearby region, a single crystal diffraction pattern was obtained as shown in Figure 10.14(b).

The diffraction patterns are taken at an accelerating voltage of 200 kV (λ = 0.0251 Å). The camera length for the spotty ring pattern is L = 500 mm, and that for the single crystal pattern is 1000 mm. The crystal structure data for CaO and HA are obtained from the Inorganic Crystal Structure Database (ICSD).

The space group for CaO is Fm-3m, and its lattice constant is a = 4.8360 Å. The d spacing values for CaO given below are calculated based on ICSD 180198.

{*hkl*}	*d* (CaO)/Å
111	2.79
002	2.42
022	1.71
113	1.46
222	1.40
004	1.21

Fundamentals of Crystallography

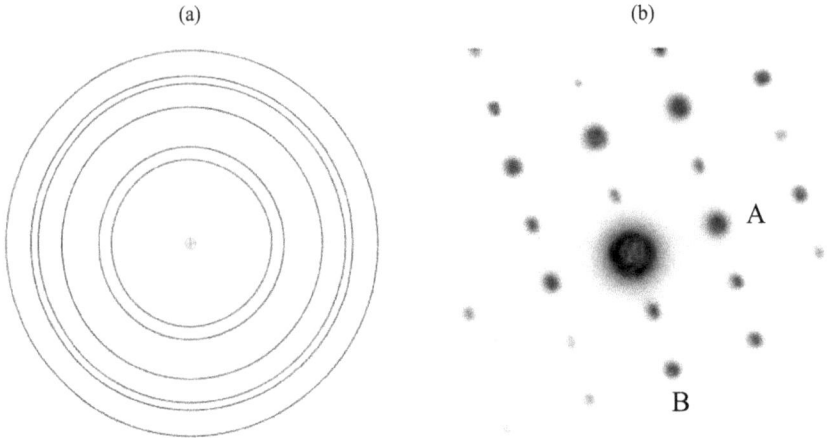

FIGURE 10.14 Selected area electron diffraction patterns (a) from a region of poly-crystals and (b) from a single grain.

In Figure 10.14(a), the measured radius values for the first five rings $R1 = 4.5$ mm $R2 = 5.2$ mm $R3 = 7.3$ mm $R4 = 8.6$ mm $R5 = 9.0$ mm	In Figure 10.14(b), the distances between the central and diffraction spots $OA = 5.3$ mm $OB = 7.3$ mm $OA \perp OB$

The space group for HA is P6$_3$/m, and its lattice constants are $a = 9.4240$ Å, $c = 6.8790$ Å. The d spacing values for HA are calculated based on ICSD 26204.

{hkl}	{hkil}	d (HA)/Å
010	01-10	8.16
011	01-11	5.26
110	11-20	4.71
020	02-20	4.08
111	11-21	3.89
021	02-21	3.51
002	0002	3.44
012	01-12	3.17
120	12-30	3.08
121	12-31	2.81

The equations for d spacing and the interplanar angle for the cubic lattice are

$$d_{hkl} = \frac{a}{\sqrt{h^2 + k^2 + l^2}}, \quad \cos\phi = \frac{h_1 h_2 + k_1 k_2 + l_1 l_2}{\sqrt{h_1^2 + k_1^2 + l_1^2}\sqrt{h_2^2 + k_2^2 + l_2^2}}$$

The equations for d spacing and the interplanar angle for the hexagonal lattice are

$$\cos\phi = \frac{h_1h_2 + k_1k_2 + \frac{1}{2}(h_1k_2 + h_2k_1) + \frac{3a^2}{4c^2}l_1l_2}{\sqrt{h_1^2 + k_1^2 + h_1k_1 + \frac{3a^2}{4c^2}l_1^2}\sqrt{h_2^2 + k_2^2 + h_2k_2 + \frac{3a^2}{4c^2}l_2^2}}$$

$$\frac{1}{d_{hkl}^2} = \frac{4}{3}\left(\frac{h^2 + hk + k^2}{a^2}\right) + \frac{l^2}{c^2}.$$

(I) Based on the crystal structure data, determine whether the ring pattern is from CaO or HA. Index the pattern.

(II) Determine whether the single crystal diffraction pattern is from CaO, HA or something else. Please index the pattern.

Solution to part (I): Based on the ring pattern obtained from the electron diffraction experiment, d spacings can be calculated using $Rd = L\lambda$ when R values are measured from the ring pattern. The results are:

$$d_1 = (L\lambda)/R_1 = (12.55/4.5)\,\text{Å} = 2.79\,\text{Å}$$

$$d_2 = (L\lambda)/R_2 = (12.55/5.2)\,\text{Å} = 2.41\,\text{Å}$$

$$d_3 = (L\lambda)/R_3 = (12.55/7.3)\,\text{Å} = 1.71\,\text{Å}$$

$$d_4 = (L\lambda)/R_4 = (12.55/8.6)\,\text{Å} = 1.46\,\text{Å}$$

$$d_5 = (L\lambda)/R_4 = (12.55/9.0)\,\text{Å} = 1.40\,\text{Å}$$

To compare the experimental and the database d spacing values, it is noticed that the ring pattern is from CaO, not from HA.

Figure 10.15 shows the indexed pattern.

Solution to part (II): Based on the electron pattern obtained from the electron diffraction experiment, the d spacings can be calculated using $Rd = L\lambda$ when the R values are measured. The results are:

$d_A = (L\lambda)/R_A = (25.1\ \text{mm Å})/(5.3\ \text{mm}) = 4.73\ \text{Å}$
$d_B = (L\lambda)/R_B = (25.1\ \text{mm Å})/(7.3\ \text{mm}) = 3.44\ \text{Å}$

The above results suggest that spot A is due to the diffraction from a plane in the {110} family, or the {11-20} family in the four-index scheme. Spot B is from a plane in the {002} family, or the {0002} family in the four-index scheme.
Using,

$$\cos\phi = \frac{h_1h_2 + k_1k_2 + \frac{1}{2}(h_1k_2 + h_2k_1) + \frac{3a^2}{4c^2}l_1l_2}{\sqrt{h_1^2 + k_1^2 + h_1k_1 + \frac{3a^2}{4c^2}l_1^2}\sqrt{h_2^2 + k_2^2 + h_2k_2 + \frac{3a^2}{4c^2}l_2^2}}$$

the interplanar angles can be calculated.

The angle between (110) and (002) is 90°. In the four-index system, the planes are (11-20) and (0002).

The zone axis calculated is [−110], or [−1100], and the indexed pattern is given in Figure 10.16.

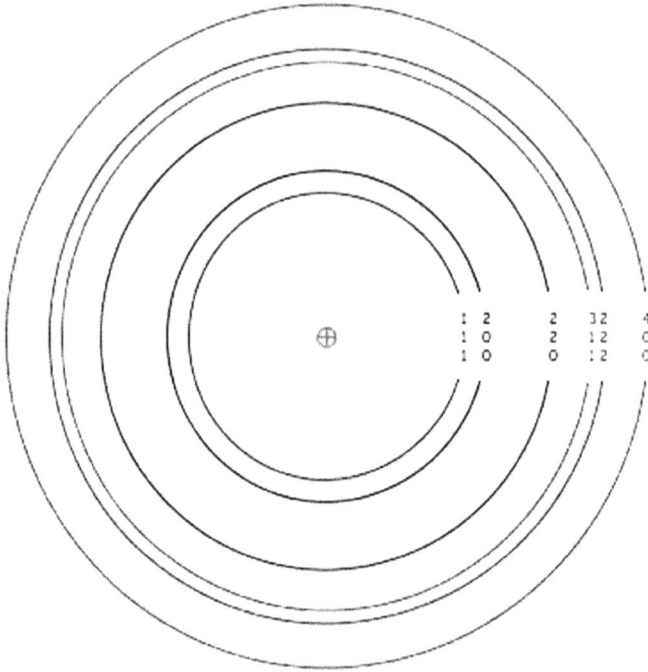

FIGURE 10.15 Indexed spotty ring pattern from CaO polycrystalline region.

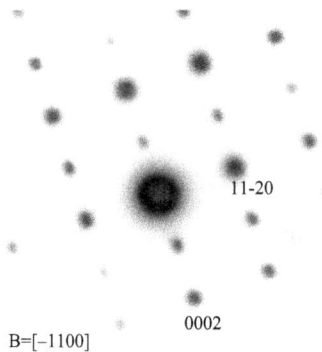

FIGURE 10.16 Electron diffraction pattern from the [10-10] zone axis of hydroxypaptite.

From Example 10.3, we notice that, for a hexagonal structure, we can do the indexing using the three-index scheme first, as the d spacing calculation and interplanar angle calculation are based on this scheme. After that, we can convert the indices from the three-index scheme to the four-index scheme (see Chapter 1). If some students prefer using the three-index scheme for hexagonal crystal systems, no conversion is needed.

In materials research, selected area electron diffraction is an effective method to obtain the orientation relationship between two phases. In a collaborative research project of an epitaxial ferroelectric tunnel junction on silicon (Li et al., 2014), selected area electron diffraction patterns were taken from an area including the strontium titanate perovskite phase and the silicon substrate (Figure 10.17). Based on the analysis of the selected area electron diffraction patterns, an HRTEM image and

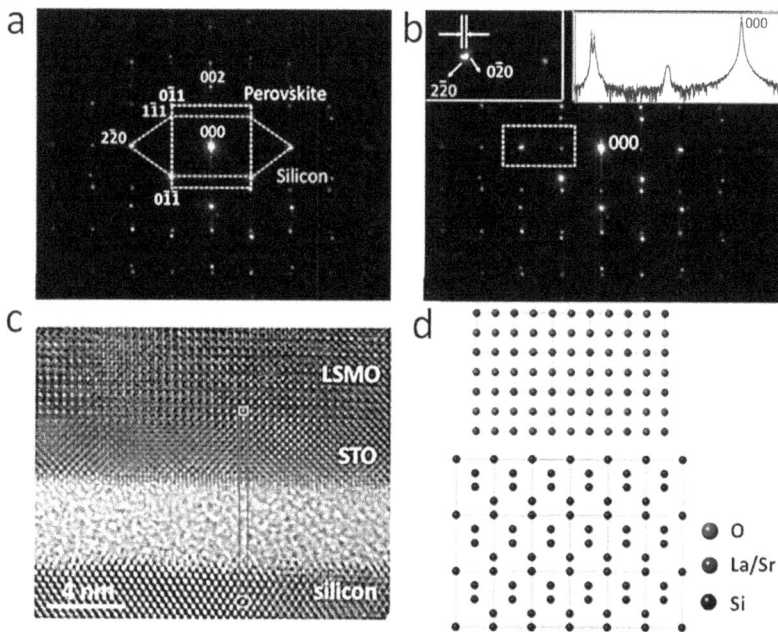

FIGURE 10.17   The orientation relationship between a silicon substrate and perovskite coating, $(010)_{perovskite}//(-110)_{Si}$, $[100]_{perovskite}//[110]_{Si}$, revealed by SAED and HRTEM images. (a) The indexed SAED pattern, showing both the diffraction spots from the perovskite crystal along the [100] zone axis, and the silicon along the [110] zone axis. (b) The SAED pattern of the silicon based FTJ junction, the dashed, rectangular region is enlarged and shown in the inset in the top left corner, the accurate distances from (000) to the silicon (2$\bar{2}$0) and perovskite (0$\bar{2}$0) spots are determined from their intensity maxima plot shown in the top right corner. (c) An HRTEM image of the perovskite crystals and silicon. The yellow square indicates a unit cell in LSMO, and the yellow hexagon similarly marks a unit cell in the silicon. (d) The crystal model which is to show the crystal relationship between perovskite and silicon is constructed based on the analysis results from the HRTEM image and SAED pattern.

(Source: Li et al. (2014), with the permission of John Wiley & Sons.)

the crystal structure data, the orientation relationship between the silicon substrate and the perovskite coating was revealed, which can be expressed as $(010)_{perovskite}//(-110)_{Si}$, $[100]_{perovskite}//[110]_{Si}$. This method is often used to analyze the orientation relationships when precipitates are formed in a matrix phase. In our $(Ti_{0.6}Al_{0.4})_{1-x}Y_xN$ multilayer coating research project, the observed orientation relationship between a rock salt type crystal and the wurtzite type secondary phase was $(1-10)_R//(10-3)_W$ and $[11-2]_R//[311]_W$ (Wang et al., 2017).

Example 10.4

$SrTiO_3$, having a primitive cubic lattice with $a = 3.905$ Å, was examined using a TEM operating at 200 kV. Figure 10.18 is a selected area electron diffraction pattern of the ZOLZ from a thin crystal of $SrTiO_3$. Given the camera constant $L\lambda = 25.1$ mm Å, index diffraction spots A, B and C, and determine the zone axis of the diffraction pattern.

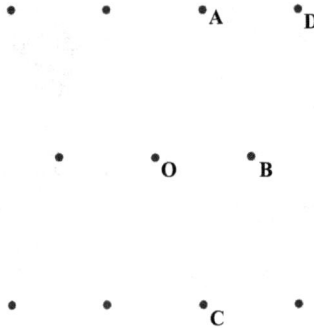

FIGURE 10.18 Selected area electron diffraction pattern from a thin film of $SrTiO_3$.

OA = 14.4 mm, OB = 9.1 mm and OC = 14.4 mm. The angle between OA and OB is 71.6°, and the angle between OB and OC is also 71.6°.

Solution: Based on the electron pattern obtained from the electron diffraction experiment, d spacings can be calculated using $Rd = L\lambda$ when R values are measured. The results are:

$$d_A = (L\lambda)/R_A = \left(25.1 mm Å\right)/(14.4 mm) = 1.74 Å$$

$$d_B = (L\lambda)/R_B = \left(25.1 mm Å\right)/(9.1 mm) = 2.76 Å$$

$$d_C = (L\lambda)/R_C = \left(25.1 mm Å\right)/(14.4 mm) = 1.74 Å$$

Using the d spacing formula $d = \dfrac{a}{\sqrt{h^2 + k^2 + l^2}}$, the interplanar spacings for different crystallographic planes can be calculated and the values are:

{hkl}	d (SrTiO₃)/Å
001	3.90
011	2.76
111	2.25
002	1.95
012	1.74
112	1.59

The above results suggest that spots A and C are due to the diffraction from planes in the {012} family, and spot B is from a plane in the {110} family.

Using $\cos\phi = \dfrac{\vec{g}_1 \cdot \vec{g}_2}{|\vec{g}_1||\vec{g}_2|} = \dfrac{h_1 h_2 + k_1 k_2 + l_1 l_2}{\sqrt{h_1^2 + k_1^2 + l_1^2}\sqrt{h_2^2 + k_2^2 + l_2^2}}$, the interplanar angles can be calculated.

The angle between $\left(10\bar{1}\right)$ and $\left(02\bar{1}\right)$ is 71.6°.

The zone axis calculated from $\left(10\bar{1}\right)$ and $\left(02\bar{1}\right)$ is [212], and the indexed pattern is given in Figure 10.19.

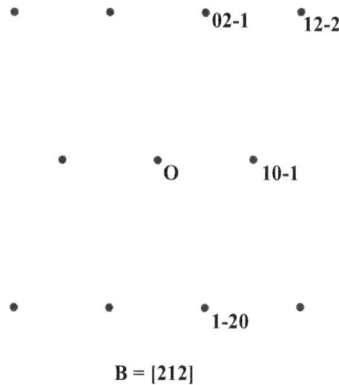

B = [212]

FIGURE 10.19 Indexed electron diffraction pattern of SrTiO₃.

The above is a correct answer.

Remarks: While resolving the problem, a student chose another pair of planes (110) and $\left(\bar{1}20\right)$, and the angle between (110) and $\left(\bar{1}20\right)$ was also 71.6°. However, when the pattern was indexed, a confusion occurred.

Figure 10.20 shows the problems, which should be avoided. The problem is explained as follows.

If we assign the B and A as (110) and $\left(\bar{1}20\right)$ reflections as shown in Figure 10.20(a), we can see that OA+OB is the (030) reflection as presented in Figure 10.20(b). In this

(a)

-120   030

O   110

2-10

B = [001]

(b)

-120   030

020

010

O   110

2-10

B = [001]

(c)

010

O   110

100

B = [001]

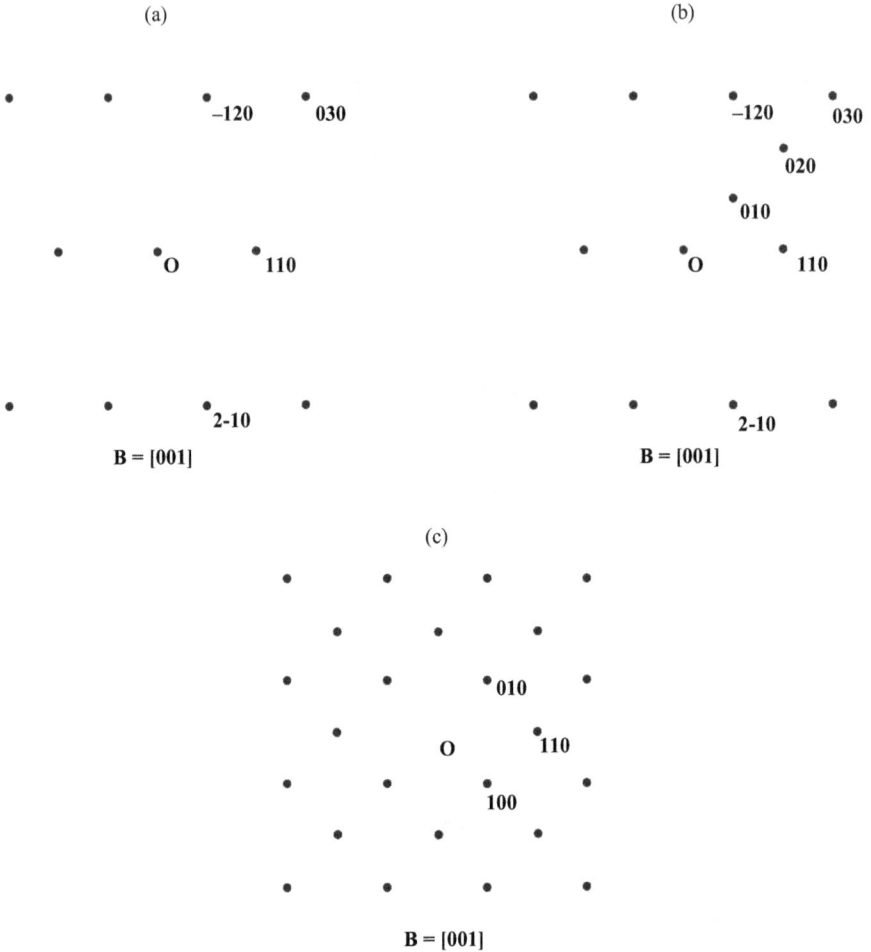

FIGURE 10.20 Explanation of a wrong indexing of the selected area electron diffraction pattern from a thin film of SrTiO$_3$. (a) ($h_1k_1l_1$) and ($h_2k_2l_2$) reflections assigned to spot A and B. (b) Spots generated through the vectorial additions, showing confusions due to the existence of the spots between O and D, which are not observed in the experiment. (c) The resultant diffraction pattern. This is from the [001] zone axis, and does not match the observed diffraction pattern shown in Figure 10.18.

case, inside (030), there should be reflections (010) and (020), and the pattern should be displayed as in Figure 10.20(c). The experimental observation shows that there were no reflections between the center spot and spot D, hence the indexing is not correct.

This can happen to crystals with high symmetries. Therefore, it is recommended to check whether the solution is reasonable even if the calculations look fine. The intensities for different spots can also give some suggestions as to whether the indexing is reasonable or not. However, due to double diffractions (or multiple diffractions), it is not easy to quantitatively analyze the spot intensities.

Once again we want to remind our students of the following. After indexing a diffraction pattern, check again whether other reflections generated by vector operations are reasonable. Check further as to whether double diffraction spots appear at the forbidden reflections. Check through the TEM software, e.g. JEMS: what are the relative intensities of different spots?

SUMMARY

Powder X-ray diffraction provides the average crystal structure of a polycrystal or powder sample, whereas electron diffraction reveals the local structure of crystals within a grain or a collection of grains. The wavelengths for electron beams are normally at the 10^{-2} angstrom level, and the radius of the Ewald sphere is much larger in comparison with the reciprocal lattice vectors. The portion of the Ewald sphere cutting the relrods near the reciprocal lattice origin O* looks very flat. The view of the diffraction pattern along the incident electron beam direction reflects the geometry of the weighted reciprocal lattice points.

During electron diffraction analysis, it is recommended to collect more diffraction patterns through tilting the crystal to different zone axes, in which case the three-dimensional structural information can be obtained.

For a very thin crystal, the selected area electron diffraction pattern may contain both ZOLZ and HOLZ information as the relrods are long. The HOLZ information is very helpful in resolving the three-dimensional crystal structure.

An important feature that electron diffraction has is double diffraction. Some forbidden reflections may be observed due to double diffraction as the diffracted beams have relatively high intensity.

Convergent beam electron diffraction is a powerful technique to study the symmetries in a crystal. This chapter has covered CBED beam geometry, which enables a better understanding of beam intensity as a function of the deviation parameter s inside the incident beam disk and inside the diffracted beam disk.

REFERENCES

Hirsch, P. B. (1977) *Electron microscopy of thin crystals/P. B. Hirsch, A. Howrie, R. B. Nicholson, D. W. Pashley and M. J. Whelan.* Malabar, FL.

Khor, K. A. et al. (2000) 'Microstructure investigation of plasma sprayed HA/Ti$_6$Al$_4$V composites by TEM', *Materials Science and Engineering A*, 281(1–2), pp. 221–228. doi:10.1016/s0921-5093(99)00717-0.

Li, Z. et al. (2014) 'An epitaxial ferroelectric tunnel junction on silicon', *Advanced Materials*, 26(42), pp. 7185–7189. doi:10.1002/adma.201402527.

Wahab, M. A. (Mohammad A.) (2009) *Essentials of crystallography/M.A. Wahab.* Oxford, UK: Alpha Science International.

Wang, J. et al. (2017) 'The yttrium effect on nanoscale structure, mechanical properties, and high-temperature oxidation resistance of (Ti0.6Al0.4)1−xYxN multilayer coatings', *Metallurgical and Materials Transactions A: Physical Metallurgy and Materials Science*, 48(9), pp. 4097–4110. doi:10.1007/s11661-017-4187-6.

Wei, Q. J. (1990) *Electron microscopy analysis of materials/Wei, Q. J.* Beijing: Chinese Metallurgical Industry Publisher.

Williams, D. B. (1996) *Transmission electron microscopy [electronic resource]: A Textbook for Materials Science/by David B. Williams, C. Barry Carter*. Edited by C. B. Carter. Boston, MA: Springer US. doi:10.1007/978-1-4757-2519-3.

Xia, J. et al. (2018) 'Morphological growth and theoretical understanding of gold and other noble metal nanoplates', *Chemistry – A European Journal*, 24(58), pp. 15589–15595. doi:10.1002/chem.201802372.

Xu, C. X. et al. (2005) 'Magnetic nanobelts of iron-doped zinc oxide', *Applied Physics Letters*, 86(17), pp. 1–3. doi:10.1063/1.1919391.

11 Diffraction Contrast

For a thin crystalline specimen, Bragg diffraction creates variations in the diffracted wave amplitude, ϕ_g, and the transmitted wave amplitude, ϕ_0, along the specimen thickness direction, or the z direction. At the exit surface of the specimen, both the direct wave amplitude and diffracted wave amplitude carry the local structural information of the specimen. When the structure of the material varies across the specimen, both transmitted beam and diffracted beam intensity vary across the specimen, or the x–y plane on the exit surface of the specimen. During the operation of TEM, choosing one of the beams using the objective aperture can obtain images with contrast caused by the intensity variation of the transmitted beam or diffracted beam across the specimen's exit surface. When the transmitted beam is chosen by the objective aperture near the back focal plane, the contrast of the image formed at the image plane is due to the variation of the direct beam intensity at the exit surface. Then the image is called a bright field image. In the case where a diffracted beam is chosen by the objective aperture, the contrast of the image formed at the image plane reflects the intensity of that diffracted beam across the exit surface of the thin specimen. Then the image is called a dark field image.

Diffraction contrast and the appearance of features in bright field and dark field images depend sensitively on how the Bragg condition is satisfied, which diffraction is active and the value of the deviation parameter. Defects in a crystal can be revealed in diffraction contrast images. If the transmitted beam or any of the diffracted beams is selected by the objective lens aperture in the back focal plane of the objective lens, only the selected beam can pass through the aperture, contributing to the image formed on the image plane of the objective lens. This image displaying diffraction contrast is enlarged by the intermediate and projector lenses, and can be observed on the viewing screen, or collected by the CCD camera. The intensity (square of the amplitude) variation of the selected beam across the specimen contains the crystal structure information of the specimen. Depending on the crystal orientation and which beam is selected by the objective aperture, either a bright field image, a dark field image or a weak beam dark field image can be obtained. Figure 11.1 is a bright field image from a hydroxyapatite coating showing diffraction contrast.

This assumes that the incident electron beam illuminates a thin crystalline specimen as shown in Figure 11.2.

From the figure we know that there are dz/V_{cell} unit cells per unit area in an element of thickness dz, each of which scatters with the structure amplitude F_g. We can use the scattered wavefront with a scattering angle $\theta_S = 2\theta$ to calculate the contribution $d\phi_g$ of the layer dz to the diffracted amplitude at the point P (Reimer, 1997). For a unit area in the wavefront, the corresponding area in the dz layer is $\dfrac{1}{\cos\theta_S}$. Therefore, for an area element dS in the wavefront, the corresponding area in the dz layer is $\dfrac{1}{\cos\theta_S}dS$, and the corresponding volume is $\dfrac{dS}{\cos\theta_S}\cdot dz$. Since θ_S is very small,

DOI: 10.1201/9780429351662-15

(a)   (b)

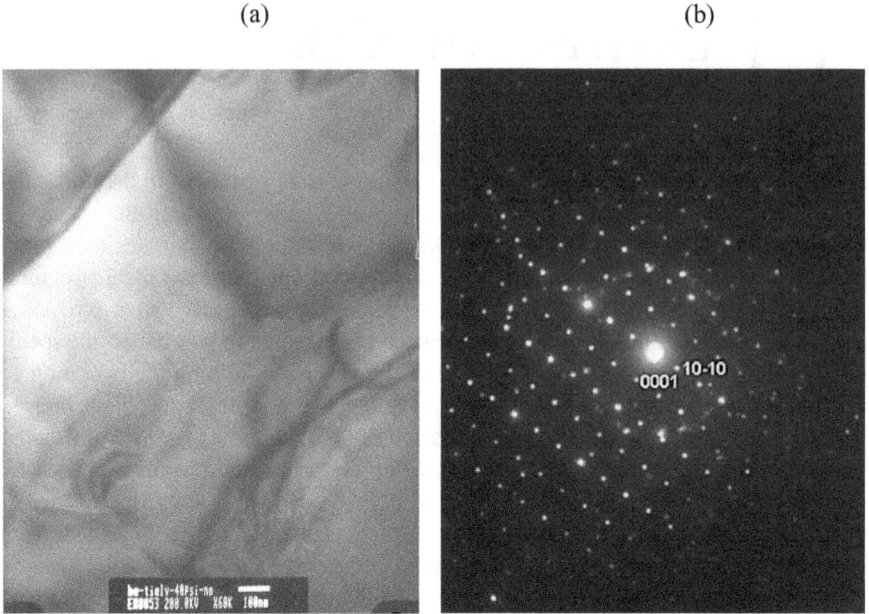

FIGURE 11.1 (a) Bright field image from a hydroxyapatite region showing diffraction contrast. Variation of intensities of different grains are due to different orientations. (b) Selected area electron diffraction pattern from the HA [-1 2-1 0] zone axis.

(Source: Khor et al. (2000), with the permission of Elsevier.)

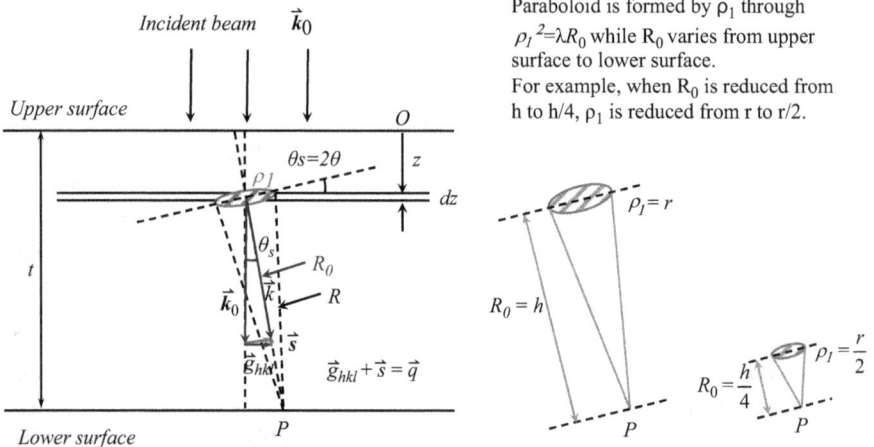

Paraboloid is formed by ρ_1 through $\rho_1{}^2=\lambda R_0$ while R_0 varies from upper surface to lower surface. For example, when R_0 is reduced from h to h/4, ρ_1 is reduced from r to r/2.

FIGURE 11.2 The first Fresnel zone located at R_0, or depth z, for calculating the amplitude of a diffracted wave at a point P at the lower surface of a crystal. The intensity of the beams at point P is influenced by all the scattering within a cone, the section of which is $\pi\rho_1{}^2$. ρ_1 varies with R_0 from the top surface to the lower surface through $\rho_1{}^2 = \lambda R_0$, forming a paraboloid. In TEM, a column is employed to replace the cone.

(Source: Reimer (1997), with the permission of Springer Nature.)

$\cos\theta_S = \cos 2\theta \approx 1$ and $\cos\theta \approx 1$. At the exact Bragg condition, $\theta_S = 2\theta_B$. When there is a deviation, the excitation error $\bar{s} \neq 0$, then $\theta \approx \theta_B$ and $\theta_S = 2\theta \approx 2\theta_B$.

Assume that the incident electron beam has an amplitude ϕ_0, then the diffracted amplitude from an element of area dS in the wavefront is $\phi_0 \dfrac{\left(\dfrac{dS}{\cos\theta_S} \cdot dz\right)}{V_{cell}} F_g$.

Hence

$$d\phi_g = \int_S \phi_0 \frac{1}{\cos\theta_S} \frac{1}{V_{cell}} F_g \frac{e^{2\pi ikR}}{R} dSdz = \phi_0 \frac{F_g dz}{\cos\theta_S V_{cell}} \int_S \frac{e^{2\pi ikR}}{R} dS$$

From Figure 11.2, we know that

$$\rho^2 = R^2 - R_0^2$$
$$S = \pi\rho^2 = \pi R^2 - \pi R_0^2$$
$$dS = 2\pi R dR$$

We also know that, when we calculate the integral, we only consider the first Fresnel zone as shown in Figure 11.2. The integral value from the wavefront at thickness z is only half of the integral value of the first Fresnel zone (Reimer, 1997). Therefore,

$$d\phi_g = \phi_0 \frac{F_g dz}{\cos\theta V_{cell}} \int_S \frac{e^{2\pi ikR}}{R} dS = \phi_0 \frac{F_g dz}{\cos\theta V_{cell}} \left(\frac{1}{2} \int_{R_0}^{R_0+\frac{\lambda}{2}} 2\pi e^{2\pi ikR} dR\right)$$

$$= i\phi_0 \frac{\lambda F_g}{\cos\theta V_{cell}} e^{2\pi ikR_0} dz \qquad\qquad (11.1)$$

$$= \frac{i\pi}{\xi_g} \phi_0 e^{2\pi ikR_0} dz$$

where ξ_g is the extinction distance; its explanation will be given later.

The first Fresnel zone with radius ρ_1 can be calculated using the geometry shown in Figure 11.2. As

$$R_0^2 + \rho_1^2 = R^2 = \left(R_0 + \frac{\lambda}{2}\right)^2$$

so the radius of the first Fresnel zone is

$$\rho_1 \approx \sqrt{\lambda R_0} \qquad\qquad (11.2)$$

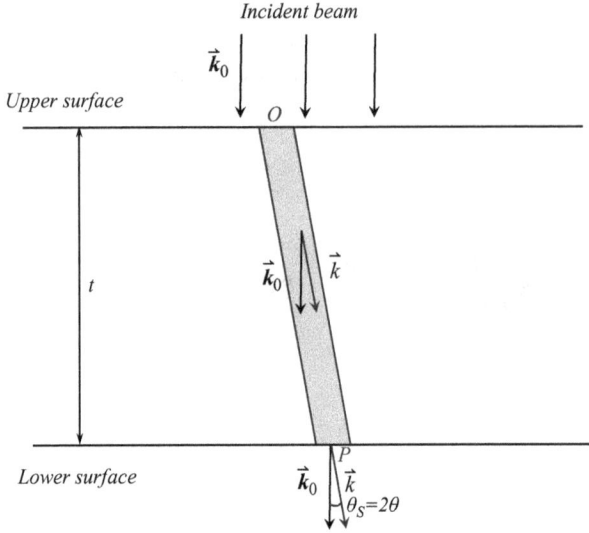

FIGURE 11.3 Schematic representation of column approximation. The diameter value depends on the thickness of the sample. In practice it is usually taken to be about 2 nm.

(Source: Williams (1996).)

Because ρ_1 is small, only a column with a diameter of 1–2 nm is contributing to the amplitude at the point P (Reimer, 1997). In TEM, column approximation is used for the discussion on diffraction contrast formation. Figure 11.3 shows the column of crystal used in calculating the diffracted beam amplitude (Hirsch et al., 1960).

In fact, the intensity of the beams at point P at the lower surface of the specimen is influenced by all the scattering within the cone of material. As an approximation, a column is employed to replace the cone. It is also known that the Bragg angles of interest are usually no more than 1° when we are forming images (Williams, 1996). As $2\,\theta_B$ is very small, the direction of the inclined column in Figure 11.3 can be considered along the thickness direction during mathematical treatment.

The formula (11.1) shows that when we assume that the amplitude of the incident beam is 1, or $\phi_0 = 1$, we get $d\phi_g = i\dfrac{\lambda F_g}{\cos\theta V_{cell}}e^{2\pi i k R_0}dz$. If we skip the $\dfrac{\pi}{2}$ phase difference, as well as the phase term $e^{2\pi i k R_0}$ in the expression, we realize that the contribution to the diffracted beam amplitude from a thickness element dz is

$$\frac{\lambda F_g}{\cos\theta V_{cell}}dz$$

The amplitude–phase diagram, a regular polygon, can be considered as a circle, as each edge of the polygon is small (Hirsch, 1977). The diameter of the circle is the maximum value of the beam amplitude, and this value is 1 as the incident beam amplitude $\phi_0 = 1$. If we assume that the amplitude–phase diagram contributed from m layers just forms half of the circle, then $m \cdot \left(\dfrac{\lambda F_g}{\cos \theta V_{cell}} dz \right) = \dfrac{circumference}{2} = \dfrac{\pi}{2}.$

The length of these m layers along the depth direction is $m \cdot dz$.

The extinction distance ξ_g is associated with twice the distance $m \cdot dz$ in the crystal, and its value is

$$2m \cdot dz = \frac{\pi \cos \theta V_{cell}}{\lambda F_g} \approx \frac{\pi V_{cell}}{\lambda F_g}$$

Hence, the extinction distance is

$$\xi_g = \frac{\pi V_c \cos \theta}{\lambda F_g} \approx \frac{\pi V_c}{\lambda F_g} \tag{11.3}$$

So, we can write $d\phi_g = i \dfrac{\lambda F_g}{\cos \theta V_{cell}} e^{2\pi ikR_0} dz = \dfrac{i\pi}{\xi_g} e^{2\pi ikR_0} dz.$ If we do not assume $\phi_0 \neq 1$, we cannot omit ϕ_0, and the contribution to the diffracted beam amplitude from the thickness element dz is $d\phi_g = \dfrac{i\pi}{\xi_g} \phi_0 e^{2\pi ikR_0} dz.$

11.1 TWO-BEAM DYNAMICAL THEORY

The kinematical theory is valid only for very thin films for which the diffraction intensity is small and the decrease of the primary-beam intensity can be neglected (Reimer, 1997). Therefore, instead of discussing the kinematical theory, we present the simplified two-beam dynamical theory. The real situation can involve many beams and the physics involved is more complicated to materials scientists. Students equipped with sound physics foundations can read Chapters 14 and 15 of Williams (1996) and Chapter 4 of Thomas and Goringe (1979). The simplified two-beam model tells us that the incident beam will be diffracted by the crystal, contributing to the amplitude of the diffracted beam. The diffracted beam can be diffracted back to the incident beam direction, contributing to the amplitude of the incident beam.

The incident beam and the diffracted beam vary their amplitudes while they travel along the depth direction, or z direction, of the specimen (see Figure 11.4). The amplitudes on the lower surface of the specimen can be calculated mathematically based on such a simplified two beam model.

I have followed some approaches presented in the textbook *Electron Microscopy of Thin Crystals* by Hirsch (1977), which can be observed in the discussion below.

The electron wave is composed of the incident wave and the diffraction wave in the assumption of the two-beam dynamical theory:

$$\psi (\bar{r}) = \phi_0 (z) \exp(2\pi i \bar{\chi}_0 \cdot \bar{r}) + \phi_g (z) \exp(2\pi i \bar{\chi} \cdot \bar{r}) \tag{11.4}$$

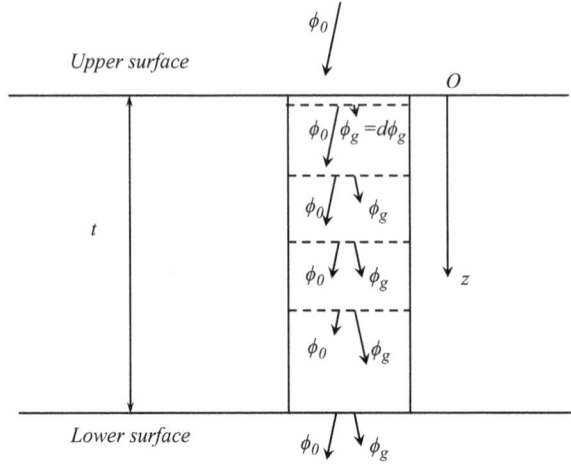

FIGURE 11.4 Schematic diagram of the incident and diffracted waves that change their amplitudes after scattering by each slice of dz thickness inside the specimen.

where ϕ_0 and ϕ_g are the incident and diffracted wave amplitudes. $\bar{\chi}_0$ and $\bar{\chi}$ are the incident and diffracted wave vectors.

In the case where we consider the diffraction of the incident beam from a thickness element dz at any position r, the phase difference with respect to the upper surface of the specimen $-2\pi q \cdot \bar{r}$ needs to be included. Therefore, at this position r:

$$d\phi_g = \left(\frac{i\pi}{\xi_g}\right)\phi_0 \exp\left(-2\pi i \bar{q} \cdot \bar{r}\right) dz = \left(\frac{i\pi}{\xi_g}\right)\phi_0 \exp\left(-2\pi i s z\right) dz \qquad (11.5)$$

where $\bar{q} = \bar{\chi} - \bar{\chi}_0 = \bar{g} + \bar{s}$.

As $\bar{g} \cdot \bar{r}$ is an integer, $\bar{s}$ and $\bar{r}$ are almost parallel or antiparallel, and $r \approx z$; so we have $\bar{q} \cdot \bar{r} = (\bar{g} + \bar{s}) \cdot \bar{r} = \bar{s} \cdot \bar{r} \approx sr \approx sz$.

Now let us consider the process in which the incident beam and the diffracted beam are propagating across the thickness direction: (i) the incident beam will be diffracted by the crystal, contributing to the amplitude of the diffracted beam; (ii) the diffracted beam can be diffracted back to the incident beam direction, contributing to the amplitude of the incident beam. Therefore, both $d\phi_0$ and $d\phi_g$ can be derived:

$$
\begin{aligned}
d\phi_0 &= \left(\frac{i\pi}{\xi_0}\right)\phi_0 \exp\left[-2\pi i (\bar{\chi}_0 - \bar{\chi}_0) \cdot \bar{r}\right] dz + \left(\frac{i\pi}{\xi_g}\right)\phi_g \exp\left[-2\pi i (\bar{\chi}_0 - \bar{\chi}) \cdot \bar{r}\right] dz \\
&= \left(\frac{i\pi}{\xi_0}\right)\phi_0 dz + \left(\frac{i\pi}{\xi_g}\right)\phi_g \exp\left[2\pi i s z\right] dz,
\end{aligned}
\qquad (11.6)
$$

The phase differences with respect to the upper surface of the specimen $-2\pi i(\bar{\chi}_0 - \bar{\chi}_0) \cdot \bar{r}$ and $-2\pi i(\bar{\chi}_0 - \bar{\chi}) \cdot \bar{r}$ for the dz element at position r are included.

ξ_g is the distinction distance $\xi_g = \dfrac{\pi V_c \cos\theta_S}{\lambda F_g} \approx \dfrac{\pi V_c}{\lambda F_g}$. ξ_0 is a quantity similar to ξ_g

(Howie and Whelan, 1961), and $\xi_0 = \dfrac{\pi V_c \cos\theta_S}{\lambda F_0} = \dfrac{\pi V_c}{\lambda F_0}$.

Similarly,

$$
\begin{aligned}
d\phi_g &= \left(\frac{i\pi}{\xi_0}\right)\phi_g \exp\left[-2\pi i\left(\bar{\chi}-\bar{\chi}\right)\cdot\bar{r}\right]dz + \left(\frac{i\pi}{\xi_g}\right)\phi_0 \exp\left[-2\pi i\left(\bar{\chi}-\bar{\chi}_0\right)\cdot\bar{r}\right]dz \\
&= \left(\frac{i\pi}{\xi_0}\right)\phi_g dz + \left(\frac{i\pi}{\xi_g}\right)\phi_0 \exp\left[-2\pi isz\right]dz
\end{aligned}
\tag{11.7}
$$

The following two expressions in (11.8) are known as the Howie–Whelan, or Darwin–Howie–Whelan equations, which are used in two-beam analysis:

$$
\begin{aligned}
\frac{d\phi_0}{dz} &= \frac{i\pi}{\xi_0}\phi_0 + \frac{i\pi}{\xi_g}\phi_g \exp\left(2\pi isz\right) \\
\frac{d\phi_g}{dz} &= \frac{i\pi}{\xi_g}\phi_0 \exp\left(-2\pi isz\right) + \frac{i\pi}{\xi_0}\phi_g
\end{aligned}
\tag{11.8}
$$

Through the transformation of variables (Hirsch, 1977),

$$
\begin{aligned}
\phi_0' &= \phi_0 \exp\left(-i\pi z/\xi_0\right) \\
\phi_g' &= \phi_g \exp\left(2\pi isz - i\pi z/\xi_0\right),
\end{aligned}
\tag{11.9}
$$

or

$$
\begin{aligned}
\phi_0 &= \phi_0' \exp\left(i\pi z/\xi_0\right) \\
\phi_g &= \phi_g' \exp\left(-2\pi isz + i\pi z/\xi_0\right),
\end{aligned}
\tag{11.10}
$$

Simplified equations are derived by the substitution of Equation (11.10) into Equation (11.8):

$$
\begin{aligned}
\frac{d\phi_0'}{dz} &= \frac{i\pi}{\xi_g}\phi_g' \\
\frac{d\phi_g'}{dz} &= \frac{i\pi}{\xi_g}\phi_0' + 2\pi is\phi_g'
\end{aligned}
\tag{11.11}
$$

and the electron wave is expressed as

$$
\psi\left(\bar{r}\right) = \phi_0'\left(z\right)\exp\left(2\pi i\bar{k}_0\cdot\bar{r}\right) + \phi_g'\left(z\right)\exp\left(2\pi i\bar{k}\cdot\bar{r}\right)
\tag{11.12}
$$

where $\bar{k} = \bar{k}_0 + \bar{g} + \bar{s}$.

Hirsch (1977) has given a clear explanation of $\bar{k}_0$ and $\bar{\chi}_0$, namely that they have the same components in the x and y directions, and different components in the z direction as $k_z = \chi + (2\xi_0)^{-1}$, which includes the effect of the refraction of the crystal.

Let us remove ′ from ϕ_0' and ϕ_g' in the following mathematical treatment:

$$\frac{d\phi_0}{dz} = \frac{i\pi}{\xi_g}\phi_g$$
$$\frac{d\phi_g}{dz} = \frac{i\pi}{\xi_g}\phi_0 + 2\pi i s \phi_g$$

(11.13)

Equation (11.13) can be further treated as follows by differentiating the first equation:

$$\frac{d^2\phi_0}{dz^2} = \frac{i\pi}{\xi_g}\cdot\frac{d\phi_g}{dz} = \frac{i\pi}{\xi_g}\left(\frac{i\pi}{\xi_g}\phi_0 + 2\pi i s \phi_g\right) = -\frac{\pi^2}{\xi_g^2}\phi_0 - \frac{2\pi^2 s}{\xi_g}\phi_g$$

$$= -\frac{\pi^2}{\xi_g^2}\phi_0 - \frac{2\pi^2 s}{\xi_g}\left(\frac{\xi_g}{i\pi}\frac{d\phi_0}{dz}\right) = -\frac{\pi^2}{\xi_g^2}\phi_0 - \frac{2\pi s}{i}\frac{d\phi_0}{dz},$$

or

$$\frac{d^2\phi_0}{dz^2} - 2\pi i s \frac{d\phi_0}{dz} + \frac{\pi^2}{\xi_g^2}\phi_0 = 0$$

(11.14)

The solution to the above equation has the form:

$$\phi_0(z) = C_0 \exp(2\pi i \gamma z)$$

By substituting this into Equation (11.14), we have:

$$\gamma^2 - s\gamma - \frac{1}{4\xi_g^2} = 0$$

By solving this equation, we get:

$$\gamma^{(1)} = \frac{1}{2}\left(s + \sqrt{s^2 + \frac{1}{\xi_g^2}}\right)$$
$$\gamma^{(2)} = \frac{1}{2}\left(s - \sqrt{s^2 + \frac{1}{\xi_g^2}}\right)$$

(11.15)

It can be seen that $\gamma^{(1)}$ is larger than $\gamma^{(2)}$, and $\gamma^{(1)} - \gamma^{(2)} = \sqrt{s^2 + \xi_g^{-2}}$. Different textbooks assign $\gamma^{(1)}$ and $\gamma^{(2)}$ differently, for instance, in Hirsch (1977), $\gamma^{(2)}$ is larger than $\gamma^{(1)}$, and $\gamma^{(2)} - \gamma^{(1)} = \sqrt{s^2 + \xi_g^{-2}}$. Some textbooks (Fultz and Howe, 2008; Williams,

1996) present the Bloch waves based on the quantum mechanics approach with an in-depth physics picture. In my lectures, I have only chosen some explanations suitable to materials scientists. Following the textbook *Transmission Electron Microscopy and Diffractometry of Materials* (Fultz and Howe, 2008), we assign $\gamma^{(1)}$ to be larger than $\gamma^{(2)}$, and $\gamma^{(1)} - \gamma^{(2)} = \sqrt{s^2 + \xi_g^{-2}}$.

Then the two independent solutions for the amplitudes of the transmitted wave are:

$$\phi_0^{(1)}(z) = C_0^{(1)} \exp\left(2\pi i \gamma^{(1)} z\right)$$
$$\phi_0^{(2)}(z) = C_0^{(2)} \exp\left(2\pi i \gamma^{(2)} z\right) \tag{11.16}$$

According to the first equation in (11.13), the corresponding amplitudes of the diffracted waves are:

$$\phi_g^{(1)}(z) = \frac{\xi_g}{i\pi}\frac{d\phi_0^{(1)}}{dz} = 2\gamma^{(1)}\xi_g C_0^{(1)} \exp\left(2\pi i \gamma^{(1)} z\right) = C_g^{(1)} \exp\left(2\pi i \gamma^{(1)} z\right)$$
$$\phi_g^{(2)}(z) = \frac{\xi_g}{i\pi}\frac{d\phi_0^{(2)}}{dz} = 2\gamma^{(2)}\xi_g C_0^{(2)} \exp\left(2\pi i \gamma^{(2)} z\right) = C_g^{(2)} \exp\left(2\pi i \gamma^{(2)} z\right) \tag{11.17}$$

It is useful to define a dimensionless parameter $w = s\xi_g = ctg\beta$ before further calculation. It can be observed from Equation (11.17) that

$$\frac{C_g^{(1)}}{C_0^{(1)}} = 2\gamma^{(1)}\xi_g = s\xi_g + \sqrt{\left(s\xi_g\right)^2 + 1} = \frac{\cos\dfrac{\beta}{2}}{\sin\dfrac{\beta}{2}}$$

$$\frac{C_g^{(2)}}{C_0^{(2)}} = 2\gamma^{(2)}\xi_g = s\xi_g - \sqrt{\left(s\xi_g\right)^2 + 1} = -\frac{\sin\dfrac{\beta}{2}}{\cos\dfrac{\beta}{2}} \tag{11.18}$$

Then the electron wave is represented by two Bloch waves:

$$b^{(1)}\left(\vec{k}^{(1)}, \vec{r}\right) = C_0^{(1)} \exp\left(2\pi i \vec{k}_0^{(1)} \cdot \vec{r}\right) + C_g^{(1)} \exp\left(2\pi i \vec{k}^{(1)} \cdot \vec{r}\right)$$
$$b^{(2)}\left(\vec{k}^{(2)}, \vec{r}\right) = C_0^{(2)} \exp\left(2\pi i \vec{k}_0^{(2)} \cdot \vec{r}\right) + C_g^{(2)} \exp\left(2\pi i \vec{k}^{(2)} \cdot \vec{r}\right) \tag{11.19}$$

$\vec{k}_0^{(1)}, \vec{k}_0^{(2)}, \vec{k}_0$ and $\vec{\chi}_0$ have the same components along x and y directions, and have different projections along the z axis. Their relationships are expressed as:

$$k_{0z}^{(1)} = k_{0z} + \gamma^{(1)}$$
$$k_{0z}^{(2)} = k_{0z} + \gamma^{(2)}$$

$$\vec{k}^{(1)} - \vec{k}_0^{(1)} = \vec{k}^{(2)} - \vec{k}_0^{(2)} = \vec{k} - \vec{k}_0 = \vec{g} + \vec{s}$$

or

$$\vec{k}^{(1)} = \vec{k}_0^{(1)} + \vec{g} + \vec{s},\ \vec{k}^{(2)} = \vec{k}_0^{(2)} + \vec{g} + \vec{s},\ \vec{k} = \vec{k}_0 + \vec{g} + \vec{s}$$

In expressions (11.19), $C_0^{(1)}$ and $C_g^{(1)}$, $C_0^{(2)}$ and $C_g^{(2)}$ can be normalized:

$$\left|C_0^{(1)}\right|^2 + \left|C_g^{(1)}\right|^2 = 1$$
$$\left|C_0^{(2)}\right|^2 + \left|C_g^{(2)}\right|^2 = 1 \tag{11.20}$$

Considering $\dfrac{C_g^{(1)}}{C_0^{(1)}} = \dfrac{\cos\dfrac{\beta}{2}}{\sin\dfrac{\beta}{2}}$ and $\dfrac{C_g^{(2)}}{C_0^{(2)}} = -\dfrac{\sin\dfrac{\beta}{2}}{\cos\dfrac{\beta}{2}}$, the expressions for $C_0^{(1)}$ and $C_g^{(1)}$, $C_0^{(2)}$

and $C_g^{(2)}$ can be obtained:

$$C_0^{(1)} = -C_g^{(2)} = \sin\left(\beta / 2\right)$$
$$C_0^{(2)} = C_g^{(1)} = \cos\left(\beta / 2\right) \tag{11.21}$$

Hence the two Bloch waves are expressed by

$$b_1\left(\vec{r}\right) = \sin\left(\frac{\beta}{2}\right)\exp\left(2\pi i \vec{k}_0^{(1)} \cdot \vec{r}\right) + \cos\left(\frac{\beta}{2}\right)\exp\left(2\pi i \vec{k}^{(1)} \cdot \vec{r}\right)$$
$$b_2\left(\vec{r}\right) = \cos\left(\frac{\beta}{2}\right)\exp\left(2\pi i \vec{k}_0^{(2)} \cdot \vec{r}\right) - \sin\left(\frac{\beta}{2}\right)\exp\left(2\pi i \vec{k}^{(2)} \cdot \vec{r}\right) \tag{11.22}$$

The electron wave can be described as:

$$\psi\left(\vec{r}\right) = \psi^{(1)}\left\{\sin\left(\beta / 2\right)\exp\left(2\pi i \vec{k}_0^{(1)} \cdot \vec{r}\right) + \cos\left(\beta / 2\right)\exp\left(2\pi i \vec{k}^{(1)} \cdot \vec{r}\right)\right\}$$
$$+ \psi^{(2)}\left\{\cos\left(\beta / 2\right)\exp\left(2\pi i \vec{k}_0^{(2)} \cdot \vec{r}\right) - \sin\left(\beta / 2\right)\exp\left(2\pi i \vec{k}^{(2)} \cdot \vec{r}\right)\right\} \tag{11.23}$$

where $\psi^{(1)}$ and $\psi^{(2)}$ are constants.

By applying the boundary condition, $\psi^{(1)}$ and $\psi^{(2)}$ can be resolved. At the upper surface of the crystal where $z = 0$ and $r = 0$, it can be seen that $\phi_0(0) = 1$ and $\phi_g(0) = 0$. Therefore,

$$\phi_0\left(0\right) = \psi^{(1)}\sin\left(\beta / 2\right) + \psi^{(2)}\cos\left(\beta / 2\right) = 1$$
$$\phi_g\left(0\right) = \psi^{(1)}\cos\left(\beta / 2\right) - \psi^{(2)}\sin\left(\beta / 2\right) = 0 \tag{11.24}$$

Hence,

$$\psi^{(1)} = \sin\left(\beta/2\right)$$
$$\psi^{(2)} = \cos\left(\beta/2\right) \tag{11.25}$$

Substituting Equation (11.25) into Equation (11.23), the electron wave can be written as:

$$\psi\left(\vec{r}\right) = \sin\left(\frac{\beta}{2}\right)\left[\sin\left(\frac{\beta}{2}\right)\exp\left(2\pi i\vec{k}_0^{(1)}\cdot\vec{r}\right) + \cos\left(\frac{\beta}{2}\right)\exp\left(2\pi i\vec{k}^{(1)}\cdot\vec{r}\right)\right]$$
$$+ \cos\left(\frac{\beta}{2}\right)\left[\cos\left(\frac{\beta}{2}\right)\exp\left(2\pi i\vec{k}_0^{(2)}\cdot\vec{r}\right) - \sin\left(\frac{\beta}{2}\right)\exp\left(2\pi i\vec{k}^{(2)}\cdot\vec{r}\right)\right] \tag{11.26}$$
$$= \sin\left(\frac{\beta}{2}\right)b_1\left(\vec{r}\right) + \cos\left(\frac{\beta}{2}\right)b_2\left(\vec{r}\right),$$

where $\sin(\beta/2)(=\psi^{(1)})$ and $\cos(\beta/2)(=\psi^{(2)})$ are the Bloch-wave excitation coefficients.

11.2 DISCUSSION OF DIFFRACTION CONTRAST

11.2.1 AMPLITUDE OF THE DIFFRACTED BEAM AND THE FORMATION OF THICKNESS FRINGES AND BEND CONTOURS

When the incident wave and the diffracted wave are considered separately as

$$\psi_0\left(z\right) = \sin^2\left(\frac{\beta}{2}\right)\exp\left(2\pi i\vec{k}_0^{(1)}\cdot\vec{r}\right) + \cos^2\left(\frac{\beta}{2}\right)\exp\left(2\pi i\vec{k}_0^{(2)}\cdot\vec{r}\right)$$
$$\psi_g\left(z\right) = \sin\left(\frac{\beta}{2}\right)\cos\left(\frac{\beta}{2}\right)\exp\left(2\pi i\vec{k}^{(1)}\cdot\vec{r}\right) - \cos\left(\frac{\beta}{2}\right)\sin\left(\frac{\beta}{2}\right)\exp\left(2\pi i\vec{k}^{(2)}\cdot\vec{r}\right)$$

simplified forms for $\psi_0(z)$ and $\psi_g(z)$ can be derived. We do this through the following example.

Example 11.1   Simplify the following expressions of the incident beam and diffracted beam:

$$\psi_0\left(z\right) = \sin^2\left(\frac{\beta}{2}\right)\exp\left(2\pi i\vec{k}_0^{(1)}\cdot\vec{r}\right) + \cos^2\left(\frac{\beta}{2}\right)\exp\left(2\pi i\vec{k}_0^{(2)}\cdot\vec{r}\right)$$
$$\psi_g\left(z\right) = \sin\left(\frac{\beta}{2}\right)\cos\left(\frac{\beta}{2}\right)\exp\left(2\pi i\vec{k}^{(1)}\cdot\vec{r}\right) - \cos\left(\frac{\beta}{2}\right)\sin\left(\frac{\beta}{2}\right)\exp\left(2\pi i\vec{k}^{(2)}\cdot\vec{r}\right)$$

Solution: Using some formulas in trigonometry, the incident beam can be simplified as:

$$\psi_0(z) = \sin^2\left(\frac{\beta}{2}\right)\exp\left(2\pi i\bar{k}_0^{(1)}\cdot\bar{r}\right) + \cos^2\left(\frac{\beta}{2}\right)\exp\left(2\pi i\bar{k}_0^{(2)}\cdot\bar{r}\right)$$

$$= \left(\frac{1-\cos\beta}{2}\right)\exp\left(2\pi i\bar{k}_0^{(1)}\cdot\bar{r}\right) + \frac{1+\cos\beta}{2}\exp\left(2\pi i\bar{k}_0^{(2)}\cdot\bar{r}\right)$$

$$= \frac{1}{2}\left\{\exp\left(2\pi i\bar{k}_0^{(1)}\cdot\bar{r}\right) + \exp\left(2\pi i\bar{k}_0^{(2)}\cdot\bar{r}\right)\right\} - \frac{\cos\beta}{2}\left\{\exp\left(2\pi i\bar{k}_0^{(1)}\cdot\bar{r}\right) - \exp\left(2\pi i\bar{k}_0^{(2)}\cdot\bar{r}\right)\right\}$$

The first term:

$$\frac{1}{2}\left\{\exp\left(2\pi i\bar{k}_0^{(1)}\cdot\bar{r}\right) + \exp\left(2\pi i\bar{k}_0^{(2)}\cdot\bar{r}\right)\right\}$$

$$= \frac{1}{2}\left\{\cos\left(2\pi\bar{k}_0^{(1)}\cdot\bar{r}\right) + i\sin\left(2\pi\bar{k}_0^{(1)}\cdot\bar{r}\right) + \cos\left(2\pi\bar{k}_0^{(2)}\cdot\bar{r}\right) + i\sin\left(2\pi\bar{k}_0^{(2)}\cdot\bar{r}\right)\right\}$$

$$= \cos\left[\pi\left(\bar{k}_0^{(1)} + \bar{k}_0^{(2)}\right)\cdot\bar{r}\right]\cos\left[\pi\left(\bar{k}_0^{(1)} - \bar{k}_0^{(2)}\right)\cdot\bar{r}\right] + i\sin\left[\pi\left(\bar{k}_0^{(1)} + \bar{k}_0^{(2)}\right)\cdot\bar{r}\right]\cos\left[\pi\left(\bar{k}_0^{(1)} - \bar{k}_0^{(2)}\right)\cdot\bar{r}\right]$$

$$= \cos\left[\pi\left(\bar{k}_0^{(1)} - \bar{k}_0^{(2)}\right)\cdot\bar{r}\right]\left\{\cos\left[\pi\left(\bar{k}_0^{(1)} + \bar{k}_0^{(2)}\right)\cdot\bar{r}\right] + i\sin\left[\pi\left(\bar{k}_0^{(1)} + \bar{k}_0^{(2)}\right)\cdot\bar{r}\right]\right\}$$

$$= \cos\left[\pi\left(\bar{k}_0^{(1)} - \bar{k}_0^{(2)}\right)\cdot\bar{r}\right]\exp\left[i\pi\left(\bar{k}_0^{(1)} + \bar{k}_0^{(2)}\right)\cdot\bar{r}\right]$$

$$= \cos\left[\pi\left(k_{0z}^{(1)} - k_{0z}^{(2)}\right)\cdot z\right]\exp\left[i\pi\left(\bar{k}_0^{(1)} + \bar{k}_0^{(2)}\right)\cdot\bar{r}\right]$$

$$= \cos[\pi\Delta k_z z]\exp\left[2\pi i\left(\frac{\bar{k}_0^{(1)} + \bar{k}_0^{(2)}}{2}\right)\cdot\bar{r}\right],$$

where $\Delta k_z = k_z^{(1)} - k_z^{(2)} = \gamma^{(1)} - \gamma^{(2)} = \sqrt{s^2 + \xi_g^{-2}}$.

The second term:

$$-\frac{\cos\beta}{2}\left\{\exp\left(2\pi i\bar{k}_0^{(1)}\cdot\bar{r}\right) - \exp\left(2\pi i\bar{k}_0^{(2)}\cdot\bar{r}\right)\right\}$$

$$= -\frac{\cos\beta}{2}\left\{\cos 2\pi\bar{k}_0^{(1)}\cdot\bar{r} + i\sin 2\pi\bar{k}_0^{(1)}\cdot\bar{r} - \cos 2\pi\bar{k}_0^{(2)}\cdot\bar{r} - i\sin 2\pi\bar{k}_0^{(2)}\cdot\bar{r}\right\}$$

$$= -\frac{\cos\beta}{2}\left\{-2\sin\left[\pi\left(\bar{k}_0^{(1)} + \bar{k}_0^{(2)}\right)\cdot\bar{r}\right]\sin\left[\pi\left(\bar{k}_0^{(1)} - \bar{k}_0^{(2)}\right)\cdot\bar{r}\right]\right.$$
$$\left. + 2i\cos\left[\pi\left(\bar{k}_0^{(1)} + \bar{k}_0^{(2)}\right)\cdot\bar{r}\right]\sin\left[\pi\left(\bar{k}_0^{(1)} - \bar{k}_0^{(2)}\right)\cdot\bar{r}\right]\right\}$$

$$= -\frac{\cos\beta}{2}(2i)\sin\left[\pi\left(\bar{k}_0^{(1)} - \bar{k}_0^{(2)}\right)\cdot\bar{r}\right]\left\{\cos\left[\pi\left(\bar{k}_0^{(1)} + \bar{k}_0^{(2)}\right)\cdot\bar{r}\right] + i\sin\left[\pi\left(\bar{k}_0^{(1)} + \bar{k}_0^{(2)}\right)\cdot\bar{r}\right]\right\}$$

$$= -i\cos\beta\sin\left[\pi\left(k_{0z}^{(1)} - k_{0z}^{(2)}\right)z\right]\left\{\exp\left[i\pi\left(\bar{k}^{(1)} + \bar{k}^{(2)}\right)\cdot\bar{r}\right]\right\}$$

$$= -i\cos\beta\sin[\pi\Delta k_z z]\left\{\exp\left[2\pi i\left(\frac{\bar{k}_0^{(1)} + \bar{k}_0^{(2)}}{2}\right)\cdot\bar{r}\right]\right\}.$$

Hence,

$$\psi_0(z) = \left\{\cos\left[\pi\Delta k_z z\right] - i\cos\beta\sin\left[\pi\Delta k_z z\right]\right\}\left\{\exp\left[2\pi i\left(\frac{\vec{k}_0^{(1)} + \vec{k}_0^{(2)}}{2}\right)\cdot\vec{r}\right]\right\}$$

and the amplitude of the incident beam is:

$$\phi_0(z) = \cos\left(\pi\Delta k_z z\right) - i\cos\beta\sin\left(\pi\Delta k_z z\right) \tag{11.27}$$

Similarly, the diffracted beam can be simplified as:

$$\psi_g(z) = \sin\left(\frac{\beta}{2}\right)\cos\left(\frac{\beta}{2}\right)\exp\left(2\pi i\vec{k}^{(1)}\cdot\vec{r}\right) - \cos\left(\frac{\beta}{2}\right)\sin\left(\frac{\beta}{2}\right)\exp\left(2\pi i\vec{k}^{(2)}\cdot\vec{r}\right)$$

$$= \sin\left(\frac{\beta}{2}\right)\cos\left(\frac{\beta}{2}\right)\left\{\exp\left(2\pi i\vec{k}^{(1)}\cdot\vec{r}\right) - \exp\left(2\pi i\vec{k}^{(2)}\cdot\vec{r}\right)\right\}$$

$$= \frac{\sin\beta}{2}\left\{\cos 2\pi\vec{k}^{(1)}\cdot\vec{r} + i\sin 2\pi\vec{k}^{(1)}\cdot\vec{r} - \cos 2\pi\vec{k}^{(2)}\cdot\vec{r} - i\sin 2\pi\vec{k}^{(2)}\cdot\vec{r}\right\}$$

$$= \frac{\sin\beta}{2}\left\{-2\sin\left[\pi\left(\vec{k}^{(1)} + \vec{k}^{(2)}\right)\cdot\vec{r}\right]\sin\left[\pi\left(\vec{k}^{(1)} - \vec{k}^{(2)}\right)\cdot\vec{r}\right]\right.$$
$$\left. + 2i\cos\left[\pi\left(\vec{k}^{(1)} + \vec{k}^{(2)}\right)\cdot\vec{r}\right]\sin\left[\pi\left(\vec{k}^{(1)} - \vec{k}^{(2)}\right)\cdot\vec{r}\right]\right\}$$

$$= \frac{\sin\beta}{2}\cdot 2i\sin\left[\pi\left(\vec{k}^{(1)} - \vec{k}^{(2)}\right)\cdot\vec{r}\right]\left\{\cos\left[\pi\left(\vec{k}^{(1)} + \vec{k}^{(2)}\right)\cdot\vec{r}\right] + i\sin\left[\pi\left(\vec{k}^{(1)} + \vec{k}^{(2)}\right)\cdot\vec{r}\right]\right\}$$

$$= i\sin\beta\sin\left[\pi\left(k_z^{(1)} - k_z^{(2)}\right)z\right]\left\{\exp\left[i\pi\left(\vec{k}^{(1)} + \vec{k}^{(2)}\right)\cdot\vec{r}\right]\right\}$$

$$= i\sin\beta\sin\left[\pi\Delta k_z z\right]\left\{\exp\left[2\pi i\left(\frac{\vec{k}^{(1)} + \vec{k}^{(2)}}{2}\right)\cdot\vec{r}\right]\right\}$$

Hence, the amplitude of the diffracted beam is:

$$\phi_g(z) = i\sin\beta\sin\left(\pi\Delta k_z z\right), \tag{11.28}$$

where $\Delta k_z = k_z^{(1)} - k_z^{(2)} = \gamma^{(1)} - \gamma^{(2)} = \sqrt{s^2 + \xi_g^{-2}}$.

If we analyze the oscillations of the diffracted beam amplitude $\phi_g(z) = i\sin\beta\sin(\pi\Delta k_z z)$, or the diffracted beam intensities $I \propto |\phi_g(z)|^2$, it is clear that the origin of the thickness oscillations is the difference in wavelengths of the two Bloch waves. It is the beating between the two Bloch waves (Williams, 1996).

Some textbooks have expressed the effective excitation error s_{eff} (Williams, 1996) as:

$$s_{eff} = \Delta k = \frac{1}{\xi_g^{eff}} = \sqrt{s^2 + \xi_g^{-2}} \tag{11.29}$$

The effective excitation error s_{eff} is never zero. When $s = 0$, $s_{eff} = \frac{1}{\xi_g}$. When s is very large, $s_{eff} \approx s$.

Then the diffracted beam amplitude is in a similar format as in the case of the kinematical approximation:

$$\phi_g(z) = i \sin \beta \sin(\pi s_{eff} z),$$ (11.30)

as $ctg\beta = s\xi_g$ and $\sin \beta = \dfrac{1}{\sqrt{1+s\xi_g^{\,2}}} = \dfrac{1}{\xi_g s_{eff}}$, the diffracted beam intensity can be expressed as:

$$I_g(t) = |\phi_g(t)|^2 = \frac{\pi^2}{\xi_g^{\,2}} \frac{\sin^2(\pi s_{eff} t)}{(\pi s_{eff})^2}$$ (11.31)

$\dfrac{\xi_g^{eff}}{\xi_g} = \dfrac{1}{s_{eff}} = \dfrac{1}{\sqrt{s^2 + \xi_g^{-2}}}$ is the effective extinction distance, and the direct beam intensity is

$$I_0(t) = 1 - I_g(t) = 1 - |\varphi_g(t)|^2$$

Expression (11.31) can be used to explain the formation of thickness fringes in the wedge of a specimen with changing thickness, and bend contours with changing s when the specimen is bent.

An image showing thickness fringes from a thin silicon specimen and an image showing the bend contours from a zinc oxide belt are presented in Figure 11.5.

(a)   (b)

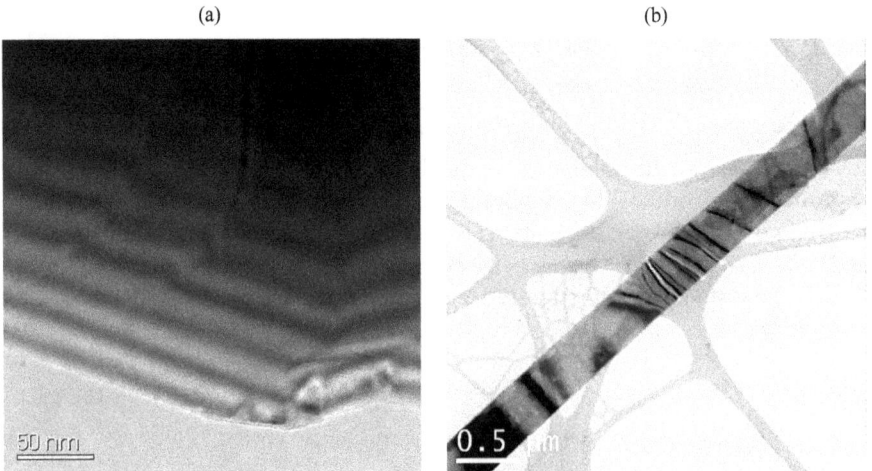

FIGURE 11.5 Bright field images of (a) a silicon crystal showing thickness fringes at the wedge and (b) a ZnO nanobelt showing bending contours due to different s values.

(Source: Xu et al. (2005), with the permission of AIP Publishing.)

11.2.2 Discussion of the Two Bloch Waves

For the two Bloch waves,

$$b^{(1)}(\vec{r}) = \sin\left(\frac{\beta}{2}\right)\exp\left(2\pi i \vec{k}_0^{(1)} \cdot \vec{r}\right) + \cos\left(\frac{\beta}{2}\right)\exp\left(2\pi i \vec{k}^{(1)} \cdot \vec{r}\right)$$

$$b^{(2)}(\vec{r}) = \cos\left(\frac{\beta}{2}\right)\exp\left(2\pi i \vec{k}_0^{(2)} \cdot \vec{r}\right) - \sin\left(\frac{\beta}{2}\right)\exp\left(2\pi i \vec{k}^{(2)} \cdot \vec{r}\right)$$

If the exact Bragg diffraction condition is satisfied, or $\theta = \theta_B$, we have

$s = 0$, $ctg\beta = s\xi_g = 0$ and $\beta = \dfrac{\pi}{2}$.

In this case, we also have

$\vec{s} = 0$, $\vec{k}^{(1)} = \vec{k}_0^{(1)} + \vec{g}$ and $\vec{k}^{(2)} = \vec{k}_0^{(2)} + \vec{g}$

Therefore,

$$\begin{aligned}b^{(1)}(\vec{r}) &= \sin\left(\frac{\beta}{2}\right)\exp\left(2\pi i \vec{k}_0^{(1)} \cdot \vec{r}\right) + \cos\left(\frac{\beta}{2}\right)\exp\left(2\pi i \vec{k}^{(1)} \cdot \vec{r}\right)\\ &= \frac{1}{\sqrt{2}}\left\{\exp\left(2\pi i \vec{k}_0^{(1)} \cdot \vec{r}\right) + \exp\left[2\pi i \left(\vec{k}_0^{(1)} + \vec{g}\right) \cdot \vec{r}\right]\right\}\\ &= \frac{1}{\sqrt{2}}\left\{\left[\exp\left(-\pi i \vec{g} \cdot \vec{r}\right) + \exp\left(\pi i \vec{g} \cdot \vec{r}\right)\right]\exp\left[2\pi i \left(\vec{k}_0^{(1)} + \frac{1}{2}\vec{g}\right) \cdot \vec{r}\right]\right\}\\ &= \sqrt{2}\cos\left(\pi \vec{g} \cdot \vec{r}\right)\exp\left[2\pi i \left(\vec{k}_0^{(1)} + \frac{1}{2}\vec{g}\right) \cdot \vec{r}\right]\end{aligned}$$

And

$$\begin{aligned}b^{(2)}(\vec{r}) &= \cos\left(\frac{\beta}{2}\right)\exp\left(2\pi i \vec{k}_0^{(2)} \cdot \vec{r}\right) - \sin\left(\frac{\beta}{2}\right)\exp\left(2\pi i \vec{k}^{(2)} \cdot \vec{r}\right)\\ &= \frac{1}{\sqrt{2}}\left\{\exp\left(2\pi i \vec{k}_0^{(2)} \cdot \vec{r}\right) - \exp\left[2\pi i \left(\vec{k}_0^{(2)} + \vec{g}\right) \cdot \vec{r}\right]\right\}\\ &= \frac{1}{\sqrt{2}}\left\{\left[\exp\left(-\pi i \vec{g} \cdot \vec{r}\right) - \exp\left(\pi i \vec{g} \cdot \vec{r}\right)\right]\exp\left[2\pi i \left(\vec{k}_0^{(2)} + \frac{1}{2}\vec{g}\right) \cdot \vec{r}\right]\right\}\\ &= -i\sqrt{2}\sin\left(\pi \vec{g} \cdot \vec{r}\right)\exp\left[2\pi i \left(\vec{k}_0^{(2)} + \frac{1}{2}\vec{g}\right) \cdot \vec{r}\right]\end{aligned}$$

If we choose $x \parallel \vec{g}$, then $\vec{g} \cdot \vec{r} = gx$. In this case

$$b^{(1)}(\vec{r}) = \sqrt{2}\cos\left(\pi gx\right)\exp\left[2\pi i \left(\vec{k}_0^{(1)} + \frac{1}{2}\vec{g}\right) \cdot \vec{r}\right]$$

$$b^{(2)}(\vec{r}) = -i\sqrt{2}\sin\left(\pi gx\right)\exp\left[2\pi i \left(\vec{k}_0^{(2)} + \frac{1}{2}\vec{g}\right) \cdot \vec{r}\right]$$

(11.32)

It is noticed that the intensities of the two Bloch waves at the Bragg condition are:

$$\left|b^{(1)}(\bar{r})\right|^2 \propto \cos^2\left(\pi g x\right)$$

$$\left|b^{(2)}(\bar{r})\right|^2 \propto \sin^2\left(\pi g x\right)$$

(11.33)

For the $b^{(1)}(\bar{r})$ wave, the beam is more concentrated in sheets in the vicinity of the atoms. For the $b^{(2)}(\bar{r})$ wave, the situation is reversed and the maxima occur between the crystallographic planes (Fultz and Howe, 2008). So, $b^{(1)}(\bar{r})$ is absorbed faster than $b^{(2)}(\bar{r})$, and $b^{(2)}(\bar{r})$ penetrates deeper than $b^{(1)}(\bar{r})$ inside a specimen.

When the exact Bragg condition is satisfied, we have $s = 0$, $ctg\beta = s\xi_g = 0$ and $\beta = \dfrac{\pi}{2}$. Hence $\Psi^{(1)}\left(= \sin\dfrac{\beta}{2}\right) = \Psi^{(2)}\left(= \cos\dfrac{\beta}{2}\right)$, suggesting that the two Bloch waves are excited by the same amplitude.

When $\theta < \theta_B$, $s < 0$, $ctg\beta = s\xi_g < 0$, we have $\beta > \dfrac{\pi}{2}$ and

$$\sin\left(\frac{\beta}{2}\right) > \cos\left(\frac{\beta}{2}\right)$$

The amplitude of $b^{(1)}(\bar{r})$ is greater and the scattering/absorption is stronger.

When $\theta > \theta_B$, $s > 0$, $ctg\beta = s\xi_g > 0$, we have $\beta < \dfrac{\pi}{2}$ and

$$\sin\left(\frac{\beta}{2}\right) < \cos\left(\frac{\beta}{2}\right)$$

The amplitude of $b^{(2)}(\bar{r})$ is greater, which means a higher penetration.

11.2.3 DISCUSSION ON THE CONTRAST OF THE CRYSTAL DEFECTS

As the defect alters the diffraction amplitude, the contrast caused by the defects can be observed. Using $\bar{r} + \bar{R}$ to represent the defect location where $\bar{R}$ is the displacement caused by a defect, the expressions (11.8) can be rewritten as

$$\frac{d\phi_0}{dz} = \frac{i\pi}{\xi_0}\phi_0 + \frac{i\pi}{\xi_g}\phi_g \exp\left(2\pi i s z + 2\pi i \bar{g}\cdot\bar{R}\right)$$

$$\frac{d\phi_g}{dz} = \frac{i\pi}{\xi_g}\phi_0 \exp\left(-2\pi i s z + 2\pi i \bar{g}\cdot\bar{R}\right) + \frac{i\pi}{\xi_0}\phi_g$$

Using the transformation of wave amplitudes (Hirsch, 1977):

$$\phi_0'' = \phi_0 \exp\left(-i\pi z / \xi_0\right)$$

$$\phi_g'' = \phi_g \exp\left(2\pi i s z - i\pi z / \xi_0 + 2\pi i \bar{g}\cdot\bar{R}\right)$$

which leads to:

$$\frac{d\phi_0^{''}}{dz} = \frac{i\pi}{\xi_g}\phi_g^{''}$$

$$\frac{d\phi_g^{''}}{dz} = \frac{i\pi}{\xi_g}\phi_0^{''} + 2\pi i\left(s + \vec{g}\cdot\frac{d\vec{R}}{dz}\right)\phi_g^{''}$$

It can be seen that the contrast of a defect is caused by the factor, $\vec{g}\cdot d\vec{R}/dz$, or caused by the changes of value of the excitation error s at the distorted area of defect, which indicates the local rotation caused by $\vec{R}$. The weak beam technique is based on the condition, $\vec{s} + \vec{g}\cdot d\vec{R}/dz = 0$, and this condition can be used to improve the resolution of the defect image.

SUMMARY

Diffraction contrast is created due to the intensity variations of the incident beam and diffracted beams at the exit surface of the thin crystalline specimen. When the transmitted beam or one of the diffracted beams is selected by the objective lens aperture at the back focal plane of the objective lens, only the selected beam can pass through the aperture, contributing to the image formed in the image plane of the objective lens.

The features observed in the bright field image and dark field image depend sensitively on how the Bragg condition is satisfied, for instance: Which diffraction is active? What is the value of the deviation parameter?

In a perfect crystal, typical features are thickness fringes and bending contours. When the specimen has some defects, such as dislocation, precipitates, grain boundaries or inclusions, the defects in a crystal can be revealed through bright field imaging or dark field imaging.

REFERENCES

Fultz, B. and Howe, J. M. (2008) *Transmission electron microscopy and diffractometry of materials/Brent Fultz, James Howe.* 3rd ed. Berlin: Springer.

Hirsch, P. B. (1977) *Electron microscopy of thin crystals/P. B. Hirsch, A. Howrie, R. B. Nicholson, D. W. Pashley and M. J. Whelan.* Malabar, FL: Krieger.

Hirsch, P. B., Howie, A. and Whelan, M. J. (1960) 'A kinematical theory of diffraction contrast of electron transmission microscope images of dislocations and other defects', *Philosophical Transactions of the Royal Society of London. Series A, Mathematical and Physical Sciences.* The Royal Society, 252(1017), pp. 499–529. Available at: http://www.jstor.org/stable/73192.

Howie, A. and Whelan, M. J. (1961) 'Diffraction contrast of electron microscope images of crystal lattice defects. II. The development of a dynamical theory', *Proceedings of the Royal Society of London. Series A, Mathematical and Physical Sciences.* The Royal Society, 263(1313), pp. 217–237. Available at: http://www.jstor.org/stable/2414111.

Khor, K. A. et al. (2000) 'Microstructure investigation of plasma sprayed HA/Ti$_6$Al$_4$V composites by TEM', *Materials Science and Engineering A*, 281(1–2), pp. 221–228. doi:10.1016/s0921-5093(99)00717-0.

Reimer, L. (1997) *Transmission electron microscopy: physics of image formation and micro-analysis/Ludwig Reimer*. 4th ed. Berlin: Springer-Verlag (Springer series in optical sciences; v. 36).

Thomas, G. and Goringe, M. J. (Michael J.) (1979) *Transmission electron microscopy of materials/Gareth Thomas, Michael J. Goringe*. New York: Wiley.

Williams, D. B. (1996) *Transmission electron microscopy [electronic resource]: A textbook for materials science/by David B. Williams, C. Barry Carter*. Edited by C. B. Carter. Boston, MA: Springer US. doi:10.1007/978-1-4757-2519-3.

Xu, C. X. et al. (2005) 'Magnetic nanobelts of iron-doped zinc oxide', *Applied Physics Letters*, 86(17), pp. 1–3. doi:10.1063/1.1919391.

12 Phase Contrast

Both mass–thickness contrast and diffraction contrast are amplitude contrasts. The amplitude variations for the incident beam and scattered beams at the exit surface of a thin specimen create contrasts in the bright field images and dark field images.

Across the lower surface of a thin specimen, if there are phase variations for the electron wave, the phase difference can be transformed to intensity differences in the image plane of the objective lens, which are depicted effectively as darker or brighter areas of the resultant image. The phase contrast enables very high resolution imaging, making it possible to distinguish features at the angstrom or sub-angstrom scale.

12.1 INTERFERENCE OF TWO BEAMS

When we consider the interference of the incident beam and the diffracted beam in the two-beam condition, we need to remove the objective lens aperture. Some students prefer using a larger objective aperture to let the two beams pass through it. When the objective lens aperture is removed from the beam path, the interference of the incident beam and the diffracted beam occurs as they have a fixed phase relationship. The incident beam and the diffracted beam are (Williams, 1996):

$$\psi = \phi_0(z)\exp 2\pi i\left(\bar{k}_0 \cdot \bar{r}\right) + \phi_g(z)\exp 2\pi i\left(\bar{k} \cdot \bar{r}\right)$$
$$\bar{k} - \bar{k}_0 = \bar{q} = \bar{g} + \bar{s}$$

If the amplitudes of the incident beam and the diffracted beam are rewritten in a simple form as $\phi_0 = A\exp(i\alpha_1)$ and $\phi_g = B\exp(i\alpha_2)$, where A and B are real numbers, and $\alpha_2 - \alpha_1 = \Delta\alpha$, then:

$$\psi = A\exp(i\alpha_1)\exp\left[2\pi i\left(\bar{k}_0 \cdot \bar{r}\right)\right] + B\exp(i\alpha_2)\exp\left[2\pi i\left(\bar{k} \cdot \bar{r}\right)\right]$$
$$= \exp(i\alpha_1)\exp\left[2\pi i\left(\bar{k}_0 \cdot \bar{r}\right)\right]\left\{A + B\exp i\left(2\pi\bar{q}\cdot\bar{r} + \Delta\alpha\right)\right\}$$

Then $I = |\psi|^2$ can be obtained:

$$I = \left\{A + B\exp\left[i\left(2\pi\bar{q}\cdot\bar{r} + \Delta\theta\right)\right]\right\}\left\{A + B\exp\left[-i\left(2\pi\bar{q}\cdot\bar{r} + \Delta\alpha\right)\right]\right\}$$
$$= A^2 + B^2 + AB\left\{\exp\left[i\left(2\pi\bar{q}\cdot\bar{r} + \Delta\alpha\right)\right] + \exp\left[-i\left(2\pi\bar{q}\cdot\bar{r} + \Delta\alpha\right)\right]\right\}$$
$$= A^2 + B^2 + 2AB\cos\left(2\pi\bar{q}\cdot\bar{r} + \Delta\alpha\right)$$

DOI: 10.1201/9780429351662-16

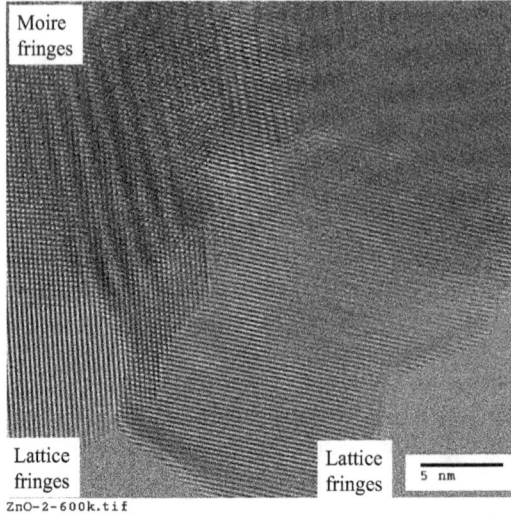

FIGURE 12.1 Lattice fringes from a wurtzite-type zinc oxide specimen.

If the Bragg condition is satisfied, or $\bar{q} = \bar{g}$, and we set $\bar{g}$ parallel to x, then:

$$I = A^2 + B^2 + 2AB\cos\left(2\pi gx + \Delta\alpha\right) = A^2 + B^2 + 2AB\cos\left(2\pi\frac{x}{d} + \Delta\alpha\right) \quad (12.1)$$

The image contrast due to the intensity variations across the exit surface of the thin specimen shows the fringes of the periodicity of the d spacing. When $\bar{s}$ is not zero, $\bar{q} = \bar{g} + \bar{s}$; the periodicity of the fringes is only an approximation of the d spacing. In transmission electron microscope (TEM) analysis, it is very difficult to accurately measure the d spacing from an interference pattern. Reliable d spacing data can be obtained from X-ray diffraction measurements.

Although lattice fringes are not direct images of the projected structure, lattice spacing information is helpful for some analyses (see Figure 12.1).

12.2 PROJECTED POTENTIAL OF THIN CRYSTALS AND HRTEM IMAGES

In the case where there is no objective lens aperture (or larger objective aperture), many electron beams can pass through and contribute to the image formation. The process is treated differently from those for electron diffraction and diffraction contrast. For electron diffraction, the geometry of the diffraction pattern and the intensity of each diffraction spot at the back focal plane of the objective lens are correlated to the crystal structure of the specimen. For diffraction contrast, the incident beam amplitude and the diffracted beam amplitude vary across the lower surface of the specimen, which is caused by structural feature variations inside the specimen. In the imaging process, only the beam(s) selected by the objective aperture contributes to the image at the image plane of the objective lens, forming the diffraction contrast and revealing the structures of the defects. In phase contrast formation, when an electron beam passes through a thin specimen, it carries the information of the projected potential of the specimen.

In this section, students need to understand two pictures of physics: (i) the wave at the exit surface of the specimen that carries information of the specimen, and (ii) the imaging process that can reveal the information of the specimen.

It is known that the relativistic wavelength is $\lambda = h/(mv) = h/[2m eE(1 + eE/2m_0c^2)]^{1/2}$. In approximation, $\lambda \approx h/\sqrt{2m_0eE}$ in a vacuum, and $\lambda'(x, y, z) \approx h/\sqrt{2m_0e\left[E+V(x, y, z)\right]}$ inside the specimen. Considering the difference of wavelengths in the vacuum and inside the specimen, the phase shift caused by an element of thickness dz of the specimen is:

$$d\chi(x,y,z) = 2\pi \frac{dz}{\lambda'} - 2\pi \frac{dz}{\lambda}$$

$$= 2\pi \frac{dz}{\lambda}\left[\frac{\sqrt{E+V(x,y,z)}}{\sqrt{E}} - 1\right]$$

$$\approx \sigma V(x,y,z)dz,$$

where $\sigma = \dfrac{\pi}{\lambda E}$.

For a specimen having a thickness t, the integral is:

$$\chi(x,y) = \sigma \int_0^t V(x,y,z)dz = \sigma V_t(x,y) \tag{12.2}$$

Under week-phase object approximation when the specimen is very thin, the wave function on the lower surface of the specimen is written as:

$$\psi_e(x,y) = \exp\left[-i\sigma V_t(x,y)\right]$$
$$\approx \left[1 - i\sigma V_t(x,y)\right], \tag{12.3}$$

or

$$\psi_e(\bar{r}) \approx \left[1 - i\sigma V_t(\bar{r})\right]$$

For the above result, it has been assumed that the incident wave amplitude $\phi_0 = 1$, and that absorption is neglected. In many textbooks, $\phi_p(x, y)$ is used to represent the projected specimen potential. In this textbook, we have been using $V(\bar{r})$ and $U(\bar{r})$ to represent the electric potential and potential energy respectively. Therefore, we use $V_t(\bar{r})$ here to express the projected specimen potential to avoid confusion.

In electron diffraction and diffraction contrast studies, we discuss lens aberration very little. However, in a phase contrast study, via Abbe's theory, we need to consider the influence of objective lens aberration while the image is formed.

In Abbe's imaging theory, three planes associated with the objective lens are considered: (i) the lower surface of the specimen, (ii) the back focal plane and (iii) the image plane (Figure III.2). For a real space vector on the lower surface of the specimen or on the image plane, we can use $\bar{r}$, or the coordinates x, y in the x–y plane, to represent it. For a vector on the back focal plane, which is associated with reciprocal space, we can use vector $\bar{u}$, or coordinates u, v in the u–v plane, to represent it. For simplicity, in many cases we can just use the one-dimensional coordinate u for mathematical treatments. As the three planes are not inside the thin specimen, or a crystal, there is no need to use the axes of the seven crystal system to define the vectors or the coordinates.

In the back focal plane of the objective lens, the wave function $\psi\left(\bar{r}\right) \approx \left[1 - i\sigma V_t\left(\bar{r}\right)\right]$ can be expressed in reciprocal space through a Fourier transform:

$$\Psi_d\left(\bar{u}\right) = \left[\delta\left(\bar{u}\right) - i\sigma FT\left[V_t\left(\bar{r}\right)\right]\right]$$

as $FT\left[1\right] = \delta\left(\bar{u}\right)$.

It can be seen that the phase shift at the back focal plane is related to the spherical aberration coefficient C_s and the defocus ε. The value of ε is negative when the lens is under-focused and it is positive when it is over-focused. The phase change is expressed as

$$\chi\left(\bar{u}\right) = \pi\varepsilon\lambda\left|\bar{u}\right|^2 + \frac{1}{2}\pi C_s\lambda^3\left|\bar{u}\right|^4 \tag{12.4}$$

Hence, the phase factor $\exp i\chi\left(\bar{u}\right) = \cos \chi\left(\bar{u}\right) + i\sin \chi\left(\bar{u}\right)$ is introduced at the back focal plane of the objective lens before the second Fourier transform. As a vector in the back focal plane can also be represented using coordinates u and v, $\chi\left(\bar{u}\right) = \pi\varepsilon\lambda\left|\bar{u}\right|^2 + \frac{1}{2}\pi C_s\lambda^3\left|\bar{u}\right|^4$ is written as

$$\chi\left(\bar{u}\right) = \pi\varepsilon\lambda\left(u^2 + v^2\right) + \frac{1}{2}\pi C_s\lambda^3\left(u^2 + v^2\right)^2$$

Introducing $\exp i\chi\left(\bar{u}\right) = \cos \chi\left(\bar{u}\right) + i\sin \chi\left(\bar{u}\right)$ at the back focal plane results in

$$\Psi_d\left(\bar{u}\right)\cdot\exp i\chi\left(\bar{u}\right) = \left[\delta\left(\bar{u}\right) - i\sigma FT\left[V_t\left(\bar{r}\right)\right]\right]\cdot\exp i\chi\left(\bar{u}\right)$$
$$= \left[\delta\left(\bar{u}\right)\cos \chi\left(\bar{u}\right) + \sigma FT\left[V_t\left(\bar{r}\right)\right]\sin \chi\left(\bar{u}\right)\right]$$
$$+ i\left[\delta\left(\bar{u}\right)\sin \chi\left(\bar{u}\right) - \sigma FT\left[V_t\left(\bar{r}\right)\right]\cos \chi\left(\bar{u}\right)\right]$$

We know that $FT\left[FT\left[V_t\left(\vec{r}\right)\right]\right]=V_t\left(-\vec{r}\right)$, and for a function $f\left(\vec{u}\right)$, we have $\int\delta\left(\vec{u}\right)f\left(\vec{u}\right)d\vec{u}=f\left(0\right)$. We also know that when $u=0$, $\chi=0$, $\cos\chi=1$ and $\sin\chi=0$. Hence, through a Fourier transform, the electron wave in the image plane is

$$\psi_i\left(\vec{r}\right)=\left\{1+\sigma V_t\left(-\vec{r}\right)\otimes FT\left[\sin\chi\left(\vec{u}\right)\right]\right\}-i\left\{\sigma V_t\left(-\vec{r}\right)\otimes FT\left[\cos\chi\left(\vec{u}\right)\right]\right\}$$

Therefore

$$I=\psi_i\left(\vec{r}\right)\cdot\psi_i^*\left(\vec{r}\right)=1+2\sigma V_t\left(-\vec{r}\right)\otimes FT\left[\sin\chi\left(\vec{u}\right)\right]$$
$$+\sigma^2\left\{V_t\left(-\vec{r}\right)\otimes FT\left[\sin\chi\left(\vec{u}\right)\right]\right\}^2+\sigma^2\left\{V_t\left(-\vec{r}\right)\otimes FT\left[\cos\chi\left(\vec{u}\right)\right]\right\}^2$$

or

$$I\approx1+2\sigma V_t\left(-\vec{r}\right)\otimes FT\left[\sin\chi\left(\vec{u}\right)\right]\tag{12.5}$$

where $\sin\chi\left(\vec{u}\right)$ is the contrast transfer function and the symbol $\otimes$ (we can also use symbol $*$) means convolution. In fact, in the back focal plane, the physical aperture $A(u)$ and the envelope damping function $E(u)$ need to be considered (Williams, 1996); they can be combined with the contrast transfer function $\sin\chi\left(\vec{u}\right)$. If we further include the constant 2, then the transfer function becomes

$$T\left(u\right)=A\left(u\right)E\left(u\right)2\sin\chi\left(u\right)$$

where $A(u)$ is the aperture function and $E(u)$ is the envelope function. A transfer function is presented in an HRTEM image simulation example in Section 12.3.

In modern HRTEM, chromatic aberration is reduced due to the stable high voltage and reduced lens current fluctuations. Beam convergence can be adjusted so the incident beam can be nearly parallel. The specimen drift and vibration can be minimized by using a suitable working environment. During the HRTEM operation, proper alignment is necessary and objective lens astigmatism should be corrected to ensure the imaging quality. With a certain C_s value, the optimum focus or Scherzer defocus condition (Williams, 1996) occurs about

$$\varepsilon_{Sch}=-1.2\left(\lambda C_s\right)^{1/2}$$

Under this condition, the contrast transfer function crosses the axis at

$$u_{Sch}=1.51C_s^{-1/4}\lambda^{-3/4}$$

and the point resolution is

$$d=\frac{1}{u_{Sch}}=0.66C_s^{1/4}\lambda^{3/4}\tag{12.6}$$

There are detailed explanations of HRTEM imaging and point resolution in Chapter 2 in Buseck et al. (1988). As phase contrast images are directly interpretable only up to the point resolution, and the information limit goes well beyond the point resolution for microscopes with field emission guns, image simulation software such as JEMS can be employed to assist in the interpretation of image details beyond the point resolution.

For microscopes with Cs correctors, the first crossover of the contrast transfer function can be extended beyond the information limit, in which case the features that are directly interpretable can reach the sub-angstrom level.

Example 12.1

Calculate the Scherzer defocus and the point resolution for (i) a TEM operating at 200 kV with Cs = 0.5 mm, and (ii) a TEM operating at 300 kV with Cs = 1.4 mm.

Solution to (i): Based on the discussion in this section, we know that the Scherzer defocus can be calculated using equation $\varepsilon_{Sch} = -1.2(\lambda C_s)^{1/2}$, and the point resolution can be calculated using $d = \dfrac{1}{u_{Sch}} = 0.66C_s^{1/4}\lambda^{3/4}$. For a microscope operating at 200 kV, the wavelength of the electron beam is 0.0251 Å. Therefore, the Scherzer defocus value is $\varepsilon_{Sch} = -1.2(\lambda C_s)^{1/2} = -425$ Å.

The point resolution is $d = \dfrac{1}{u_{Sch}} = 0.66C_s^{1/4}\lambda^{3/4} = 1.9$ Å.

Solution to (ii): For a microscope operating at 300 kV, the wavelength of the electron beam is 0.0197 Å. Therefore, the Scherzer defocus value is $\varepsilon_{Sch} = -1.2(\lambda C_s)^{1/2} = -630$ Å

The point resolution is $d = \dfrac{1}{u_{Sch}} = 0.66C_s^{1/4}\lambda^{3/4} = 2.1$ Å

While we operate a TEM, we can choose not to insert the objective aperture, even the largest one. In this case the aperture function $A(u)$ is not included in the transfer in the back focal plane and there is no cutoff by the aperture function. However, the damping of the envelope function always exists. To impose the damping envelope functions on the normal phase-contrast transfer function curve was reported earlier by J. Frank in 1973, and further discussion on the envelope function and information limit can be found in O'Keefe (1992) and O'Keefe et al. (2001). The brief explanation below is based on O'Keefe (1992) and will help students understand the physics of the envelope function and information limit. The envelope function for incident beam convergence is

$$E_\alpha(u) = \exp\left\{-\pi^2\alpha^2\left(\varepsilon + \lambda^2 C_S u^2\right)^2 u^2\right\}$$

where ε is the objective lens defocus, and the value of ε is negative when the lens is under-focused and positive when it is over-focused; C_S is the spherical aberration coefficient at an electron wavelength λ; α is the standard deviation of a Gaussian over a convergent cone, equal to 0.77 times the measured semi-angle.

Since $E\alpha(u)$ has a cutoff that increases with an increase in under-focusing, the effect of convergence does not constitute an information limit for a microscope.

The envelope function for a spread of focusing is

$$E_\Delta(u) = \exp\left\{-\frac{1}{2}\pi^2\lambda^2\Delta^2 u^4\right\}$$

where Δ is the standard deviation of a Gaussian spread-of-focus. The cutoff frequency for $E_\Delta(u)$ imposes an information limit on the microscope, at a level of $\exp(-2)$, or 13.5%.

From $\exp\left\{-\frac{1}{2}\pi^2\lambda^2\Delta^2 u^4\right\} = \exp(-2)$, we have

$$|u|_\Delta = (2/\pi\lambda\Delta)^{1/2} \tag{12.7}$$

and the information limit is

$$d_\Delta = \frac{1}{|u|_\Delta} = \sqrt{(\pi\lambda\Delta/2)} \tag{12.8}$$

The spread of focus is expressed as (O'Keefe, 1992):

$$\Delta = C_C\sqrt{(\delta V/V)^2 + (2\delta I/I)^2 + (\delta E/E)^2}$$

where C_c is the chromatic aberration coefficient for the objective lens, $\delta V/V$ is the fractional change in voltage over the time scale of image acquisition, $\delta I/I$ is the fractional change in the lens current, and $\delta E/E$ is the energy spread in the electron beam as a function of the total energy.

We can analyze $T(u) = A(u) E(u) 2\sin\chi(u)$ again, and only consider the combination of $E(u)$ and $\sin\chi(u)$. For a high resolution TEM without a spherical aberration corrector, the first crossover of $\sin\chi(u)$ is at a u value less than $1 \overset{\circ}{A}{}^{-1}$ and the point resolution is over 1 Å. With a Cs corrector, the first crossover of the contrast transfer function $\sin\chi(u)$ can be extended beyond $|u|_\Delta = (2/\pi\lambda\Delta)^{1/2}$ and which defines the information limit. In this situation we use the information limit $d_\Delta = \frac{1}{|u|_\Delta} = \sqrt{(\pi\lambda\Delta/2)}$ to define the resolution of the microscope, and for microscopes with the highest resolution so far, the information limit approaches 0.5 Å.

In summary: (i) the wave at the exit surface of a thin specimen is related to the projected potential of the specimen, and (ii) phase shift is introduced at the back focal plane of the objective lens, and the image formed by the objective lens can reveal the projected potential of the thin specimen.

12.3 HRTEM IMAGE SIMULATION

The projected potential for thin crystals can be revealed when weak-phase object approximation with kinematic scattering is valid. For a thick specimen, the observed image cannot be interpreted directly as the projected crystal structure. During image simulation, the specimen can be considered to have many slices, which are normal to the incident beam. For each of the slices, weak phase approximation is still valid. From one slice to the slice below it, Fresnel diffraction occurs.

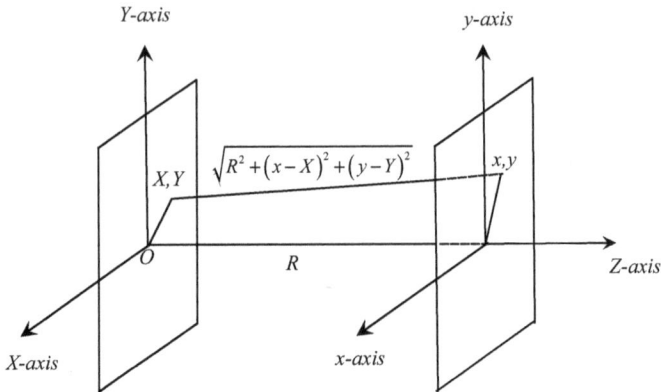

FIGURE 12.2 Schematic diagram showing transmission of a wave from a plane with coordinates X, Y to a parallel plane with coordinates x, y through an optical system.

The textbook *High-Resolution Transmission Electron Microscopy and Associated Techniques* (Buseck et al., 1988) explains the Fresnel diffraction when a wave propagates through one plane to another (Figure 12.2).

In most physics textbooks, the wave number is expressed as $k = \dfrac{2\pi}{\lambda}$, and 2π will not appear inside the phase term of the wave. In order to maintain consistency throughout our XRD and TEM discussions and avoid confusion, we use $k = \dfrac{1}{\lambda}$ as the wave number, and keep 2π inside the phase term. The expression below shows the wave function on one plane, given the wave function on a previous plane:

$$\psi\left(xy\right) = \frac{\exp\left(2\pi ikR\right)}{i\lambda R} \int \psi_1\left(xy\right)\exp\left\{2\pi ik\frac{\left[\left(x-X\right)^2+\left(y-Y\right)^2\right]}{2R}\right\}dXdY$$

or

$$\psi\left(xy\right) = \frac{\exp\left(2\pi ikR\right)}{i\lambda R}\left\{\psi_1\left(xy\right)\otimes\exp\left[\frac{\pi i\left(x^2+y^2\right)}{R\lambda}\right]\right\}$$

In the above expression, the convolution $f\left(x\right)\otimes g\left(x\right)=\displaystyle\int_{-\infty}^{\infty} f\left(X\right)g\left(x-X\right)dX$ is used.

Explanation of the Fresnel propagator can be found in other textbooks as well, for instance in Williams (1996). Williams has illustrated the methods for performing image simulation, which include: (i) reciprocal-space formalism, (ii) FFT formalism, (ii) the real-space approach and (iv) the Bloch-wave approach. In the real-space approach, for example, the wave function can be expressed by:

$$\Psi_{n+1}\left(\bar{r}\right) = \left[\Psi_n\left(\bar{r}\right)\otimes P_{n+1}\left(\bar{r}\right)\right]\cdot q_{n+1}\left(\bar{r}\right) \tag{12.9}$$

where $\Psi_{n+1}(\vec{r})$ is the wave function in real space at the exit of the $n+1$ slice; $\Psi_n(\vec{r})$ is the wave function in real space at the exit of the n slice; $p_{n+1}(\vec{r})$ is the Fresnel propagator for the $n+1$ slice in a near-field calculation in real space; and $q_{n+1}(\vec{r})$ is the phase grating (the specimen) in real space.

During image simulations, both the above multi-slice treatments for the specimen, and the operating conditions and lens aberrations for the microscope, need to be considered. More detailed atomistic structural information can be revealed by HRTEM image simulations (Kirkland, 1998; Stadelmann, 1987).

In an apatite research project, hydroxyapatite whiskers were hydrothermally synthesized in ETI labs, and the crystal structure data were obtained through Rietveld refinement. Both HRTEM observation and multi-slice simulation were conducted to assist local crystal structure analysis. The TEM specimen was prepared by dispersing the synthetic hydroxyapatite whiskers with ethanol in an ultrasonic bath and a drop of suspension deposited onto a holey-carbon coated copper grid. A JEM-3010 (Cs = 1.4 mm) with a LaB$_6$ gun operating at 300 kV was employed for the high resolution image observation. Figure 12.3 shows the materials obtained when hydrothermal synthesis was conducted at 225 °C for 6 hours. The observation was conducted from the [10-10] direction. As the hydroxyapatite material is electron beam sensitive, quick collection of the HRTEM images before structure alteration is critical.

The simulated HRTEM images were obtained through a multi-slice function using JEMS (Figure 12.4), which can assist in HRTEM image interpretation.

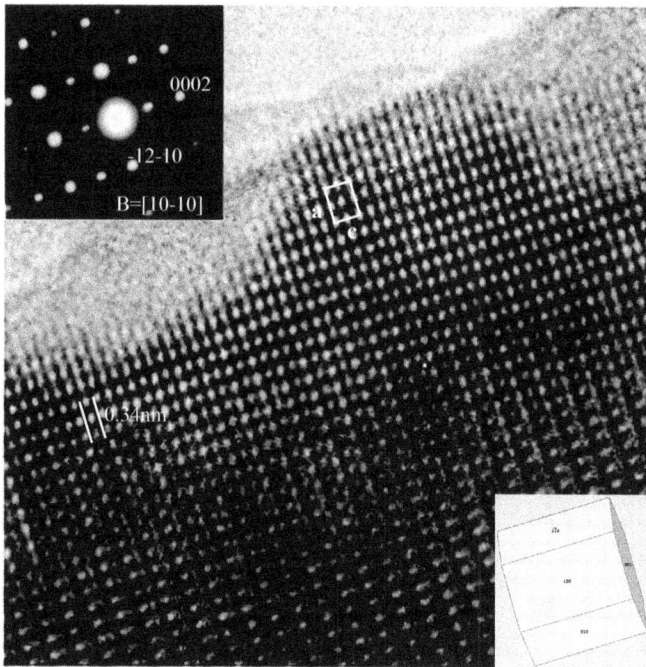

FIGURE 12.3 HRTEM image of hydroxyapatite whiskers obtained after the precursor was hydrothermally treated at 225 °C for 6 hours.

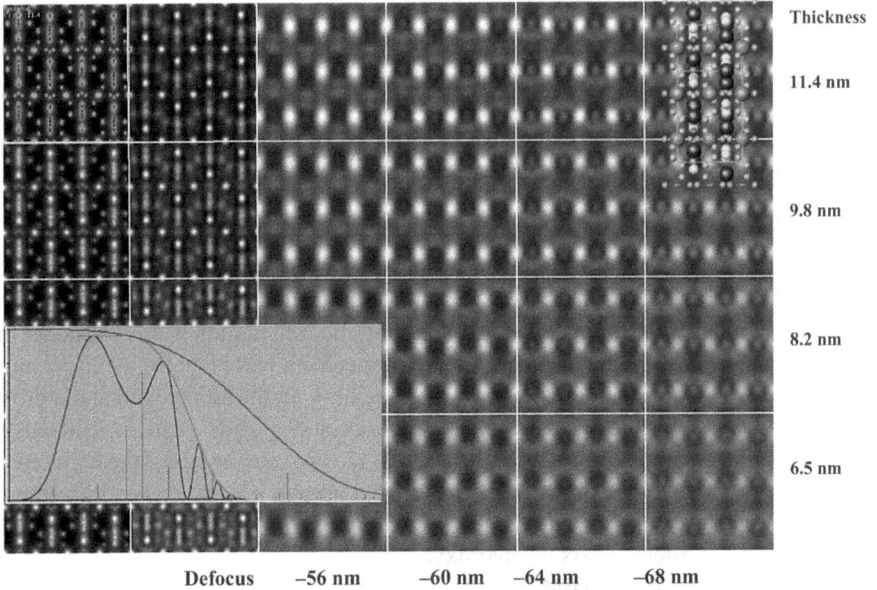

FIGURE 12.4 Image simulation of hydroxyapatite from the [10-10] zone axis. The transfer function and projected atom positions are shown.

The materials obtained are pure hydroxyapatite whiskers without an impurity phase. It is believed that the hydroxyapatite whiskers can be used as strengthening and bioactive components in composite bio-materials. However, TEM observations indicate that the sizes of the whiskers are not uniform. The length and width of the HA whiskers need to be further controlled, which would subsequently influence the strength of the bio-composites if they are used as the reinforcement phase.

12.4 LENS ABERRATIONS AND ABERRATION CORRECTIONS

In Section 12.2, we introduced the phase shift caused by aberrations in the back focal plane of the objective lens. The influence of the spherical aberration of the objective lens and the defocus is introduced in the back focal plane, which is expressed as

$B(\bar{u}) = \exp\left[i\chi(\bar{u})\right]$, where $\chi(\bar{u}) = \frac{1}{2}\pi C_s \lambda^3 |\bar{u}|^4 + \pi \varepsilon \lambda |\bar{u}|^2$.

For students from the materials science and engineering area, the analysis by Reimer will help you to better understand the influence of spherical aberration and defocus (Reimer, 1997).

In optics, we have the following lens equation

$$\frac{1}{focal\ length} = \frac{1}{object\ distance} + \frac{1}{image\ distance}$$

And we can simply express it using $\frac{1}{f} = \frac{1}{u} + \frac{1}{v}$ or $\frac{1}{f} = \frac{1}{a} + \frac{1}{b}$ as used in Reimer (1997).

Based on Figure 12.5, for a lens with spherical aberration coefficient Cs, the corresponding angular deviation is

$$\alpha_S \approx \frac{\Delta r}{b} = \frac{C_S \theta^3 M}{b}$$

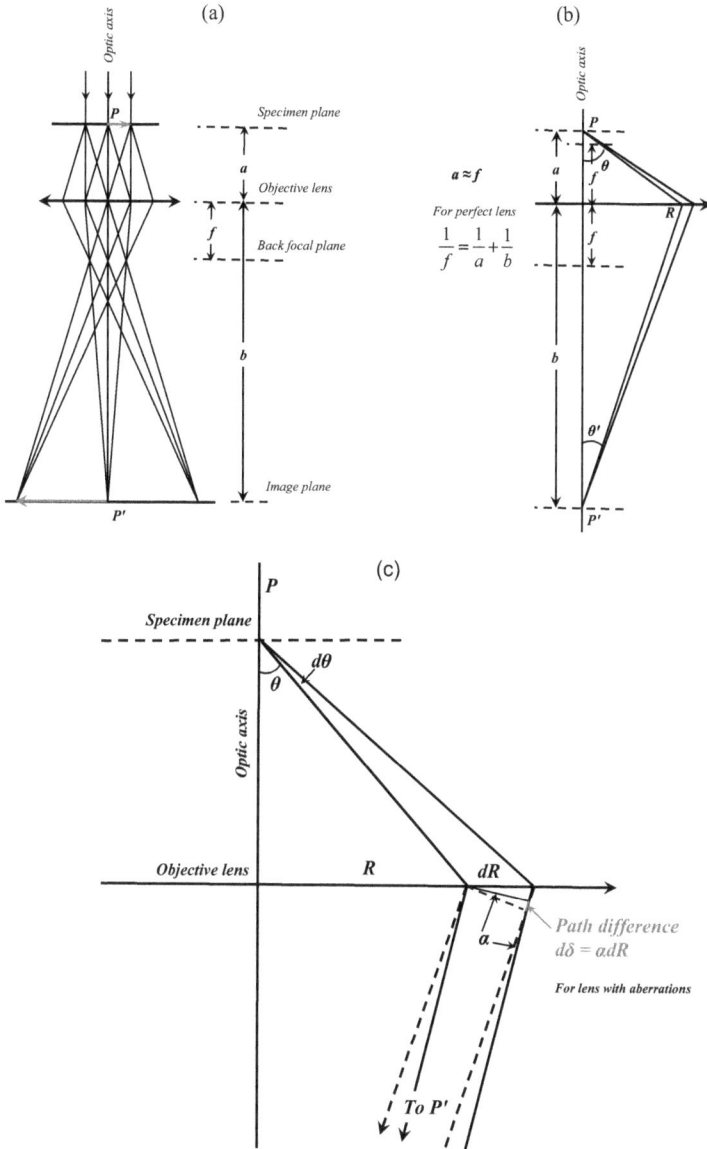

FIGURE 12.5 (a) Ray diagram of the objective lens showing image formation of an object. (b) Ray diagram of a perfect lens showing the relationships between object distance a, image distance b and focal length f. (c) Part of the outer zone of the objective lens at a distance R from the optic axis showing the angular deviation α and the optical path difference δ.

(Source: Reimer (1997), with the permission of Springer Nature.)

Δr is the spread of electron beams at the image plane of the lens when the electrons are scattered in the specimen through angles within θ from the optic axis, which is due to the existence of spherical aberration.

As $\theta \approx \dfrac{R}{a}$, $M = \dfrac{b}{a}$ and $a \approx f$, we have

$$\alpha_S = \frac{C_S R^3}{f^4} \tag{12.10}$$

For defocus ε, its value is $-\Delta f$. This means that when the focal length change Δf is positive or the focal length is increased, the defocus value ε is negative, which is called underfocus. Overfocus has a positive εvalue.

For the change of specimen position Δa, we can just keep the first order in the Taylor series, as $\Delta a/a$ and $\Delta a/b$ are small. We can write

$$\frac{1}{f} = \frac{1}{a+\Delta a} + \frac{1}{b+\Delta b} = \frac{1}{a}\left(1+\frac{\Delta a}{a}\right)^{-1} + \frac{1}{b}\left(1+\frac{\Delta b}{b}\right)^{-1}$$
$$= \frac{1}{a}\left(1-\frac{\Delta a}{a}+\dots\right) + \frac{1}{b}\left(1-\frac{\Delta b}{b}+\dots\right)$$

Solving for Δb and using $\dfrac{1}{f} = \dfrac{1}{a} + \dfrac{1}{b}$ yield

$$\Delta b = -\Delta a b^2 / a^2$$

We know that $\theta' \approx \dfrac{R}{b}$ and $a \approx f$. Therefore, the angular deviation is expressed as (Reimer, 1997):

$$\alpha_a = \left(-\Delta b \theta'\right)/b = \Delta a R / f^2 \tag{12.11}$$

Using the equation $\dfrac{1}{f} = \dfrac{1}{a} + \dfrac{1}{b}$, we can further obtain the image distance change Δb caused by the focal length change Δf. As discussed previously, we just keep the first order in the Taylor series, as $\Delta f/f$ and $\Delta b/b$ are small.

$$\frac{1}{f+\Delta f} = \frac{1}{a} + \frac{1}{b+\Delta b}$$

$$\frac{1}{f}\left(1+\frac{\Delta f}{f}\right)^{-1} = \frac{1}{a} + \frac{1}{b}\left(1+\frac{\Delta b}{b}\right)^{-1}$$

$$\frac{1}{f}\left(1-\frac{\Delta f}{f}+\dots\right) = \frac{1}{a} + \frac{1}{b}\left(1-\frac{\Delta b}{b}+\dots\right)$$

Solving for Δb and using $\dfrac{1}{f} = \dfrac{1}{a} + \dfrac{1}{b}$ yield

$$\Delta b = \frac{\Delta f b^2}{f^2}$$

As $\theta' \approx R/b$, the corresponding angular deviation is expressed as (Reimer, 1997):

$$\alpha_f = \frac{\left(-\Delta b \theta'\right)}{b} = -\frac{\Delta f R}{f^2} \tag{12.12}$$

The increase of f or $\Delta f > 0$, and the decrease of a or $\Delta a < 0$ have a similar effect. Combining Equations (12.10), (12.11) and (12.12), the total angular deviation is written as

$$\alpha_{tot} = \alpha_s + \alpha_a + \alpha_f = C_S\left(\frac{R^3}{f^4}\right) + \left(\Delta a - \Delta f\right)\frac{R}{f^2} \tag{12.13}$$

Based on Reimer's analysis as presented in Figure 12.5(b) (Reimer, 1997), the phase shift is

$$\chi(\theta) = \frac{2\pi}{\lambda}\delta = \frac{2\pi}{\lambda}\int_0^R \alpha_{tot} dR = \frac{2\pi}{\lambda}\left[\frac{1}{4}C_S\frac{R^4}{f^4} + \frac{1}{2}\left(\Delta a - \Delta f\right)\frac{R^2}{f^2}\right] \tag{12.14}$$

If we use $\dfrac{R}{f} \approx \dfrac{R}{a} \approx \theta$ and defocusing $\varepsilon = \Delta a - \Delta f$, the phase shift can be rewritten as

$$\chi(\theta) = \frac{2\pi}{\lambda}\left[\frac{1}{2}\varepsilon\theta^2 + \frac{1}{4}C_s\theta^4\right] \tag{12.15}$$

As $u = \dfrac{1}{\lambda}\theta$, we can also express the phase shift as

$$\chi(u) = 2\pi\left(\frac{1}{2}\varepsilon\lambda u^2 + \frac{1}{4}C_s\lambda^3 u^4\right) \tag{12.16}$$

In our previous discussion in Section 12.2, the specimen position change was not included. In this section, the change of specimen position and the change of focus are combined together as explained in Reimer (1997). Students should remember that the ε value is negative at the underfocus condition. The ε value is positive when the objective lens is over-focused. Therefore, we will use $\varepsilon = \Delta a - \Delta f$ here.

The above discussions show the influence of the defocus and the third-order spherical aberration. In modern TEMs with aberration correctors, Cs and some higher order aberrations can also be corrected.

The wave aberration coefficients can be represented using different notations as reported by Uhlemann and Haider (1998), Kirkland and Meyer (2004) and Erni (2010). Here we use the descriptions from Uhlemann and Haider (1998), where the phase shift is presented as follows.

$$
\begin{aligned}
\text{Phase shift} = \frac{2\pi}{\lambda} \, \mathrm{Re} \Bigg\{ &\frac{1}{2}\omega\bar{\omega}C_1 + \frac{1}{2}\bar{\omega}^2 A_1 + \omega^2\bar{\omega}B_2 + \frac{1}{3}\bar{\omega}^3 A_2 \\
&+ \frac{1}{4}(\omega\bar{\omega})^2 C_3 + \omega^3\bar{\omega}S_3 + \frac{1}{4}\bar{\omega}^4 A_3 + \omega^3\bar{\omega}^2 B_4 \\
&+ \omega^4\bar{\omega}D_4 + \frac{1}{5}\bar{\omega}^5 A_4 + \frac{1}{6}(\omega\bar{\omega})^3 C_5 \\
&+ \frac{1}{6}\bar{\omega}^6 A_5 + \dots \Bigg\}.
\end{aligned}
\tag{12.17}
$$

where $\omega = \theta_x + i\theta_y$ is the complex scattering angle and $\bar{\omega} = \theta_x - i\theta_y$ is its complex conjugate (Erni, 2010) $|\omega| = \dfrac{|\bar{u}|}{k} = \lambda|\bar{u}|$. Equation (12.17) includes many aberration coefficients. C_1 or $-\Delta f$ is the defocus, A_1 is the twofold astigmatism coefficient, A_2 is the threefold astigmatism coefficient, B_2 is the axial coma coefficient, C_3 or C_S is third-order spherical aberration coefficient, A_3 is the fourfold astigmatism coefficient, S_3 is the star aberration coefficient, A_4 is the fivefold astigmatism coefficient, B_4 is the axial coma coefficient, D_4 is the three lobe aberration coefficient, C_5 is the five-order spherical aberration coefficient, and A_5 is the sixfold astigmatism coefficient.

When the spherical aberration is corrected, or $Cs = 0$, the resolution of a TEM can reach the information limit. At certain conditions, the TEM image obtained can be correlated with the charge density projection.

Based on Spence (2013), the exit-face wave amplitude for a phase object is

$$
\psi_e(x,y) = \exp\left[-i\sigma V_t(x,y)\right]
$$

If the focusing error ε is very small and there is no spherical aberration, the amplitude distribution in the back focal plane can be expressed via Fraunhofer diffraction and the aberration function:

$$
\psi_d(u,v) = FT\left\{\exp\left[-i\sigma V_t(x,y)\right]\right\}\exp\left(i\pi\varepsilon\lambda\left(u^2+v^2\right)\right) \approx \Phi(u,v)\left[1+i\pi\varepsilon\lambda\left(u^2+v^2\right)\right]
$$

where $\Phi(u,v)$ is the Fourier transform of $\psi_e(x,y)$.

The amplitude distribution in the image plane at unit magnification and without rotation is given by inverse transformation as

$$
\psi_i(x,y) = \exp\left[-i\sigma V_t(x,y)\right] + i\pi\varepsilon\lambda FT^{-1}\left\{\left(u^2+v^2\right)\Phi(u,v)\right\}
\tag{12.18}
$$

Since $\psi_e(x,y)$ and $\Phi(u,v)$ are a Fourier transform pair, we have

$$FT^{-1}\left(\left(u^2+v^2\right)\Phi\left(u,v\right)\right)=-\frac{1}{4\pi^2}\nabla^2\psi_e\left(x,y\right)$$

(12.19)

To approve the relationship (12.19), we calculate $-\frac{1}{4\pi^2}\nabla^2\psi_e(x,y)$, and see whether it is equal to $FT^{-1}((u^2+v^2)\Phi(u,v))$. The derivation is:

$$-\frac{1}{4\pi^2}\nabla^2\psi_e\left(x,y\right)=-\frac{1}{4\pi^2}\nabla^2 FT^{-1}\left(\Phi\left(u,v\right)\right)$$

$$=-\frac{1}{4\pi^2}\nabla^2\left(\iint\Phi\left(u,v\right)\exp\left(2\pi iux+2\pi ivy\right)dudv\right)$$

$$=\iint\left(u^2+v^2\right)\Phi\left(u,v\right)\exp\left(2\pi iux+2\pi ivy\right)dudv$$

$$=FT^{-1}\left(\left(u^2+v^2\right)\Phi\left(u,v\right)\right)$$

Using this result gives the image amplitude as

$$\psi_i\left(x,y\right)=\exp\left[-i\sigma V_t\left(x,y\right)\right]-\left(i\pi\varepsilon\lambda/4\pi^2\right)\nabla^2\left\{\exp\left[-i\sigma V_t\left(x,y\right)\right]\right\}$$

$$=\exp\left[-i\sigma V_t\left(x,y\right)\right]+\left(i\pi\varepsilon\lambda\sigma/4\pi^2\right)\exp\left[-i\sigma V_t\left(x,y\right)\right]$$

$$\times\left\{\sigma\left(V_t\left(x,y\right)\right)^2+i\nabla^2 V_t\left(x,y\right)\right\}$$

(12.20)

Hence, the image intensity obtained, with approximation to first order, is

$$I\left(x,y\right)\approx1-\left(\varepsilon\lambda\sigma/2\pi\right)\nabla^2 V_t\left(x,y\right)$$

(12.21)

Using Poisson's equation,

$$\nabla^2 V_t\left(x,y\right)=-\rho_p\left(x,y\right)/\varepsilon_0\varepsilon_r$$

We have

$$I\left(x,y\right)=1+\left(\varepsilon\lambda\sigma/2\pi\varepsilon_0\varepsilon_r\right)\rho_p\left(x,y\right)$$

(12.22)

Figure 12.6 shows HRTEM images of an apatite specimen collected using an aberration corrected TEM with a resolution at the sub-angstrom level.

(a)

(b)

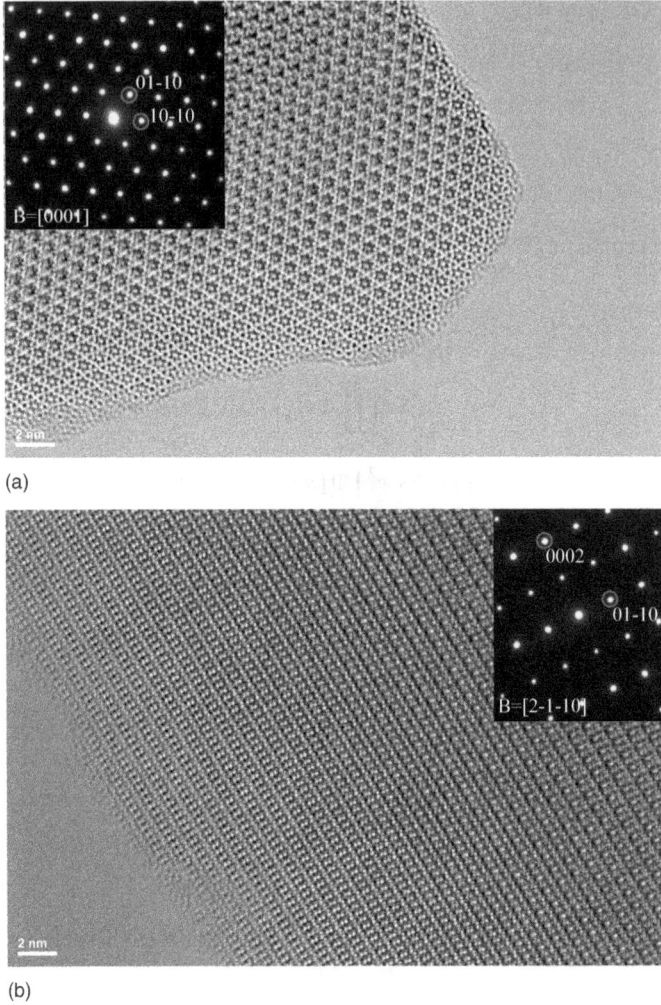

FIGURE 12.6 HRTEM images of synthetic $Nd_8Sr_2(SiO_4)_6O_2$ apatite collected using a 200kV Cs-corrected TEM. (a) Along the [0001] zone axes, (b) along the [2-1-10] zone axis.

12.5 EXAMPLE OF APPLICATION OF HRTEM TO MATERIALS RESEARCH

In this section, we use a perovskite-type coating as an example and show how to analyze materials using HRTEM techniques in combination with EELS analysis. In a Ministry of Education of Singapore (MOE) Tier 2 collaborative project "Nano-structured Ferroelectromagnetic Superlattices by Laser MBE for Spintronics", the oxygen pressure-mediated cation stoichiometry, microstructure and properties of epitaxial perovskite thin films were investigated (Li et al., 2012, 2014). One of the objectives of the research was to find out the effect of oxygen vacancies, which can be tuned by oxygen pressure, on the formation of an interfacial and surface insulating dead layer in lanthanum strontium manganite (LSMO) films. The samples are schematically shown in Figure 12.7. Furthermore, $LSMO/BaTiO_3$ multiferroic superlattices with robust ferroelectricity and ferromagnetism at room temperature by mediating oxygen pressure were fabricated.

In the HRTEM and EELS analysis, two samples were analyzed. (I) Annealed sample: Grown under lower oxygen pressure (1 Pa) first and then fully oxygen annealed to exclude the oxygen vacancies. (II) Reference sample: Grown under higher oxygen pressure (26 Pa), then oxygen annealed.

The cross-sectional TEM specimens were prepared in a method described in Li et al. (2012). The cross-sectional view of a TEM is shown in Figure 12.8. EELS analysis and composition variations are shown in Figures 12.9 and 12.10.

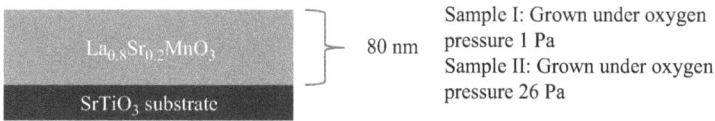

FIGURE 12.7   Pulse laser deposited $La_{0.8}Sr_{0.2}MnO_3$ coating on an $SrTiO_3$ substrate.

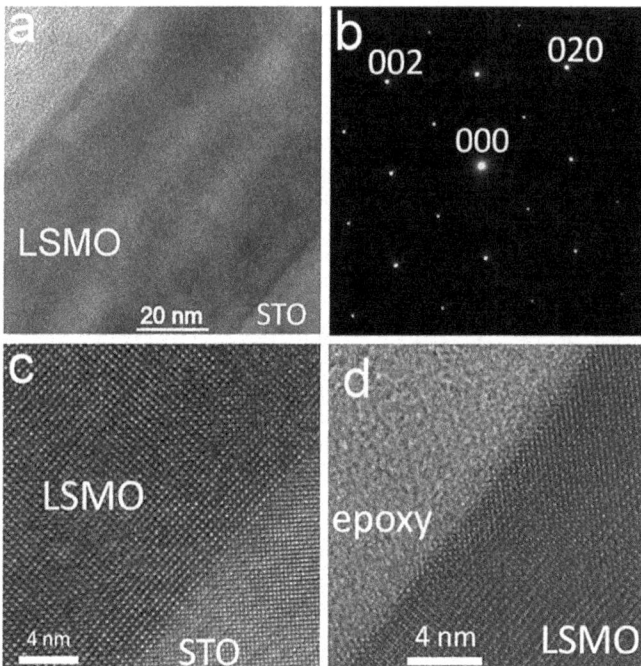

FIGURE 12.8   Morphology of the annealed LSMO thin film. (a) An overview at low magnification. (b) Selected-area electron diffraction pattern of the LSMO film at the [100] zone axis. (c) HRTEM image of the interface of LSMO with the $SrTiO_3$ substrate. (d) HRTEM image of the LSMO surface.

(Source: Li et al. (2012), with the permission of John Wiley and Sons.)

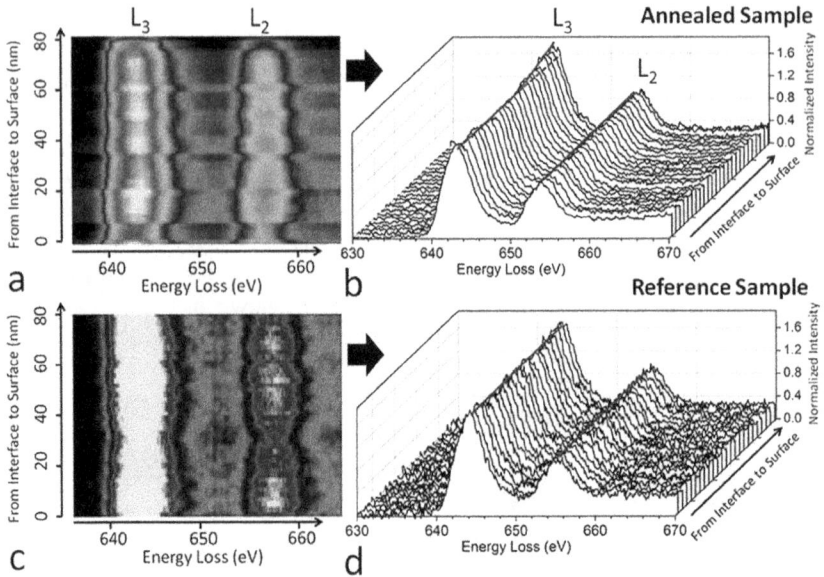

FIGURE 12.9 EELS spectrum line scan of the Mn $L_{2,3}$ edges. (a) and (c) EELS images of the Mn $L_{2,3}$ edges in annealed sample and reference sample. (b) and (d) The side view of Mn $L_{2,3}$ edges with L_2 peak intensity normalized in an annealed sample and reference sample. The red line shows the variation of the L_3/L_2 peak intensity ratio.

(Source: Li et al. (2012), with the permission of John Wiley and Sons.)

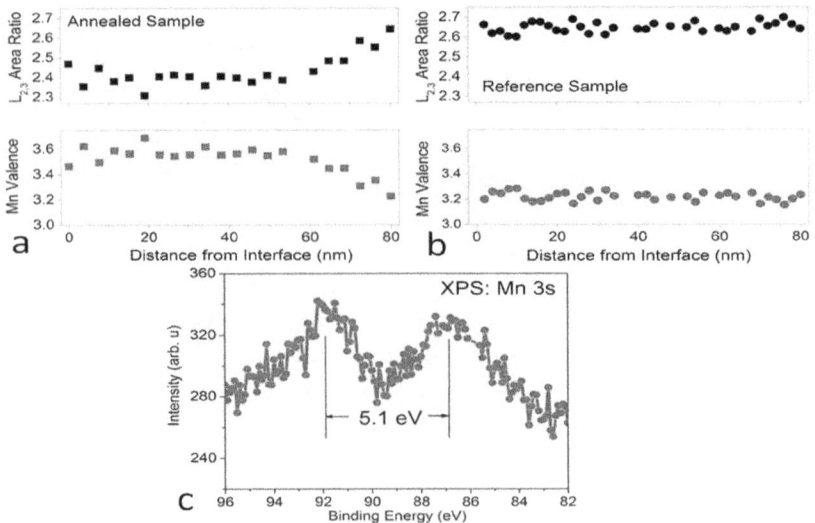

FIGURE 12.10 Variation of Mn nominal valence in LSMO films. (a) and (b) The exacted $L_{2,3}$ ratio and corresponding Mn valence vs. the distance from the interface in an annealed sample and the reference sample, respectively. (c) The XPS peak of the Mn 3s band in an annealed sample. The energy separation between two peaks indicates ΔE_{3s}.

(Source: Li et al. (2012), with the permission of John Wiley and Sons.)

Through analyzing crystal structures, the HRTEM images and the EELS analysis results, some conclusions were obtained. In the sample grown at low oxygen pressure, the lattices of LSMO were expanded due to tensile strain from the STO substrate, hence there was enough space for the La and Sr cations to be accommodated in Mn sites. At the initial growth stage, there was an excess of La and an excess of Sr in comparison with Mn. At a later stage, strain was relaxed, and there was not enough space for La/Sr to sit at Mn sites (Li et al., 2012). Obviously, the oxygen vacancies during film growth at the lower oxygen pressure enhanced the A-site cation excess and Mn deficiency, resulting in decreased saturation magnetization.

To better understand the structure–property relationship for a material system, TEM plays an important role. A modern analytical TEM is equipped with EELS and EDS, therefore the analysis based on imaging, diffraction and spectroscopy in a nano area provide key guidance of how to improve the material's structure and obtain the desired properties.

The spectroscopy analyses on the electronic structures of crystals are not the focus of this textbook. To help students read some related physics textbook while conducting spectroscopy analysis, an appendix on band structure is provided (see Appendix 2).

SUMMARY

In the early 1970s direct observation of crystal structures of inorganic compounds and organic compounds, for example chlorinated copper phthalocyanine, were realized.

With the new development of Cs-corrected HRTEM in the 1990s, sub-angstrom resolution microscopes emerged. Using these new microscopes, light atoms such as oxygen and lithium columns can be resolved. These new types of microscopes extend the first crossover of the contrast transfer function, and slow down the dropping of the envelope function at the diffraction plane.

Other than spherical aberration, other high order aberrations can be corrected for the further improvement of the resolution of TEMs.

The HRTEM images for very thin crystals are often interpreted as the projected potential of a thin specimen. However, in many cases, we need to use software to calculate and interpret the HRTEM images based on the crystal structure model.

REFERENCES

Buseck, P., Cowley, J. M. (John M.) and Eyring, L. (1988) *High-resolution transmission electron microscopy and associated techniques/editors, Peter R. Buseck, John M. Cowley, Leroy Eyring.* New York: Oxford University Press.

Erni, R. (2010) *Aberration-corrected imaging in transmission electron microscopy.* Singapore: World Scientific.

Kirkland, A. I. and Meyer, R. R. (2004) '"Indirect" high-resolution transmission electron microscopy: aberration measurement and wavefunction reconstruction', *Microscopy and Microanalysis*, 10, pp. 401–413.

Kirkland, E. J. (1998) *Advanced computing in electron microscopy [electronic resource]/by Earl J. Kirkland.* Boston, MA: Springer US. doi:10.1007/978-1-4757-4406-4.

Li, Z. et al. (2012) 'Interface and surface cation stoichiometry modified by oxygen vacancies in epitaxial manganite films', *Advanced Functional Materials*, 22(20), pp. 4312–4321. doi:10.1002/adfm.201200143.

Li, Z. et al. (2014) 'An epitaxial ferroelectric tunnel junction on silicon', *Advanced Materials*, 26(42), pp. 7185–7189. doi:10.1002/adma.201402527.

O'Keefe, M. A. (1992) '"Resolution" in high-resolution electron microscopy', *Ultramicroscopy*, 47, pp. 282–297.

O'Keefe, M. A. et al. (2001) 'Sub-Ångstrom high-resolution transmission electron Sub- A microscopy at 300 keV', *Ultramicroscopy*, 89, pp. 215–241.

Reimer, L. (1997) *Transmission electron microscopy: physics of image formation and micro-analysis/Ludwig Reimer*. 4th ed. Berlin: Springer (Springer series in optical sciences, v. 36).

Spence, J. C. H. (2013) *High-resolution electron microscopy/John C.H. Spence, Department of Physics and Astronomy, Arizona State University/LBNL California*. 4th ed. Oxford: Oxford University Press.

Stadelmann, P. A. (1987) 'EMS – a software package for electron diffraction analysis and HREM image simulation in materials science', *Ultramicroscopy*, 21(2), pp. 131–145. doi:10.1016/0304-3991(87)90080-5.

Uhlemann, S. and Haider, M. (1998) 'Residual wave aberrations in the first spherical aberration corrected transmission electron microscope', *Ultramicroscopy*, 72(3), pp. 109–119. doi:10.1016/S0304-3991(97)00102-2.

Williams, D. B. (1996) *Transmission electron microscopy [electronic resource]: A textbook for materials science/by David B. Williams, C. Barry Carter*. Edited by C. B. Carter. Boston, MA: Springer US. doi:10.1007/978-1-4757-2519-3.

Appendix 1
Fourier Series, Fourier Transforms, and Associated Equations

In one dimension, if a function $f(x)$ can be expressed by the following trigonometric series

$$f(x) = \frac{a_0}{2} + \sum_{n=1}^{\infty} a_n \cos nx + b_n \sin nx \tag{A1.1}$$

it is called the Fourier series of $f(x)$(Jeffrey, 2005), where x is within the range $[-\pi, \pi]$, and the coefficients are

$$a_n = \frac{1}{\pi} \int_{-\pi}^{\pi} f(x) \cos nx \, dx \quad (n = 0,1,2,3,\ldots)$$

$$b_n = \frac{1}{\pi} \int_{-\pi}^{\pi} f(x) \sin nx \, dx \quad (n = 1,2,3,\ldots) \tag{A1.2}$$

Dirichlet conditions must be satisfied if a Fourier series truly represents the function $f(x)$. The $\int |f(x)| dx$ over a complete period must be finite, and $f(x)$ may have no more than a finite number of discontinuities in any finite interval (Croft and Davison, 2015).

In fact, some mathematical calculations used in this textbook are only true under some conditions. As we have just introduced the ideas on how to derive more simplified expressions, we have not focused on how to present the conditions based on mathematics textbooks.

As $\{1, \cos x, \sin x, \cos 2x, \sin 2x, \ldots \cos nx, \sin nx, \ldots\}$ are orthogonal over the interval $[-\pi, \pi]$, the coefficients can be obtained in the following way. From

$$\int_{-\pi}^{\pi} f(x) dx = \frac{a_0}{2} \cdot 2\pi = a_0 \pi$$

we obtain

$$a_0 = \frac{1}{\pi} \int_{-\pi}^{\pi} f(x) dx$$

From

$$\int_{-\pi}^{\pi} f(x) \cos mx dx = \frac{a_0}{2} \int_{-\pi}^{\pi} \cos mx dx + \sum_{n=1}^{\infty} \left(a_n \int_{-\pi}^{\pi} \cos nx \cos mx dx + b_n \int_{-\pi}^{\pi} \sin nx \cos mx dx \right)$$

$$= \int_{-\pi}^{\pi} a_m \cos^2 mx dx = a_m \pi$$

we obtain

$$a_m = \frac{1}{\pi} \int_{-\pi}^{\pi} f(x) \cos mx dx$$

Similarly, we have

$$b_m = \frac{1}{\pi} \int_{-\pi}^{\pi} f(x) \sin mx dx$$

The above coefficients are derived based on the fact that $\{1, \cos x, \sin x, \cos 2x, \sin 2x, \ldots \cos nx, \sin nx, \ldots\}$ are orthogonal within the range $[-\pi, \pi]$, or the range $[0, 2\pi]$, or any other range $[c, c+2\pi]$, which can be proved as follows.

First, we can calculate the integral of the product of any two different functions from the sequence, and the integral is zero. All these integrals are shown below:

$$\int_{c}^{c+2\pi} \cos nx dx = \int_{0}^{2\pi} \cos nx dx = 0$$

$$\int_{c}^{c+2\pi} \sin nx dx = \int_{0}^{2\pi} \sin nx dx = 0$$

$$(n = 1, 2, \ldots)$$

and

$$\int_{c}^{c+2\pi} \sin nx \cos mx\,dx = 0$$

$$\int_{c}^{c+2\pi} \sin nx \sin mx\,dx = 0$$

$$\int_{c}^{c+2\pi} \cos nx \cos mx\,dx = 0$$

$$(n \neq m)$$
$$(n = 1, 2, \ldots)$$
$$(m = 1, 2, \ldots)$$

During the above calculations, we used the following formulas

$$2\sin nx \cos mx = \sin(nx + mx) + \sin(nx - mx)$$
$$-2\sin nx \sin mx = \cos(nx + mx) - \cos(nx - mx)$$
$$2\cos nx \cos mx = \cos(nx + mx) + \cos(nx - mx)$$

Second, we can calculate the integral of the product of any function with itself, and the integral is non-zero. All these integrals are shown below:

$$\int_{c}^{c+2\pi} 1^2\,dx = 2\pi$$

$$\int_{c}^{c+2\pi} \cos^2 nx\,dx = \int_{0}^{2\pi} \cos^2 nx\,dx = \int_{0}^{2\pi} \frac{1 + \cos 2nx}{2}\,dx = \pi$$

$$\int_{c}^{c+2\pi} \sin^2 nx\,dx = \int_{c}^{2\pi} \sin^2 nx\,dx = \int_{0}^{2\pi} \frac{1 - \cos 2nx}{2}\,dx = \pi$$

$$(n = 1, 2, \ldots)$$

In the case where the period for $f(x)$ is λ, but not 2π, or the Fourier series is in the range of $\left[-\dfrac{\lambda}{2}, \dfrac{\lambda}{2}\right]$, we have

$$f(x) = \frac{a_0}{2} + \sum_{n=1}^{\infty} a_n \cos nkx + b_n \sin nkx \qquad (A1.3)$$

$$a_n = \frac{2}{\lambda} \int_{-\lambda/2}^{\lambda/2} f(x) \cos nkx\, dx \qquad \left(k = \frac{2\pi}{\lambda}, n = 0, 1, 2, \dots \right)$$

$$b_n = \frac{2}{\lambda} \int_{-\lambda/2}^{\lambda/2} f(x) \sin nkx\, dx \qquad \left(k = \frac{2\pi}{\lambda}, n = 1, 2, 3, \dots \right)$$

(A1.4)

In diffraction analysis, we often use the exponential form of the Fourier series. This form within the rage of $[-\pi, \pi]$ is

$$f(x) = \sum_{n=-\infty}^{\infty} c_n e^{inx} \qquad (A1.5)$$

$$c_n = \frac{1}{2\pi} \int_{-\pi}^{\pi} f(x) e^{-inx}\, dx$$

$$(n = 0, \pm 1, \pm 2, \dots)$$

(A1.6)

The exponential form of the Fourier series can be deduced as follows.

If we employ the Euler formula, $e^{ix} = \cos x + i \sin x$, the Fourier series can be expressed in the exponential form.

We know that

$$\cos x = \frac{1}{2}\left(e^{ix} + e^{-ix} \right)$$

$$\sin x = \frac{1}{2i}\left(e^{ix} - e^{-ix} \right) = -\frac{i}{2}\left(e^{ix} - e^{-ix} \right)$$

Therefore,

$$f(x) = \frac{a_0}{2} + \sum_{n=1}^{\infty} a_n \cos nx + b_n \sin nx$$

$$= \frac{a_0}{2} + \sum_{n=1}^{\infty} \left(\frac{a_n - ib_n}{2} e^{inx} + \frac{a_n + ib_n}{2} e^{-inx} \right)$$

$$= c_0 + \sum_{n=1}^{\infty} \left(c_n e^{inx} + c_{-n} e^{-inx} \right)$$

$$= \sum_{n=-\infty}^{\infty} c_n e^{inx}$$

where

$$\frac{a_0}{2} = c_0$$

$$\frac{a_n - ib_n}{2} = c_n$$

$$\frac{a_n + ib_n}{2} = c_{-n}$$

$$(n = 1, 2, ...)$$

Now let us analyze the expression for c_n:

$$c_0 = \frac{a_0}{2} = \frac{1}{2\pi} \int_{-\pi}^{\pi} f(x) \, dx$$

$$c_n = \frac{a_n - ib_n}{2} = \frac{1}{2\pi} \int_{-\pi}^{\pi} f(x) \cos nx \, dx - i \frac{1}{2\pi} \int_{-\pi}^{\pi} f(x) \sin nx \, dx$$

$$= \frac{1}{2\pi} \int_{-\pi}^{\pi} f(x) e^{-inx} \, dx$$

$$c_{-n} = \frac{a_n + ib_n}{2} = \frac{1}{2\pi} \int_{-\pi}^{\pi} f(x) \cos nx \, dx + i \frac{1}{2\pi} \int_{-\pi}^{\pi} f(x) \sin nx \, dx$$

$$= \frac{1}{2\pi} \int_{-\pi}^{\pi} f(x) e^{inx} \, dx$$

$$(n = 1, 2, ...)$$

And we can just use one common expression for the coefficients

$$c_n = \frac{1}{2\pi} \int_{-\pi}^{\pi} f(x) e^{-inx} \, dx$$

$$(n = 0, \pm 1, \pm 2, ...)$$

The exponential form of the Fourier series we have just presented is in the range of $[-\pi, \pi]$.

In the case where we use the period λ, or discuss the Fourier series in the range of $\left[-\dfrac{\lambda}{2}, \dfrac{\lambda}{2} \right]$, we have

$$f(x) = \sum_{n=-\infty}^{\infty} c_n e^{inkx} \tag{A1.7}$$

$$c_n = \frac{1}{\lambda}\int_{-\lambda/2}^{\lambda/2} f(x)e^{-inkx}dx$$

$$\left(k = \frac{2\pi}{\lambda}, n = 0, \pm 1, \pm 2, \ldots\right)$$

(A1.8)

In the case where we use the period T, or discuss the Fourier series in the range of $\left[-\dfrac{T}{2}, \dfrac{T}{2}\right]$, we have

$$f(t) = \sum_{n=-\infty}^{\infty} c_k e^{in\omega t}$$

(A1.9)

$$c_n = \frac{1}{T}\int_{-T/2}^{T/2} f(t)e^{-in\omega t}dt$$

$$\left(\omega = \frac{2\pi}{T}, n = 0, \pm 1, \pm 2, \ldots\right)$$

(A1.10)

From the Fourier series obtained, we can explain further, and informally derive the Fourier transforms. This will help students understand and use the Fourier transforms properly.

It has been shown that the Fourier series in the range of $\left[-\dfrac{\lambda}{2}, \dfrac{\lambda}{2}\right]$ is

$$f(x) = \sum_{n=-\infty}^{\infty} c_n e^{inkx}$$

$$c_n = \frac{1}{\lambda}\int_{-\lambda/2}^{\lambda/2} f(x)e^{-inkx}dx$$

$$\left(k = \frac{2\pi}{\lambda}, n = 0, \pm 1, \pm 2, \ldots\right)$$

From the Fourier series, $f(x)$ can be written as

$$f(x) = \sum_{n=-\infty}^{\infty} c_n e^{inkx} = \sum_{n=-\infty}^{\infty}\left(\frac{1}{\lambda}\int_{-\lambda/2}^{\lambda/2} f(x)e^{-inkx}dx\right)e^{inkx}$$

$$= \sum_{n=-\infty}^{\infty}\left(\frac{k}{2\pi}\int_{-\pi/k}^{\pi/k} f(x)e^{-inkx}dx\right)e^{inkx}$$

(A1.11)

Let $nk = k_n$.

Then $\Delta k = k_n - k_{n-1} = nk - (n-1)k = k$ and $\int_{-\pi/k}^{\pi/k} f(x)e^{-inkx}dx = \int_{-\pi/k}^{\pi/k} f(x)e^{-ik_nx}dx = F(k_n)$.

Hence (A1.11) can be rewritten as

$$f(x) = \sum_{n=-\infty}^{\infty} \frac{1}{2\pi} F(k_n)e^{ik_nx}\Delta k$$

When $\lambda \to \infty$, or $\dfrac{\pi}{k} = \dfrac{\lambda}{2} \to \infty$,

$$F(k_n) = \int_{-\pi/k}^{\pi/k} f(x)e^{-ik_nx}dx = \int_{-\infty}^{\infty} f(x)e^{-ik_nx}dx \Rightarrow F(k) = \int_{-\infty}^{\infty} f(x)e^{-ikx}dx$$

When $\lambda \to \infty$, or $\Delta k = k = \dfrac{2\pi}{\lambda} \to 0$, summation $f(x) = \sum_{n=-\infty}^{\infty} \dfrac{1}{2\pi} F(k_n)e^{ik_nx}\Delta k$ can

be replaced by integration, or

$$\sum_{n=-\infty}^{\infty} \frac{1}{2\pi} F(k_n)e^{ik_nx}\Delta k$$

$$= \frac{1}{2\pi} \sum_{n=-\infty}^{\infty} F(k_n)e^{ik_nx}\Delta k$$

$$\to \frac{1}{2\pi} \int_{-\infty}^{\infty} F(k)e^{ikx}dk$$

Therefore,

$$F(k) = \int_{-\infty}^{\infty} f(x)e^{-ikx}dx$$

$$f(x) = \frac{1}{2\pi} \int_{-\infty}^{\infty} F(k)e^{ikx}dk$$

(A1.12)

We call $F(k) = \int_{-\infty}^{\infty} f(x)e^{-ikx}dx$ the Fourier transform of $f(x)$, and

$f(x) = \dfrac{1}{2\pi} \int_{-\infty}^{\infty} F(k)e^{ikx}dk$ the inverse Fourier transform.

The Fourier transform exists if the Dirichlet conditions are satisfied. It is sufficient for the application of the Fourier transform if $f(x)$ is piecewise continuous, of bound variation on any finite interval and absolutely integrable (Antimirov et al., 1993):

$$\int_{-\infty}^{\infty} |f(x)|\, dx < \infty$$

In one-dimension descriptions, the Fourier transform and inverse Fourier transform can be expressed in different ways. For example,

$$g(k) = FT\{f(x)\} = \int_{-\infty}^{+\infty} f(x)\exp\{-2\pi ikx\}\, dx$$
$$f(x) = FT^{-1}\{g(k)\} = \int_{-\infty}^{+\infty} g(k)\exp\{2\pi ikx\}\, dk$$

(A1.13)

In physics, we can use $\exp\{-ikx\}$ and $\exp\{ikx\}$ to replace $\exp\{-i2\pi kx\}$ and $\exp\{i2\pi kx\}$, which means "2π" is incorporated into "k". In this case, we can write

$$g(k) = FT\{f(x)\} = \int_{-\infty}^{+\infty} f(x)\exp\{-ikx\}\, dx$$
$$f(x) = FT^{-1}\{g(k)\} = \frac{1}{2\pi}\int_{-\infty}^{+\infty} g(k)\exp\{ikx\}\, dk$$

or

$$g(k) = FT\{f(x)\} = \frac{1}{\sqrt{2\pi}}\int_{-\infty}^{+\infty} f(x)\exp\{-ikx\}\, dx$$
$$f(x) = FT^{-1}\{g(k)\} = \frac{1}{\sqrt{2\pi}}\int_{-\infty}^{+\infty} g(k)\exp\{ikx\}\, dk$$

or

$$g(k) = FT\{f(x)\} = \frac{1}{2\pi}\int_{-\infty}^{+\infty} f(x)\exp\{-ikx\}\, dx$$
$$f(x) = FT^{-1}\{g(k)\} = \int_{-\infty}^{+\infty} g(k)\exp\{ikx\}\, dk$$

There are many associated equations, and the following are just presented as examples.

We can calculate the Fourier transform of a delta function:

$$FT\{\delta(x)\} = \int_{-\infty}^{+\infty} \delta(x)\exp\{-2\pi ikx\}\,dx = \int_{-\infty}^{+\infty} \delta(x)\exp\{-2\pi ik(0)\}\,dx = \int_{-\infty}^{+\infty} \delta(x)\,dx = 1$$

Therefore, the Fourier transform pairs are

$$FT\{\delta(x)\} = \int_{-\infty}^{+\infty} \delta(x)\exp\{-2\pi ikx\}\,dx = 1$$

$$FT^{-1}\{1\} = \int_{-\infty}^{+\infty} 1\cdot\exp\{2\pi ikx\}\,dk = \delta(x)$$

(A1.14)

In quantum mechanics,

$$\int_{-\infty}^{+\infty} \exp\left[\frac{i}{\hbar}(p_x - p_x')x\right]dx = (2\pi\hbar)\int_{-\infty}^{+\infty} \exp\left[2\pi i(p_x - p_x')\left(\frac{x}{2\pi\hbar}\right)\right]d\left(\frac{x}{2\pi\hbar}\right) = 2\pi\hbar\delta(p_x - p_x')$$

is a very important conclusion for the wave functions expressed in momentum. We can also write

$$FT^{-1}\{\delta(x)\} = \int_{-\infty}^{+\infty} \delta(x)\exp\{2\pi ikx\}\,dx = 1$$

$$FT\{1\} = \int_{-\infty}^{+\infty} 1\cdot\exp\{-2\pi ikx\}\,dk = \delta(x)$$

The following are important equations associated with convolution and Fourier transforms.

$$FT\{f(x)\otimes g(x)\} = FT\{f(x)\}\cdot FT\{g(x)\}$$

$$FT\{f(x)\cdot g(x)\} = FT\{f(x)\}\otimes FT\{g(x)\}$$

(A1.15)

The convolution of $f(x)$ and $g(x)$ is defined as

$$f(x)\otimes g(x) = \int_{-\infty}^{+\infty} f(x')g(x-x')\,dx'$$

$$= \int_{-\infty}^{+\infty} f(x-x')g(x')\,dx' = g(x)\otimes f(x)$$

(A1.16)

The convolution of $f(x)$ with a delta function is expressed as

$$f(x) \otimes \delta(x - x_0) = f(x - x_0) \tag{A1.17}$$

Therefore, in the crystallography part, a crystal is written as a lattice $\otimes$ basis.

REFERENCES

Antimirov, M. Y., Kolyshkin, A. A. and Vaillancourt, R. (1993) *Applied integral transforms/M. Ya. Antimirov, A.A. Kolyshkin, Rémi Vaillancourt.* Providence, R.I., USA: American Mathematical Society (CRM monograph series, v. 2).

Croft, T. and Davison, R. (2015) *Mathematics for engineers/Anthony Croft, Robert Davison.* 4th ed. Harlow: Pearson Prentice Hall.

Jeffrey, A. (2005) *Mathematics for engineers and scientists/Alan Jeffrey.* 6th ed. Boca Raton, FL: Chapman & Hall/CRC.

Appendix 2
Nearly Free Electron Approximation and the Band Structure of Crystals

In solid state physics, nearly free electron approximation and the tight binding model are two simplified extreme conditions for deriving energy band structures in crystals and representing band gap formation. Under the nearly free electron approximation, interactions between electrons are ignored. This approximation allows use of Bloch's theorem which describes the wave functions of electrons in the periodic potential field of a crystal. It can well characterize the crystal's electronic properties, such as the differences between metals and insulators, the band gap formation and the density of states. The nearly free electron model works well in metals where distances between neighboring atoms are small. The tight binding model is another extreme, and works well in materials with limited overlap between atomic orbitals and potentials on neighboring atoms. In investigations of the electronic structure of crystals, X-ray absorption spectroscopy and electron energy loss spectroscopy may be employed, and some basic knowledge presented here will assist materials scientists and engineers in their spectroscopy analysis.

Bloch's theorem shows that the wave functions for electrons subjected to periodic potential take the form

$$\psi_{\bar{k}}(\vec{r}) = u_{\bar{k}}(\vec{r}) e^{i\vec{k} \cdot \vec{r}}$$
$$u_{\bar{k}}(\vec{r} + \vec{R}_n) = u_{\bar{k}}(\vec{r}) \tag{A2.1}$$

or

$$\psi_{\bar{k}}(\vec{r} + \vec{R}_n) = e^{i\vec{k} \cdot \vec{R}_n} \psi_{\bar{k}}(\vec{r}) \tag{A2.2}$$

where

$\vec{R}_n = n_1 \vec{a}_1 + n_2 \vec{a}_2 + n_3 \vec{a}_3$ is a direct lattice vector, and n_1, n_2 and n_3 are integers.

For a crystal with a parallelepiped shape, the edge lengths are

$$L_1 = N_1 a_1$$
$$L_2 = N_2 a_2$$
$$L_3 = N_3 a_3$$

The boundary conditions are

$$\psi_{\bar{k}}\left(\vec{r}+N_i\vec{a}_i\right)=\psi_{\bar{k}}\left(\vec{r}\right)$$

$$i=1,2,3$$

(A2.3)

Based on Equation (A2.2), we also know that

$$\psi_{\bar{k}}\left(\vec{r}+N_i\vec{a}_i\right)=e^{i\bar{k}\cdot\left(N_i\vec{a}\right)_i}\psi_{\bar{k}}\left(\vec{r}\right)$$

$$i=1,2,3$$

(A2.4)

Comparing Equations (A2.3) and (A2.4), we obtain

$$e^{iN_i\bar{k}\cdot\vec{a}_i}=1$$

$$i=1,2,3$$

and

$$N_i\bar{k}\cdot\vec{a}_i=2\pi m_i$$

$$i=1,2,3$$

where m_i is an integer.

Consider that $\bar{k}=u_1\vec{b}_1+u_2\vec{b}_2+u_3\vec{b}_3$ is a wave vector in the reciprocal lattice; then we have

$$\bar{k}\cdot\vec{a}_i=2\pi u_i$$

$$i=1,2,3$$

Therefore, we can write

$$N_i\bar{k}\cdot\vec{a}_i=2\pi N_i u_i$$

$$i=1,2,3$$

And we have

$$N_i u_i=m_i$$

$$i=1,2,3$$

Hence we rewrite the wave vector $\bar{k}=u_1\vec{b}_1+u_2\vec{b}_2+u_3\vec{b}_3$ in the reciprocal lattice as

$$\bar{k}=\frac{m_1}{N_1}\vec{b}_1+\frac{m_2}{N_2}\vec{b}_2+\frac{m_3}{N_3}\vec{b}_3.$$

The volume associated with each state in $\vec{k}$ space is

$$\frac{\vec{b}_1}{N_1} \cdot \left(\frac{\vec{b}_2}{N_2} \times \frac{\vec{b}_3}{N_3} \right) = \frac{(2\pi)^3}{V} \tag{A2.5}$$

where $V = N_1 N_2 N_3 V_{cell}$ is the volume of the crystal.

To simply the discussion, we just consider a one-dimensional crystal with periodic potentials.

From the boundary condition and $\psi_k(x + N_1 a_1) = e^{iN_1 ka_1} \psi_k(x)$, we know that $N_1 k a_1 = 2\pi l_1$ and that l_1 is an integer. Hence

$$k = l_1 \frac{2\pi}{N_1 a_1} \tag{A2.6}$$

Under the empty lattice approximation, $U(x) = 0$, $\left[-\frac{\hbar^2}{2m} \frac{d^2}{dx^2} + U(x) \right] \psi^{(0)}(x) = E\psi^{(0)}(x)$; the wave function and energy are:

$$\psi_k^{(0)}(x) = \frac{1}{\sqrt{L_1}} e^{ikx}$$

$$E_k^{(0)} = \frac{\hbar^2 k^2}{2m}$$

$$k = \frac{l_1}{N_1 a_1}(2\pi), \ L_1 = N_1 a_1$$

In the following discussions, we just use a to replace a_1. The periodic potential experienced by an electron is expressed by $U(x)$, and it is known that $U(x) = U(x + na)$. The potential $U(x)$ includes the interaction of the electron with all atoms in the crystal, as well as its interaction with other electrons (Omar, 1993), and it can be expressed using a Fourier series:

$$U(x) = \Sigma U_n e^{i\frac{2\pi}{a}nx} = U_0 + \sum_{n \neq 0} U_n e^{i\frac{2\pi}{a}nx}$$

where

$$U_0 = \frac{1}{L_1} \int_0^{L_1} U(x)dx = \bar{U}$$

$$U_n = \frac{1}{L_1} \int_0^{L_1} U(x) \exp\left(-i\frac{2\pi}{a}nx \right) dx$$

Let $U_0 \left(= \bar{U} \right) = 0$, as assuming the average potential to be zero will not affect the discussion.

Because the potential is a real term, and not complex, $U(x) = U^*(x), U_n^* = U_{-n}$.

To resolve the Schrödinger equation

$$\left[-\frac{\hbar^2}{2m}\frac{d^2}{dx^2} + U(x) \right] \psi(x) = E\psi(x)$$

The periodic potential felt by the electron is considered as a perturbation; non-degenerate perturbation theory is employed below.

$$H\psi_k = E(k)\psi_k$$

where

$$H = -\frac{\hbar^2}{2m}\frac{d^2}{dx^2} + U(x) = H_0 + H'$$

$$H_0 = -\frac{\hbar^2}{2m}\frac{d^2}{dx^2} + U_0$$

The perturbation is

$$H' = \sum_{n \neq 0} U_n e^{i\frac{2\pi}{a}nx} = \Delta U$$

The energy of an electron and its wavefunction can be written as

$$E(k) = E^{(0)}(k) + E^{(1)}(k) + E^{(2)}(k) + \cdots$$
$$\psi_k(x) = \psi_k^{(0)}(x) + \psi_k^{(1)}(x) + \psi_k^{(2)}(x) + \cdots$$

As an approximation, the corrections higher than second order are not included. Based on the non-degenerate perturbation theory (Zhou, 1979; Liboff, 2003),

$$\left(H_0 - E^{(0)}(k) \right)\psi_k^{(0)}(x) = 0 \tag{A2.7}$$

$$\left(H_0 - E^{(0)}(k) \right)\psi_k^{(1)}(x) = -\left(H' - E^{(1)}(k) \right)\psi_k^{(0)}(x) \tag{A2.8}$$

$$\left(H_0 - E^{(0)}(k) \right)\psi_k^{(2)}(x) = -\left(H' - E^{(1)}(k) \right)\psi_k^{(1)}(x) + E^{(2)}(k)\psi_k^{(0)}(x) \tag{A2.9}$$

From Equation (A2.7)

$$H_0 \psi_k^{(0)}(x) = E^{(0)}(k) \psi_k^{(0)}(x)$$

The energy for an electron, without considering the perturbation, is obtained as follows:

$$E^{(0)}(k) = \frac{\hbar^2 k^2}{2m} + U_0$$

Let $U_0 = 0$, then

$$E^{(0)}(k) = \frac{\hbar^2 k^2}{2m}$$

The corresponding wave function is

$$\psi_k^{(0)}(x) = \frac{1}{\sqrt{L_1}} e^{ikx}$$

The first-order and second-order energy corrections can be obtained from Equations (A2.8) and (A2.9):

$$E^{(1)}(k) = H'_{kk} = \int_0^{L_1} \psi_k^{(0)*}(x) \sum_{n \neq 0} U_n(x) e^{i\frac{2\pi}{a}nx} \psi_k^{(0)}(x) dx = \overline{\Delta U} = 0$$

$$E^{(2)}(k) = \sum_{k' \neq k} \frac{|H'_{k'k}|^2}{E^{(0)}(k) - E^{(0)}(k')}$$

where

$$H'_{kk'} = H'_{k'k}{}^*$$
$$|H'_{kk'}|^2 = |H'_{k'k}|^2$$

$|H'_{kk'}|$ can be calculated as follows:

$$|H'_{k'k}| = \int_0^{L_1} \psi_{k'}^{(0)*}(x) \sum_{n \neq 0} U_n(x) e^{i\frac{2\pi}{a}nx} \psi_k^{(0)}(x) dx$$

$$= \frac{1}{L_1} \int_0^{L_1} \sum_{n \neq 0} U_n(x) e^{i\frac{2\pi}{a}nx} e^{-i(k'-k)x} dx$$

Consider the $k' - k$ values; note that

$$\left|H'_{k'k}\right| = \begin{cases} U_n & \text{if} \quad k'-k=\dfrac{2\pi}{a}n \\ 0 & \text{if} \quad k'-k\neq\dfrac{2\pi}{a}n \end{cases} \tag{A2.10}$$

Therefore, the second-order energy shift is

$$E^{(2)}(k) = \sum_{k'\neq k} \frac{\left|H'_{kk'}\right|^2}{E^{(0)}(k)-E^{(0)}(k')}$$

$$= \sum_{n\neq 0} \frac{\left|U_n\right|^2}{\dfrac{\hbar^2}{2m}\left[k^2 - \left(k+\dfrac{2\pi}{a}n\right)^2\right]} \tag{A2.11}$$

The perturbed energy up to the second order is expressed as

$$E(k) = E^{(0)}(k) + E^{(1)}(k) + E^{(2)}(k) + \cdots$$

$$= \frac{\hbar^2 k^2}{2m} + \sum_{n\neq 0} \frac{2m\left|U_n\right|^2}{\hbar^2 k^2 - \hbar^2\left(k+\dfrac{2\pi}{a}n\right)^2} + \cdots \tag{A2.12}$$

The wave function is

$$\psi_k(x) = \psi_k^{(0)}(x) + \psi_k^{(1)}(x) + \cdots$$

$$= \psi_k^{(0)}(x) + \sum_{k'\neq k} \frac{H'_{k'k}}{E^{(0)}(k)-E^{(0)}(k')}\psi_{k'}^{(0)}(x) + \cdots$$

Since

$$\psi_k^{(0)}(x) = \frac{1}{\sqrt{L_1}}e^{ikx}$$

$$\psi_{k'}^{(0)}(x) = \frac{1}{\sqrt{L_1}}e^{i\left(k+\frac{2\pi}{a}n\right)x},$$

$\psi_k(x)$ can be expressed as

$$\psi_k(x) = \psi_k^{(0)}(x) + \sum_{k' \neq k} \frac{H'_{k'k}}{E^{(0)}(k) - E^{(0)}(k')} \psi_{k'}^{(0)}(x)$$

$$= \frac{1}{\sqrt{L_1}} e^{ikx} + \sum_{n \neq 0} \frac{2mU_n}{\hbar^2 k^2 - \hbar^2 \left(k + \dfrac{2\pi}{a} n\right)^2} \frac{1}{\sqrt{L_1}} e^{i\left(k + \frac{2\pi}{a} n\right)x} \qquad (A2.13)$$

$$= \frac{1}{\sqrt{L_1}} e^{ikx} \left[1 + \sum_{n \neq 0} \frac{2mU_n e^{i\frac{2\pi}{a} nx}}{\hbar^2 k^2 - \hbar^2 \left(k + \dfrac{2\pi}{a} n\right)^2}\right]$$

or

$$\psi_k(x) = \frac{1}{\sqrt{L_1}} e^{ikx} u_k(x) \qquad (A2.14)$$

where

$$u_k(x) = 1 + \sum_{n \neq 0} \frac{2mU_n e^{i\frac{2\pi}{a} nx}}{\hbar^2 k^2 - \hbar^2 \left(k + \dfrac{2\pi}{a} n\right)^2} \qquad (A2.15)$$

It can be seen that $u_k(x) = u_k(x + na)$, therefore $\psi_k(x) = \dfrac{1}{\sqrt{L_1}} e^{ikx} u_k(x)$ is a Bloch wave.

Equations (A2.12) and (A2.13) are used to express the energy and wave function not near the boundaries of the Brillouin zone.

When $E^{(0)}(k) = E^{(0)}(k')$, it can be seen that the denominator of the perturbation term in (A2.11) vanishes, and $E^{(2)}(k) \to \pm \infty$.

In this case

$$\frac{\hbar^2 k^2}{2m} = \frac{\hbar^2 \left(k + \dfrac{2\pi}{a} n\right)^2}{2m}$$

$$k^2 = \left(k + \frac{2\pi}{a} n\right)^2 = (k + g_n)^2$$

or

$$k = -\frac{\pi}{a}n, \qquad k' = \frac{\pi}{a}n \tag{A2.16}$$

Which means the wave vectors end at the Brillouin zone boundaries.

Since the perturbation theory presumes the smallness of the correction, the above treatments are no longer valid at or near the zone boundaries. At or near the boundary of the Brillouin zone, we must adopt degenerate perturbation treatment. Consider the situation at the boundary of the Brillouin zone:

$$k = -\frac{\pi}{a}n, \qquad k' = \frac{\pi}{a}n$$

or near the boundary of the Brillouin zone:

$$\begin{cases} k = -\frac{\pi}{a}n(1-\Delta) \\ k' = k + \frac{2\pi}{a}n = \frac{\pi}{a}n(1+\Delta) \quad |\Delta| \ll 1 \end{cases} \tag{A2.17}$$

The following quantum mechanical treatment is employed (Huang, 1988; Liboff, 2003; Wahab, 2015).

Consider the combination of the unperturbed wave functions $\psi_{k'}^{(0)}$ and $\psi_{k}^{(0)}$:

$$\psi = a\psi_k^{(0)} + b\psi_{k'}^{(0)} \tag{A2.18}$$

Based on the Schrödinger equation:

$$\left[-\frac{\hbar^2}{2m}\frac{d^2}{dx^2} + U(x) \right]\psi(x) = E\psi(x) \tag{A2.19}$$

As well as

$$\left[-\frac{\hbar^2}{2m}\frac{d^2}{dx^2} + \overline{U(x)} \right]\psi_k^{(0)}(x) = E_k^{(0)}\psi_k^{(0)}(x) \tag{A2.20}$$

$$\left[-\frac{\hbar^2}{2m}\frac{d^2}{dx^2} + \overline{U(x)} \right]\psi_{k'}^{(0)}(x) = E_{k'}^{(0)}\psi_k^{(0)}(x) \tag{A2.21}$$

The calculation (A2.19) − [(A2.20) × a + (A2.21) × b] cancels out the $-\frac{\hbar^2}{2m}\frac{d^2}{dx^2}$ terms. Then substituting $\psi = a\psi_k^{(0)} + b\psi_{k'}^{(0)}$ into it yields

$$a\left(E_k^{(0)} - E + \Delta U \right)\psi_k^{(0)} + b\left(E_{k'}^{(0)} - E + \Delta U \right)\psi_{k'}^{(0)} = 0 \tag{A2.22}$$

If we multiply Equation (A2.22) either by $\psi_k^{(0)*}$ or $\psi_{k'}^{(0)*}$ and integrate, we obtain the following equations:

$$\left\{ \begin{array}{l} \left(E_k^{(0)} - E \right) a + U_n^* b = 0 \\ U_n a + \left(E_{k'}^{(0)} - E \right) b = 0 \end{array} \right. \tag{A2.23}$$

In the above calculations, the following two relations are used

$$\left\langle k \left| \Delta U \right| k \right\rangle = \left\langle k' \left| \Delta U \right| k' \right\rangle = 0$$

$$\left\langle k \left| \Delta U \right| k' \right\rangle = \left\langle k' \left| \Delta U \right| k \right\rangle^* = U_n^*$$

To employ Cramer's rule and resolve the equation, we have

$$\begin{vmatrix} E_k^{(0)} - E & U_n^* \\ U_n & E_{k'}^{(0)} - E \end{vmatrix} = 0$$

or

$$\left(E_k^{(0)} - E \right)\left(E_{k'}^{(0)} - E \right) - \left| U_n \right|^2 = 0 \tag{A2.24}$$

The solutions are

$$E_\pm = \frac{1}{2}\left\{ \left(E_k^{(0)} + E_{k'}^{(0)} \right) \pm \sqrt{\left(E_k^{(0)} - E_{k'}^{(0)} \right)^2 + 4\left| U_n \right|^2} \right\} \tag{A2.25}$$

(I) At the boundary of the Brillouin zone

$$k = -\frac{\pi}{a}n, \qquad k' = \frac{\pi}{a}n$$

$$E_k{}^{(0)} = E_{k'}{}^{(0)} = \frac{\hbar^2}{2m}\left(\frac{\pi n}{a} \right)^2$$

$$\begin{array}{l} E_+ = E_k^{(0)} + \left| U_n \right| \\ E_- = E_k^{(0)} - \left| U_n \right| \end{array} \tag{A2.26}$$

And the band gap is

$$E_g = E_+ - E_- = 2|U_n| \tag{A2.27}$$

(II) Near the boundary of the Brillouin zone

$$\begin{cases} k = -\dfrac{\pi}{a}n(1-\Delta) \\ k' = k + \dfrac{2\pi}{a}n = \dfrac{\pi}{a}n(1+\Delta) \end{cases} \qquad |\Delta| << 1$$

$$E_k^{(0)} = \bar{U} + \frac{\hbar^2 k^2}{2m} = \bar{U} + \frac{\hbar^2}{2m}\left(\frac{\pi}{a}n\right)^2 (1-\Delta)^2$$

$$E_{k'}^{(0)} = \bar{U} + \frac{\hbar^2 k'^2}{2m} = \bar{U} + \frac{\hbar^2}{2m}\left(\frac{\pi}{a}n\right)^2 (1+\Delta)^2$$

($\bar{U}$ can be omitted as we assume $\bar{U} = 0$)

As Δ is small,

$$\left|E_k^{(0)} - E_{k'}^{(0)}\right| << |U_n|$$

To employ the Taylor series, and proceed up to the second order, we have

$$f(x) = f(0) + \frac{f'(0)}{1!}x + \frac{f''(0)}{2!}x^2 + \cdots$$

Let $x = \dfrac{E_{k'}^{(0)} - E_k^{(0)}}{|U_n|}$, then

$$\sqrt{\left(E_k^{(0)} - E_{k'}^{(0)}\right)^2 + 4|U_n|^2} = 2|U_n|\left(1 + \frac{1}{4}x^2\right)^{\frac{1}{2}} = 2|U_n| + \frac{\left(E_{k'}^{(0)} - E_k^{(0)}\right)^2}{4|U_n|}. \text{ Therefore,}$$

$$E_\pm \approx \frac{1}{2}\left\{ \left(E_k^{(0)} + E_{k'}^{(0)}\right) \pm \left[2|U_n| + \frac{\left(E_{k'}^{(0)} - E_k^{(0)}\right)^2}{4|U_n|}\right]\right\} \tag{A2.28}$$

If we use $T_n = \dfrac{\hbar^2 k^2}{2m} = \dfrac{\hbar^2}{2m}\left(\dfrac{\pi}{a}n\right)^2$ to express the kinetic energy of an electron at

$k = \dfrac{\pi}{a}n$, then near the boundary of the Brillouin zone

$$E_k^{(0)} = \bar{U} + \frac{\hbar^2 k^2}{2m} = \bar{U} + \frac{\hbar^2}{2m}\left(\frac{\pi}{a}n\right)^2 (1-\Delta)^2 = \bar{U} + T_n(1-\Delta)^2$$

$$E_{k'}^{(0)} = \bar{U} + \frac{\hbar^2 k'^2}{2m} = \bar{U} + \frac{\hbar^2}{2m}\left(\frac{\pi}{a}n\right)^2 (1+\Delta)^2 = \bar{U} + T_n(1+\Delta)^2$$

(A2.29)

Substituting Equations (A2.29) into (A2.28) yields

$$E_{\pm} = \left\langle \begin{array}{l} \bar{U}+T_n+|U_n|+T_n\left(\dfrac{2T_n}{|U_n|}+1\right)\Delta^2 \\[2ex] \bar{U}+T_n-|U_n|-T_n\left(\dfrac{2T_n}{|U_n|}-1\right)\Delta^2 \end{array} \right.$$

(A2.30)

The above analysis tells that when $\Delta \to 0$, or is near the edge of the Brillouin zone, the curves shown in Figure A2.1 differ from free-electron dispersion because of the interaction with the periodic weak potential.

When the Bragg condition is satisfied concerning the wave vector, or the end of the wave vector is on the boundary of a Brillouin zone, the wave traveling to the right is Bragg-reflected to travel to the left, and vice versa, forming standing waves. The two standing waves $\psi(+)$ and $\psi(-)$ pile up electrons at different regions, and therefore the two waves have different values of potential energy, which is the origin of the band gap (Kittel, 1996). The characteristics of the energy curve represented in reciprocal space (Figure A2.2) can well explain many behaviors of crystals. For instance, in the Introduction, we used the energy band structure to explain why some materials are metals, insulators or semiconductors.

For electrons, the number of states in a certain energy range can be obtained from the energy–wave vector expressions in three dimensions.

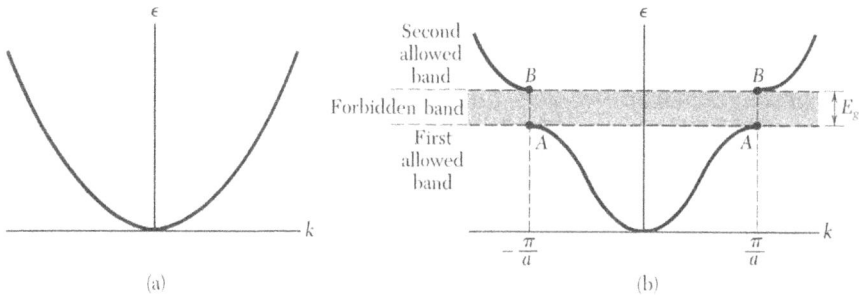

FIGURE A2.1   (a) Plot of energy versus wave vector for a free electron. (b) Plot of energy versus wave vector for an electron in a monoatomic linear lattice of lattice constant a.

(Source: Kittel (1996), with the permission of John Wiley & Sons.)

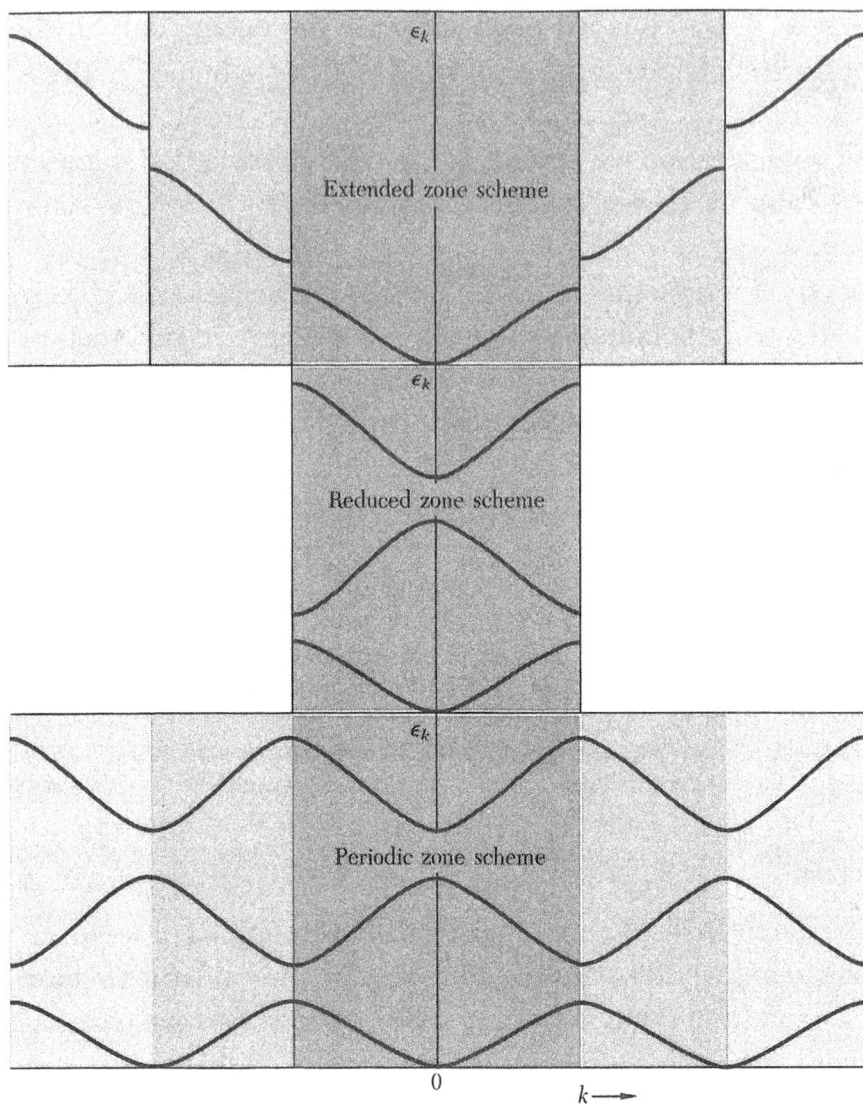

FIGURE A2.2 Three energy bands of a linear lattice plotted in the extended, reduced and periodic zone schemes.

(Source: Kittel (1996), with the permission of John Wiley & Sons.)

From Equation (A2.5), we know that the volume associated with each state in $\vec{k}$ space is $\dfrac{(2\pi)^3}{V}$. Therefore, the number of states can be written as

$$\frac{V}{(2\pi)^3} \int dS dk \qquad\qquad (A2.31)$$

We know that $\Delta E = dk|\nabla_k E|$, or $dk = \dfrac{\Delta E}{|\nabla_k E|}$, hence the number of states can be expressed as

$$\frac{V}{(2\pi)^3}\int \frac{dS}{|\nabla_k E|}\Delta E$$

Which suggests that the density of states, the number of states per unit energy range, is

$$\frac{V}{(2\pi)^3}\int \frac{dS}{|\nabla_k E|} \tag{A2.32}$$

The concept of the density of states is very useful in spectroscopy study, for example in X-ray absorption fine structure analysis and in electron energy loss spectroscopy analysis.

The tight-binding model is presented in detail in Chapter 5 in Omar (1993), and students are encouraged to read that chapter.

The focus of this textbook is on crystal structure descriptions and crystal structure determinations. Therefore, the electronic structures of crystals are only briefly explained in this appendix.

REFERENCES

Huang, K. (1988) *Solid state physics/Huang, K., Han, R. Q.* Beijing: Chinese Higher Education Publisher.

Kittel, C. (1996) *Introduction to solid state physics/Charles Kittel.* 7th ed. New York: Wiley.

Liboff, R. L. (2003) *Introductory quantum mechanics/Richard L. Liboff.* 4th ed., *Quantum mechanics.* 4th ed. San Francisco: Addison-Wesley.

Omar, M. A. (1993) *Elementary solid state physics: principles and applications/M.A. Omar.* Rev. print. Reading, Mass: Addison-Wesley Pub. Co. (Addison-Wesley series in solid state sciences).

Wahab, M. A. (Mohammad A.) (2015) *Solid state physics: structure and properties of materials/M.A. Wahab.* 3rd ed. Oxford U.K: Alpha Science International.

Zhou, S. X. (1979) *Quantum mechanics/Zhou, S. X.* Beijing: Chinese Higher Education Publisher.

Index

Page numbers in *italic* indicate figures, page numbers in **bold** indicate tables

For Product Safety Concerns and Information please contact our EU
representative GPSR@taylorandfrancis.com
Taylor & Francis Verlag GmbH, Kaufingerstraße 24, 80331 München, Germany

www.ingramcontent.com/pod-product-compliance
Lightning Source LLC
Chambersburg PA
CBHW060345220326
41598CB00023B/2817